T0302344

Introduction to Structural Analysis

Introduction to Structural Analysis

Debabrata Podder and Santanu Chatterjee

CRC Press
Taylor & Francis Group
Boca Raton London New York

CRC Press is an imprint of the
Taylor & Francis Group, an **informa** business

First edition published 2022
by CRC Press
6000 Broken Sound Parkway NW, Suite 300, Boca Raton, FL 33487-2742

and by CRC Press
2 Park Square, Milton Park, Abingdon, Oxon, OX14 4RN

© 2022 Taylor & Francis Group, LLC

CRC Press is an imprint of Taylor & Francis Group, LLC

ISBN: 978-0-367-53272-7 (hbk)
ISBN: 978-0-367-53273-4 (pbk)
ISBN: 978-1-003-08122-7 (ebk)

DOI: 10.1201/9781003081227

Typeset in Times
by KnowledgeWorks Global Ltd.

Contents

Preface

Structural analysis, as a subject, is a very important and conceptual subject in its own right among all other subjects that constitute the field of civil engineering. Starting from ancient times to present day, this subject has flourished depending upon the need and complexity of the structures coming from the advancement of human civilization. Structures are mostly classified into two categories, determinate and indeterminate structures. Determinate structures are the preliminary basic structures that need to be understood well first. Basic principles like elastic limit, Hook's law, the principle of superposition, and equilibrium conditions need to be studied very meticulously before jumping into more complex indeterminate structures. Structures that we encounter in everyday life are mostly indeterminate structures. However, the analysis process and basic compatibility conditions like force and moment equilibrium equations and superposition principle within elastic limit remain the same as those for determinate structures. That is why we have organized this book into two parts. In the first part, students will get acquainted with the basic important theories and analysis procedures for determinate structures. Having understood the basic principles, in the second part of this book, the indeterminate structural analysis will be presented, and various methods of solving the problems are described in detail.

This book is primarily aimed at students taking degree courses in civil engineering. However, students enrolled in master's degree programs in structural engineering may use this book to brush up on concepts that were taught during their undergrad study.

No matter how much theory we study, it is very important to go through as many examples as possible to get hands-on experience toward solving the problems with theoretical background that can only be understood by going through the chapters. That is why we have also included solved examples on various problems in different chapters as well as in the Appendix section, so that students can practice on their own by studying the text first and then understanding the examples.

This book is mostly self-contained. Preliminary ideas about elastic properties and some ideas of basic calculus and partial derivatives that are mostly taught at the first year or second year of degree courses in universities and colleges are assumed to be known. Apart from that, this book can be studied without taking references to other books or materials. However, for more solved examples and more advanced ideas on these topics, students are advised to go through any book provided in the bibliography section as they wish.

The authors will be highly open to suggestions regarding further development of this book and inclusion of any specific topics that may enhance the applicability to a broader audience.

We are sincerely thankful to all our colleagues, friends, and family members for standing by our side to complete this book and convert this hardworking effort into a memorable journey.

We thank you all for choosing this book for your course work, and we wish all the best to everyone on their future endeavors and successful careers in this exciting field.

Debabrata Podder
Santanu Chatterjee
December 2021

Authors

Dr. Debabrata Podder is currently working as an Assistant Professor (CE) at National Institute of Technology Meghalaya, India. He has completed his engineering education from Jadavpur University and Indian Institute of Technology Kharagpur. After a brief stint as a civil engineer at Shapoorji Pallonji & Co. Ltd., Mumbai, he took to academics. He has published the book, *Residual Stresses, Distortions and Their Mitigation for Fusion Welding* (ISBN: 978-3-659-94213-6) and papers in several peer-reviewed journals. His research interests include finite element modelling and simulation of engineering structures, theoretical and computational solid mechanics, structural analysis and design, and welding-induced deformations and residual stresses.

Mr. Santanu Chatterjee has a BE from Jadavpur University, Kolkata in construction engineering. He has more than 14 years' research and industrial experience in the field of civil/structural design engineering. He has also been involved in civil design of several solar power plants. Mr. Chatterjee also takes interest in concepts of string theory and quantum mechanics. Recently, he got his paper selected for the International Conference on Mathematical Modeling in Physical Sciences, and another of his papers on solar engineering was awarded as Best Scientific Paper at the National Institute of Technology Durgapur and the International Conference on Renewable Energy (ICCARE-2019). Another paper on pile design and its relevance (with IS-2911 part 1 section 2 for solar projects) was published at the 37th European Photovoltaic Solar Energy Conference and Exhibition (EUPVSEC 2020). In addition, a paper on quantization of classical string was also published in the *International Journal of Physics (IOP)* (vol. 1391, 2020).

Part I

Introduction to Structural
Analysis, Loads, Material,
and Section Properties

1 Introduction to Structural Analysis

1.1 INTRODUCTION

This chapter provides a general introduction to the subject of structural analysis and development of this subject from ancient ages to modern days with the help of various engineers, scientists, and philosophers. Some historical structures with their distinct characteristic features are also discussed in this chapter. One should have some knowledge of the historical backgrounds and time-to-time development of this rich and important subject. This chapter will increase interest and zeal in learning this subject by heart and setting deeper insight that may open up new avenues of the analysis process to enhance and enrich this subject to the next level.

1.2 HISTORICAL BACKGROUND

Through the contribution from various civilizations, structural analysis as we know it today evolved over several thousand years. From living under different natural habitats like tree shelters and caves, humans started colonizing beside various rivers throughout the world. There they started making their shelters using stones, clay, bricks, and numerous cementitious materials by organizing them into different geometrical shapes as per their needs. Initially, they adopted these shapes based on various thumb rules that evolved from their past experiences. They slowly modified their building types/constructions according to their diverse socioeconomic, cultural, religious, and security needs. Among these countless ancient structures, the 'Great Bath' of Mohenjo-daro (Figure 1.1) is believed to be the world's oldest public pool, built during the Indus Valley Civilization (3300–1300 BCE). The pool measures approximately 12-m long and 7-m wide, with a maximum depth of 2.4 m with two wide staircases lead down into the pool. The pool was watertight, covered by finely fitted bricks laid on edge with gypsum plaster. Most historians agree that, this would have been used for special religious occasions where its water was believed to purify and renew the well-being of the bathers.

Ancient Egyptian (3150–323 BCE) builders are mainly known for their astounding pyramid-building capabilities. They also knew the techniques of post and lintel constructions (e.g., Karnak Temple, 2000–1700 BCE). Pharaoh Djoser's official Imhotep, a famous architect, and scholar, designed Egypt's first Step Pyramid, a pharaoh's tomb at Saqqara that looks like a stairway to heaven in about 2600 BCE. Imhotep is referred to as the world's first structural engineer. Unbaked mud-brick and stone were two principal building materials in ancient Egypt.

DOI: 10.1201/9781003081227-2

FIGURE 1.1   The 'Great Bath' of Mohenjo-daro, 3300–1300 BCE.

Source: https://www.harappa.com/slide/great-bath-mohenjo-daro-0.

For determining sizes of structural members, the Egyptians, the Indus Valley Civilians, and other ancient builders surely had some kinds of empirical rules drawn from their previous experiences because, regarding the development of any theory of structural analysis, there is, however, no evidence found from their civilizations.

The ancient Greeks (1200–323 BCE) are famously known for their Post and Lintel-type constructions. Their magnificent Doric, Ionic, and Corinthian order temples were of such types. The example par excellence is undoubtedly the Parthenon (Figure 1.2) of Athens, built in the mid-5th century BCE. The temple was built to house the gigantic statue of Athena and advertise to the world the glory of Athens, which still stands majestically on the city's acropolis.

Although the ancient Greeks were blessed with many famous mathematicians, physicists, inventors, philosophers, scientists, and they also built some magnificent structures, their contributions to structural theory and analysis were few and far between. Pythagoras of Samos (about 582–500 BCE) is famous for the right-angle theorem that bears his name. However, a Babylonian clay tablet confirms that this theorem was known by the Sumerians in about 2000 BCE. Archimedes of Syracuse (287–212 BCE) developed some fundamental principles of statics and introduced the term center of gravity.

The ancient Romans (753 BCE–476 CE) were also excellent builders. Alongside various Post and Lintel constructions, they mastered the art of building arches, vaults, and domes, which helped them to cover wider space more easily. Their mastery was further enhanced by the development of concrete (*opus caementicium*), which was typically made from a mixture of lime mortar, water, sand, pozzolana,

FIGURE 1.2   Parthenon of Athens, 447–432 BCE.

Source: http://employees.oneonta.edu/farberas/arth/arth200/politics/parthenon.html.

tuff, travertine, brick, and rubbles. Some of the unusual additives were also mixed with this concrete, such as horsehair, which made the concrete less prone to cracking; animal blood that increased the resistance to frost damage. By implementing their methods, Romans built various temples, basilicas, pantheons, theatres, amphitheaters, public baths, triumphal arches, bridges, aqueducts, roads, lighthouses, etc. But like Greeks, they too had very less knowledge of structural analysis and made even less scientific progress in structural theory. They built their majestic structures from an artistic point of view based on various empirical rules gained from their past experiences. If those rules got clicked, the structures would have been survived, or else it got collapsed.

Karl-Eugen Kurrer in his book, 'The History of the Theory of Structures: Searching for Equilibrium', divided the evolution history of structural analysis into some particular periods and broke those down further into phases as shown in Figure 1.3.

The **preparatory period** (1575–1825) of the development and evolution of structural analysis stretches over around 250 years and is characterized by the direct application of mathematics and mechanics of that time to simple load-bearing elements in structures. During this time, buildings and structures were designed mostly based on empirical knowledge and theory. The theory was evident primarily in the form of geometrical design and dimensioning rules. This period mainly focused on the formulation of beam theory. The *orientation phase* (1575–1700) is characterized by the sciences (mathematics and mechanics) of this new age discovering the building industry. In the middle of this phase (1638), the final book of Galileo, *'Discorsi e Dimostrazioni Matematiche, intorno a due nuove scienze* (Dialogue Concerning Two New Sciences)', was published, which is a scientific testament covering much of

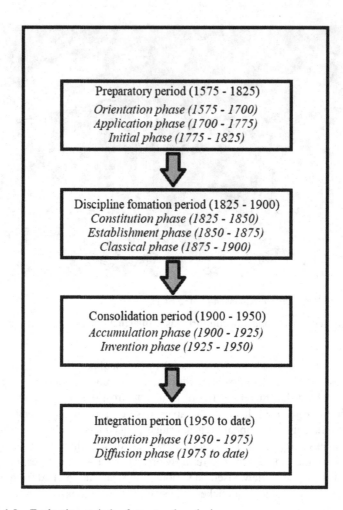

FIGURE 1.3   Evaluation periods of structural analysis.

his work in Physics over the preceding 30 years. Galileo's Dialogue contributed elements of strength of materials in the form of first beam theory to the menu, though it was erroneous. Robert Hooke (1635–1703) took the next step and discovered the law of elasticity in 1660, which later becomes as Hooke's law. As for the first time, the differential and integral calculus appeared around 1700; they found their place as a tool in the applications of astronomy, theoretical mechanics, geodesy, and construction for very obvious reasons and thus started the *application phase* (1700–1775). Mathematicians and natural scientists such as Leibniz (1646–1716), Bernoulli, and Leonhard Euler (1707–1783) made progress on the beam theory and the theory of elastic curve. In the first half of the 18th century, engineering schools were developed for the very first time in France where a scientific self-conception based on the applications of differential and integral calculus was established. Bernard Forest de Bélidor (1697–1761) in his book (1729), which was based on the calculus, dealt with earth pressure, arches, and beams in great detail. Mathematics became a useful tool for budding civil engineers in the application phase. At the beginning of the *initial phase* (1775–1825), Coulomb (1736–1806) in his paper (1776) first applied differential and integral calculus to beam, arch, and earth pressure theories in a coherent form and thus provided a method for knowledge of structural analysis. In this phase, the statics of solid bodies was the only indirectly applied mathematics unlike the application phase and calculus, which became an integral part of higher technical education.

In 1826, Louis Henri Navier (1785–1836) published his work where he discussed the practical bending theory and thus initiated the ***discipline formation period*** (1825–1900). Karl Culmann (1821–1881) expanded his trussed framework theory (1851) to form the graphical statics (1864/1866). His work was an attempt to give the structural analysis a mathematical legitimacy through projective geometry. James Clerk Maxwell (1831–1879), Emil Winkler (1835–1888), Otto Mohr (1835–1918), Alberto Castigliano (1847–1884), Heinrich Müller-Breslau (1851–1925), and Viktor Lvovich Kirpichev (1845–1913) consequentially created the linear elastic theory of trusses. Müller-Breslau with his force method – a general method for calculating statically indeterminate trusses – rounded off the discipline formation period. Navier's practical bending theory formed the basic foundation of structural analysis in the *constitution phase* (1825–1850). Due to his work, civil and structural engineers now no longer have to rely solely on their experience-based constructional knowledge rather they can create and optimize structural models through an iterative design process. Structural analysis got established in the *establishment phase* (1850–1875) in continental Europe as iron bridges became common after 1850. Culmann's trussed framework theory and graphical statics became the incarnation of iron bridge-building in search of economic use of materials. In the *classical phase* (1875–1900), Müller-Breslau developed (1886) general theory of linear elastic trusses based on the principle of virtual forces as Culmann's graphical statics was less suitable for analyzing statically indeterminate systems.

Structural analysis experienced a significant expansion of its scientific area of study in the ***consolidation period*** (1900–1950). Around 1915, the growth in reinforced concrete in the construction industry led to the development of the theory of framed structures and 20 years later to the development of the theory of plates and shells. When the existing methods of analysis reached their limits during the skyscraper boom (the 1920s), Hardy Cross (1885–1959) provided (1930) a comparatively easier iterative method, known as the moment distribution method, that could solve the internal forces with a high degree of static indeterminacy quickly. The *accumulation phase* (1900–1925) consisted of introducing statically indeterminate primary systems instead of statically determinate basic systems and thus more attention being paid to the deformations of statically indeterminate systems. Consideration of the secondary stresses in trussed frameworks and analyzing load-bearing systems of rigid frames became important issues in this phase. Due to the increase in the use of reinforced concrete, plate and shell structures became an area of study in the middle of the accumulation phase. The coherent and consistent arrangement of structural analysis arose out of the principle of virtual displacements at the end of this phase. The contents of structural analysis became tested and consolidated from the inputs/challenges of the multiple disciplines such as reinforced concrete construction, mechanical and plant engineering, crane-building, and, finally, aircraft engineering. In the *invention phase* (1925–1950), structural analysis was characterized by several new developments, such as the theory of plates and shell structures, development of displacement method alongside the force method, inclusion of nonlinear phenomena (second-order theory, plasticity), and formation of numerical methods.

The aircraft industry also reached their limits due to the continuous demands of rationalizing the calculations of airplane structures in the ***integration period*** (from 1950 to date). To make the airplanes lightweight and stable under the action of dynamic loads, engineers were lacking some reliable numerical tools where the whole body can be subdivided into some finite number of elements, considering them individually in the mechanical sense and then again putting them back together choosing the right boundary conditions. What is exactly the creator of the finite element method – Turner, Clough, Martin, and Topp – did in 1956. In the *innovation phase* (1950–1975), modern structural mechanics emerged into a theoretical level and practical level automation of the structural calculations was initiated. Various numerical methods mostly the finite element method gained more and more popularity in this phase. In 1960, Ray William Clough (1920–2016) gave this name, and in 1967, Olgierd Cecil Zienkiewicz (1921–2009) and Yau Kai Cheung outlined it in a monograph for the first time. In the *diffusion phase* (from 1975 to date), the introduction of desktop computers, computer networks, and lastly the Internet revolutionized computer-assisted structural calculations into everyday reality.

1.3   IMPORTANCE OF STRUCTURAL ANALYSIS

Structural engineering is the method of mathematical analysis and art of scheduling, designing, and constructing safe and economical structures without jeopardizing the overall integrity and which will serve their intended purposes in their anticipated lifetime. Structural analysis is the first major step of any structural engineering project, its function being the prediction of the performance of the planned structure. Without proper analysis, the critical force, and moments and the corresponding critical stresses will not be generated. Without the critical loads and their effect on the structure, a structural engineer cannot properly fix the size and types of structural members or elements so that it would not collapse under any circumstances in its entire lifetime. Most of the structures are designed to have a life span of 50 years on an average. However, as per international codes and standards design life span may

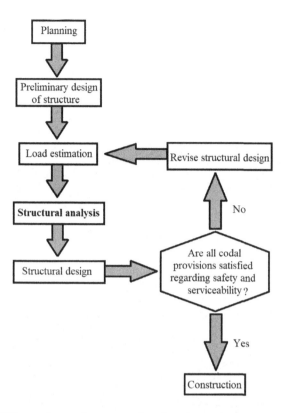

FIGURE 1.4   Different phases of a structural engineering project.

be long enough than this, and to meet higher design life, necessary coefficients for load increment and material degradation need to be considered in the design steps. However, for all structural engineers, necessary analysis steps need to be mastered well before carrying out any real-life projects. Without complete understanding the nature of critical loads and their combinations, a structure cannot be declared safe for human use. The various phases of a structural engineering project are shown in Figure 1.4.

Part II

---

*Analysis of Statically
Determinate Structures*

2 Types of Structures and Loads

2.1 INTRODUCTION

This chapter provides a general introduction of different types of loads and their combinations in structural analysis. We introduce five common types of structures: tension structures, compression structures, trusses, shear structures, and bending structures. Finally, we consider the development of the simplified models of real structures for the purpose of analysis. At the very least, loads and determination of load intensity play an important role in structural analysis. Without a good idea of the loads acting on the structure, we will not be able to analyze and design to satisfy the criticality criteria in structural design, leading to improper member selection, which will ultimately jeopardize structural integrity and quality. Thereby students are encouraged to go through this chapter without skipping to the next sections where the main topics and different analysis procedures are introduced.

2.2 STRUCTURAL CLASSIFICATIONS

It is to be understood that the most crucial decision made by a structural engineer in implementing an engineering project is the choice of the kind of structure to be used for supporting or transferring loads to the foundation systems. Most commonly used structures can be classified into five basic types, depending on the primary loads and stresses that may develop in their members under active design actions. However, it should be realized that any two or more of such basic structural types described in this section below may be combined in a single structure, such as a building or a bridge, to meet the structure's functional requirements. Depending on the joint type, the structures can also be classified into two broad categories, i.e., pin jointed structures or trusses and rigid jointed structures or frames. In the case of pin jointed structures, as members are connected through frictionless pins, reacting bending moment gets released or does not accumulate at the joints, but it is not so in case of rigid jointed structures. Selecting a proper type of structure or combination of several fundamental units to create the final complex structure profoundly impacts the overall safety and serviceability.

2.2.1 Tension Structures

Tension structures undergo purely tensile force and subsequently uniform tensile stress across their cross sections under the action of external loads. As tensile stress is distributed uniformly over the cross-sectional areas of tension members, the material of such a structure is utilized in the most effective manner. Tension structures

DOI: 10.1201/9781003081227-4

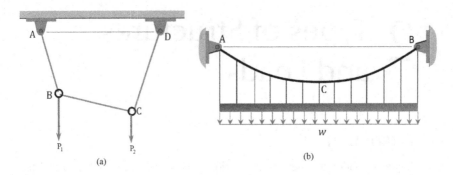

(a)

(b)

FIGURE 2.1 Tension structure – (a) Cable subjected to point load and (b) uniformly distributed load (udl).

composed of flexible steel cables are most generally preferred than other steel sections to support bridges and large-span roofs and domes. Due to their flexibility, cables have negligible bending stresses and thus subject to only tensile forces. Under imposed loads, a cable assumes a shape (the most common shape of laded cable is known as catenary) that enables it to support the load by tensile forces alone. In other words, the shape of a cable transforms/changes according to the loads (magnitude and direction) acting on it. For example, the shapes that a single cable may attain under the application of two different concentrated loads (P_1 and P_2) are provided in Figure 2.1 (a) for the ease of understanding, whereas the cable transforms its shape under the action of uniformly distributed load (w) all over it as shown in Figure 2.1 (b).

Besides cable structures, other types of tension structures used most frequently elsewhere are vertical rods as hangers (to support balconies or tanks or any overhead structures) and membrane structures like tents and roofs of large-span domes.

2.2.2 COMPRESSION STRUCTURES

Compression structures mainly develop compressive stresses under the action of externally imposed loads. The most common examples of compression structures are columns (Figure 2.2) and arches. Columns are straight structural elements subjected to axially compressive loads. Although in most of the cases, there are moments as well as axial compressive loads that act on the column, which produces uniaxial or biaxial bending stresses in the column. For an axially loaded column in which the load path has no eccentricity with respect to the central axis of column, generates compressive stress only. In the case of columns acted upon by compressive as well as bending stresses, such structural elements are called as beam-column elements. Normally beam-column terminology is used for steel structures. For RCC structures, columns are designed including uniaxial or biaxial stresses if applicable, so that failure modes do not occur during their entire life span.

An arch is a curved structure that looks like an inverted cable. These kinds of structures generally support roofs or long-span bridges. Arches develop mainly compressive forces under the action of external loads. They are designed in such a

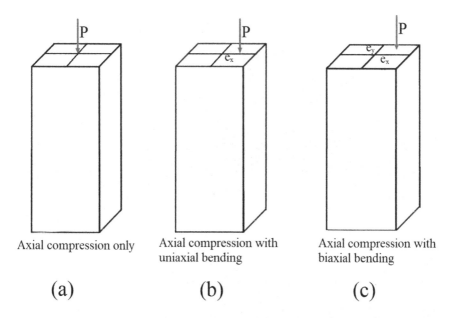

Axial compression only	Axial compression with uniaxial bending	Axial compression with biaxial bending
(a)	(b)	(c)

FIGURE 2.2 Column subjected to (a) axial compressive load only, (b) axial compression with uniaxial bending, and (c) axial compression with biaxial bending.

way so that they develop compression forces under the action of main design forces. As Arches are rigid structures, unlike cables, they produce secondary bending and shear stresses if the loading condition changes. If these secondary effects are significant, they should also be considered during the design phase.

The compression structures are susceptible to buckling failure or instability. So, they should be adequately designed considering the bracing arrangements if necessary.

2.2.3 TRUSSES

Ideally, trusses are composed of straight members connected at their ends by frictionless hinged connections to form a stable configuration. As the external loads are applied only to the joints of a truss, its members either get elongated or shortened. That means for an ideal truss, its members are either subjected to pure tension or pure compression. In the actual scenario, the truss members are connected through gusset plates using bolting or welding, which produces rigid joints instead of frictionless hinges. These rigid joints generate secondary bending stresses when the external load is applied or due to the self-weight of individual members. In most of the cases, these secondary bending stresses are negligible, and the assumption of frictionless hinges yields satisfactory results.

As the trusses are light weight and have high strength, they are the most commonly used types of structures supporting roofs of buildings. Please refer Chapter 7 for details of trusses and their analysis procedure.

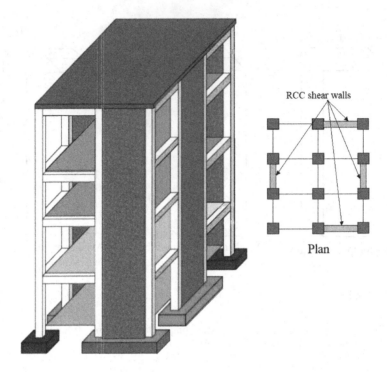

RCC shear walls

Plan

FIGURE 2.3 Multistory framed building with shear walls.

2.2.4 SHEAR STRUCTURES

Shear structures are the structural elements that develop mainly in-plane shear, with relatively small bending stresses under the application of external loads. Due to the external loads, shear is developed between two adjacent sections of a structure, which tries to slide the sections opposite each other. Usually, in concrete structures, shear walls are being designed and constructed in high-rise structures (as shown in Figure 2.3) to absorb any sudden horizontal shear load due to earthquake or wind. For shearing-resistant structural design, engineers need to take special design caution as per guidelines presented in various international codes and standards to ensure structural integrity and safety even under severe accidental loads.

2.2.5 BENDING STRUCTURES

Bending structures are structural elements subject to transverse loads and develop bending stresses as a result. As discussed in compression members, sometimes the bending moment is created due to eccentricity on the applied external load, even in columns. In such cases, the column needs to be designed against both vertical compression and uniaxial or biaxial bending effects. Typically, structures subject to pure bending are called beam elements. Beams axis is oriented along its length direction, and its cross-section is perpendicular to it. The beam is a structural element whose one dimension is much larger than its other two dimensions. In RCC and steel, beams subject to pure bending stress need to be taken care of by proper design calculation and selecting an adequate section to withstand the effect under bending stress. Beams subject to pure bending only are called Euler-Bernoulli beams and beams subject to bending as well as shear are called shear deformable beams or Timoshenko beams. If the beam's wavelength is longer than six times its height, shear deformation and rotational inertia do not play any significant role. This type of beam can be treated as an Euler-Bernoulli beam; otherwise, they are treated as Timoshenko beams. Both the beams have different stress-strain characteristics. As per various international codes and standards, analysis procedures and guidelines must be followed meticulously to ensure structural integrity even under adverse loading conditions.

2.3 STRUCTURAL SYSTEMS FOR TRANSMITTING LOADS AND LOAD PATH

In most buildings, bridges, walkover platforms, and other civil engineering facilities, two or more of the previous section's primary structural types are combined to form a structural system that can transfer the imposed loads to the ground through the foundation system. Such structures are also called framing systems or frameworks, and the components or elements of such an assembly are called structural members.

The buildings and bridges are analyzed and designed as per various international codes and standards to withstand loads in both the vertical and horizontal directions or any other general direction. Loads acting in any generalized direction can be transformed into vertical and horizontal components. These components are applied separately to the structural system during the analysis process to calculate effective net stress developed in various structural elements comprising the system. The vertical loads, generated mainly to the occupancy, self-weight, dust, and snow or rain, are commonly referred to as gravity loads or dead loads. The horizontal loads, imposed mainly due to wind and earthquakes, are called lateral loads. The term load path is considered to describe how a load acting on the structure as a whole is transmitted, through various structural elements of the structural system, to the ground. Depending on the type of load to be transferred, there are two primary load paths: gravity load path and lateral load path.

The gravity load path of a single-story building is mainly from slab to beam, beam to column, column to foundation system, and finally from foundation system to the soil. Any vertical distributed area load, such as snow, dust, or rain, acting on the roof slab is first transferred to the secondary beams as a distributed line load (force per unit length of the member). As primary beams support the secondary beams, the secondary beam reactions become concentrated forces on the supporting points of the primary beams. Similarly, the primary beams supported by columns transmit the load, via their support reactions, to the columns as axial compressive forces and/or moments. The columns transmit the load to the foundation system underneath it, which finally disperses the load to the ground soil.

The load transfer mechanism described above is yet much more straightforward than it happens. Actual load transfer from slab to primary and/or secondary beams is determined effectively by applying yield line theory. According to this theory,

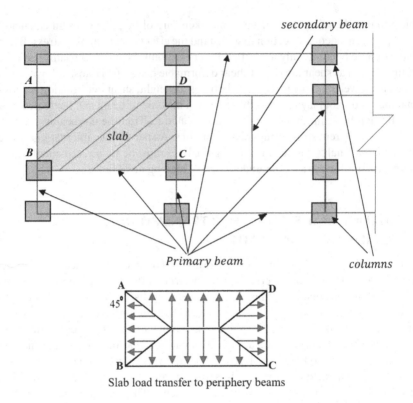

Slab load transfer to periphery beams

FIGURE 2.4   Slab to beam (primary) load transfer mechanism.

the loads transferred from slab to surrounding beams depend on many factors like the slab panel position (i.e., interior or exterior panel), edge fixity conditions of the slab, surrounding beams arrangement, and dimensions. Let us consider a typical roof and beam arrangement as shown in Figure 2.4.

The above triangular and trapezoidal distributed load will act on the corresponding primary beams, as shown in the figure. Ultimately, these loads will be transferred to columns as axial compressive loads and/or moments from the primary beams connected with the columns as per the tributary area of that particular column. This is the primary load path for gravity load transfer in all civil engineering structures. The tributary area, as shown in Figure 2.5, is related to the load path and is used to determine the amount of loads that beams, girders, columns, and walls carry. For the lateral load path, horizontal loads coming from wind or seismic force are passed to the periphery beams of the supporting slabs through shear action to the columns and progressively to the foundation systems. Horizontal loads can also be transferred through lateral load resisting elements such as shear walls to the foundation directly. In general, horizontal actions produce more complex load distribution patterns, and the analysis procedure is also very much complicated compared to gravity loads. Hence, more detailed analysis as per codal provisions and dynamic theories must be studied for detailed horizontal load analysis and its effect on the structures.

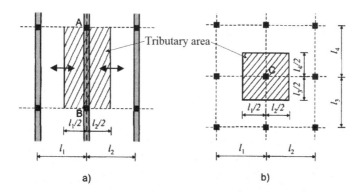

FIGURE 2.5 Tributary area for (a) beam 'AB' and (b) column 'C'.

Modern-day computer programs are primarily employed for such analysis. However, for smaller structural elements, different analysis procedures for horizontal loads are required to be studied for a basic understanding of the process.

2.4 DEAD LOADS

Dead loads are gravity loads of constant magnitudes and fixed positions that act permanently on the structure. Such loads consist of the structural system's weights and all other material and equipment permanently attached to the structural system. For example, the dead loads for a building structure include the weights of frames, framing and bracing systems, floors, roofs, ceilings, walls, stairways, heating and air-conditioning systems, plumbing, electrical systems, and other permanent miscellaneous items. The weight of the structure and all various internal equipment are not known in advance of design and is usually assumed based on experience. After the structure has been designed and the member sizes ascertained, the actual weight is computed by using the member sizes and the unit weights of materials. The actual weight is then compared to the assumed weight, and the design is reviewed and revised accordingly if required. The unit weights of some common construction materials are given in Table 2.1. The weights of permanent service equipment, such

TABLE 2.1
Unit Weights of Various Common Structural Materials

Material	Unit Weights (KN/m³)
Concrete	25
Steel	78.5
Wood	6.3
Aluminum	25.9
Brick	18.8
Cement plaster	2.4

as heating and air-conditioning systems, are usually obtained from the suppliers of respective items and from manual books of the same.

2.5 LIVE LOADS

Live loads are the transient loads, i.e., loads having variable magnitudes with respect to time and/or locations caused by the use and occupancy of the building or other structure. More specifically, the term live loads is applied to address all loads and actions on the structure which do not include construction or environmental loads such as snow load, wind load, rain load, earthquake load, floor load, or dead load. However, since the probabilities of occurrence for environmental loads are different from those due to the use of structures, the current codes use the term live loads to refer only to those variable loads caused by the use of the structure. Any such variable loads are referred to as live loads i.e., the weight of people, furniture, appliances, cars and other vehicles, and equipment. The intensity of live loads that need to be included during structural analysis and design are generally specified in various international building codes and standards. As the point of application of live loads may change, each member of the structure must be designed for the load location that will produce the maximum most severe stress conditions in members. Different members of a structure may reach their maximum allowable stress intensities at different positions for the given applied load. The method for calculating a live load location at which a particular response characteristic, such as stress resultant, or a deflection, of a structure is maximum (or minimum) is discussed in influence lines in Chapters 11 and 17.

2.6 WIND LOADS

Wind loads are produced when a structure is subjected to wind flow toward and around the structure. The intensity of wind loads that may act on a structure depends on the geographical position of the structure, wind obstructions in its vicinity, such as nearby houses, forestry, or any kind of wind obstructions present all around the structure, the plan and elevational geometry of the structure, and the natural vibrational characteristics of the structure itself. Distance from the seashore is also a major factor for the intensity of wind load that will act on the structure. Various international codes provide necessary guidelines for wind analysis that need to be incorporated during the design analysis stage, but most of the codes apply the same fundamental relationship between the wind speed V and the dynamic pressure p induced on a flat surface normal to the wind flow. This can be formulated by applying Bernoulli's principle and is given by the following equation:

$$p = \frac{1}{2}mV^2$$

in which m is the mass density of air. Using the unit weight of air as 12.02 N/m^3 for the standard atmosphere (at sea level, with a temperature of 15°C) and expressing the wind speed V in m/s, the dynamic pressure p in N/m^2 or pascal is given by:

$$p = \frac{1}{2} \times \left(\frac{12.02}{9.81} \right) V^2 = 0.613V^2$$

However, IS:875 (Part 3):2015 uses the following equation for net wind pressure due to design wind speed:

$$p_z = 0.6 \ V_z^2$$

where:

p_z = wind pressure at height z, in N/m^2
V_z = design wind speed at height z, in m/s
V_z is determined from the following equation:

$$V_z = V_b k_1 k_2 k_3 k_4$$

where:

k_1 = probability factor (risk coefficient)
k_2 = terrain roughness and height factor
k_3 = topography factor
k_4 = importance factor for the cyclonic region
V_b = basic wind speed at the site location

For all the above-mentioned factors, IS:875 (Part 3): 2015 provides detailed clauses and tables with values that need to be incorporated depending on the site location.

On the other hand, Uniform Building Code, i.e., UBC 1997 (Chap 16, Div. III, Sec. 1620) code uses the following formula for wind load analysis:

$$P = C_e C_q q_s I_w$$

where:

P = design wind pressure
C_e = combined height, exposure, and gust factor coefficient
C_q = pressure coefficient for the structure or portion of the structure under consideration
q_s = wind stagnation pressure at the standard height of 33 ft (10,000 mm)
I_w = importance factor

To determine the above values, UBC 1997 provided various design tables and guidelines to pick appropriate values as per site condition and build structural details.

2.7 SNOW LOADS

If the site location is such that there is snowfall during every winter season, snow load needs to be considered while designing and analyzing structures. The design snow load intensity on a structure is based on the ground snow thickness as per its geographical location. Various international codes and standards provide a necessary guideline for the snow load analysis process. In particular, ASCE 7 provides a clear guideline for calculating snow load intensity at any elevated structural level based on the ground peak snow thickness and meteorological data at the site location.

At first, the ground snow load was required to be calculated. Then the design snow load intensity at the roof level or at an elevated level of the structure is calculated by using factors as the structure's exposure to wind and its thermal, geometric, and functional properties and characteristics. As per ASCE 7, the following equation is used for calculating snow load intensity at the flat roof level of a building,

$$p_f = 0.7 \ C_e C_t I_s p_g$$

where:

$\quad p_f$ = snow load on flat roofs ("flat" = roof slope $\leq 5°$), in kN/m^2
$\quad p_g$ = ground snow load intensity, in kN/m^2
$\quad C_e$ = exposure factor
$\quad C_t$ = thermal factor
$\quad I_s$ = importance factor

The design snow load for a sloped roof can be determined by multiplying the corresponding flat-roof snow load by a roof slope factor C_s:

$$p_s = C_s p_f$$

There are several clauses and guidelines provided in ASCE 7 to determine the snow load intensity at the roof level using the above formulas.

2.8   EARTHQUAKE LOADS

An earthquake is a sudden vibration and movement of a portion of the earth's surface. Although the ground surface vibrates in both horizontal and vertical directions during the quake, the magnitude of the vertical component of ground vibration is usually very small and does not have an effect at a considerable scale on most structures. The horizontal component of ground vibration is the severe one that induces large dynamic stresses in the structural element that causes damage and collapse of a particular structure. Due to severity, earthquake force must be considered in designs of structures located in high earthquake prone locations. During an earthquake, the foundation of the structure vibrates with the ground, following Newton's law of inertia. On the other hand, the above-ground portion of the structure (generally called superstructure), because of the inertia of mass, always tries to resist the motion, thereby causing the structure and all its elements or members to vibrate in the horizontal direction as well. These vibrations induce horizontal shear forces in the structural elements. To predict the resulting stresses accurately in the structural element due to earthquake, dynamic analysis, considering the mass (generally called seismic mass of the structure) and stiffness characteristics of the structure, must be applied. However, for small-to-medium height buildings, most international codes provide a guideline of what is popularly known as equivalent static forces analysis or equivalent static method. The equivalent static method starts from the base shear force that will be applied due to the earthquake at the foundation level of the structure. Then this foundation shear force is required to be distributed along with various levels of the building. International codes provide all necessary equations and tables of empirical values to distribute the horizontal base shear force to various parts of the building. ASCE 7 provides quick guidelines for equivalent static method using the following equation for base shear force analysis.

$$V = C_S W$$

where:

$W =$ seismic mass of building comprising total dead load and a prescribed percentage of live loads

$C_S =$ seismic response coefficient. This coefficient is a function of several other coefficients as stated below:

$$C_S = \frac{S_{DS}}{R/I_e}$$

where:

$S_{DS} =$ design spectral response acceleration in the short period range

$R =$ response modification coefficient

$I_e =$ importance factor

However, IS:1893 (Part 1):2016 provides the following equation for calculating base shear, V_B for a structure.

$$V_B = A_h W = \frac{\left(\dfrac{Z}{2}\right)\left(\dfrac{S_a}{g}\right)}{\left(\dfrac{R}{I}\right)} \times W$$

where:

$A_h =$ design horizontal earthquake acceleration coefficient for a structure

$W =$ seismic weight of building

$Z =$ seismic zone factor

$I =$ importance factor

$\frac{S_a}{g} =$ design acceleration coefficient for different soil types, normalized with peak ground acceleration, corresponding to natural period T of the structure (considering soil-structure interaction if required)

$R =$ response reduction factor

IS:1893 (Part 1):2016 provides detailed tables of empirical values for the coefficients mentioned above based on seismic zones and structure material type. After carrying out the base shear value, it needs to be distributed among various floor levels of the building proportionate to that particular floor's seismic mass to the total seismic mass of the structure.

2.9 HYDROSTATIC AND SOIL PRESSURE

Structures constructed to retain water, such as dams and tanks, as well as structures in coastal areas partially or fully submerged in water, need to be analyzed and designed to resist water pressure. This water is the most stagnant type and does not create any dynamic effects on the structure. Hence hydrostatics is the key principle while designing these types of structures. Hydrostatic or water pressure acts normal to the structure's submerged surface as per Pascal's law, with its magnitude varying linearly with height, as shown in Figure 2.6. Thus, the hydrostatic pressure on the structure at a depth h, from the top surface, and the resultant thrust exerted on the structure can be written by the following equations, respectively.

$$p = \gamma g h$$

$$P_w = \frac{1}{2}\gamma g H^2$$

where:
p = hydrostatic pressure on the structure at a depth h
γ = density of the water or liquid retained by the structure
g = acceleration due to gravity at the location
h = depth at which pressure need to be measured
P_w = resultant thrust on the structure
H = total depth of water

Underground structures like basement walls and floors, and retaining walls need to be designed to resist soil pressure. Generally, for designing purposes, active soil pressure gives the most conservative result. The vertical soil pressure is given by the same equation as that for hydrostatic pressure with change in γ only, representing the unit weight of the soil or the earth. The lateral soil pressure is calculated based on specific parameters known as active and passive earth pressure coefficients.

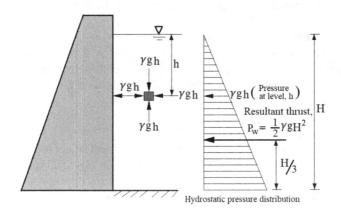

FIGURE 2.6 Hydrostatic pressure on water retaining structures.

These coefficients depend on the type of soil, and one needs to multiply the vertical pressure with these coefficients only to get the horizontal soil pressure acting on the structure. Sometimes, part of the structure remains on the earth, and partly it is submerged, or sometimes the earth is completely submerged in the water itself. In those cases, both water and soil pressures need to be applied to the structure separately. Due to water and earth, the net destabilizing force needs to be added together to proportionate and design sections for the structures correctly.

2.10   THERMAL AND OTHER EFFECTS

Statically indeterminate structures are prone to be affected due to unusual stresses induced from temperature changes, shrinkage of material, fabrication errors, and uneven settlements of supports. These effects significantly impact the increase in overall stress conditions in the structural elements, ultimately causing the unusual collapse of a structure partly or wholly. International codes and standards provide important guidelines toward these forces and their inclusion into design calculations. In most cases, one or more expansion joints are provided if the structure's length exceeds 45 m to cater to thermal expansion effects due to a temperature change. Fabrication errors induce inappropriate induction of moments and faulty joint functioning, which will cause untimely failure of the structure.

2.11   LOAD COMBINATIONS

All the load cases discussed above are first applied to the structure, and support reactions, member forces, and member stresses are preferably carried out. After that, all loads need to be combined systematically in such a way to produce the most severe load cases, which the structure needs to withstand during its entire life span. This process of combining all loads to produce a resultant effect on the structural elements and check overall stability and integrity is called load combinations. Various factors, such as an increase in magnitude and probability of occurrence, need to be incorporated while combining the loads. Hence during load combinations, various factors are multiplied with the individual load cases called safety factors. But one needs to follow specific rules as per the design code being selected for the intended job. Hence, we provide a few most commonly used design load combinations as per different international codes and standards for reference in Table 2.2.

2.12   ANALYTICAL MODEL

An analytical model is a simplified pictorial or graphical representation of the original structure for the purpose of analysis. The purpose of the model is to simplify the design and analysis of a complicated structure. The analytical model depicts or represents, as accurately as possible, the natural characteristics of the structure of interest to the designer and engineers while removing much of the details about the members, connections, and so on, that is believed to have little effect on the desired overall characteristics of the original structure under the application of prescribed loads. Formation of the analytical model is one of the most crucial steps of the design

TABLE 2.2
Load Combinations used in Different Countries

American ACI 318-11	Australian AS 3600-2009	Mexican RCDF 2004	New Zealand NZS 3101-06	Canadian CSA A23.3-04	European EURO 2-2004	Indian IS 456:2000
$1.4DL$	$1.35DL$	$1.4(DL \pm LL)$	$1.35DL$	$1.4DL$	-	$1.5(DL + LL)$
$0.9DL \pm WL$	$0.9DL \pm WL$	$0.9DL \pm WL$	$0.9DL \pm WL$	$1.25DL + 1.5LL \pm 0.4WL$	$DL \pm 1.5WL$	$1.5(DL \pm WL)$
$1.2DL + LL + 0.5L_r \pm WL$	$1.2DL + 0.4LL \pm WL$	$1.1DL + 1.1LL \pm WL$	$1.2DL + 0.4LL \pm WL$	$1.25DL + 1.5S \pm 0.5LL$	$1.35DL \pm 1.5WL$	$0.9DL \pm 1.5WL$
$1.2DL + 1.6LL \pm 0.5WL$	$DL + 0.4LL \pm EQ$	$0.9DL \pm EQ$	$DL + 0.4LL \pm EQ$	$1.25DL + 0.5LL \pm 1.4WL$	$1.35DL + 1.5LL \pm 0.9WL$	$1.2(DL + LL \pm WL)$
$0.9DL \pm EQ$	$1.2DL + 0.6LL$	$1.1DL + 1.1LL \pm EQ$	$1.2DL + 0.6LL$	$DL + 0.5LL + EQ \pm S$	$1.35DL + 1.5WL \pm 0.9LL$	$1.5(DL \pm EQ)$
$1.2DL + 1.6S + 0.5WL$	$1.2DL + 1.5LL$	-	$1.2DL + 1.5LL$	-	$DL \pm EQ$	$0.9DL \pm 1.5EQ$
$1.2DL + LL + 0.5S \pm WL$	-	-	-	-	-	$1.2(DL + LL \pm EQ)$
$1.2DL + LL + 0.2S \pm EQ$	-	-	-	-	$DL + 0.3LL \pm EQ$	-

Note: DL: Dead load; LL: Live load; EQ: Earthquake load; L_r: Roof live load; WL: Wind load; S: Snow load.

process; it demands experience and knowledge of design practices and thorough knowledge and understanding of the behavior of the original structures. Based on the structural modeling, we do mathematical modeling, enabling us to analyze the structure most effectively to get the most severe stress concentrations and its location in the structure. Based on this critical load, designers can analyze and design the structure on such a scale to resist structural failure and ensure structural integrity throughout the entire life span of the structure. In chapters following this, students will apply this method to the detailed analysis of several determinate and indeterminate structures by line diagrams/analytical models acted upon with prescribed imposed loads.

3 Material and Section Properties

3.1 INTRODUCTION

This chapter provides a general overview of different types of material and their properties. In this chapter, we will briefly introduce several important definitions and terminologies related to material properties, which directly impact determining bending stress and member deflection parameters. Important parameters like different elastic modulus and stress-strain relationships will be discussed in this chapter. Students are encouraged to brush up on their previous knowledge on strength of materials by going through this chapter in detail.

3.2 SIMPLE STRESS-STRAIN RELATIONSHIP

As per Hook's law, within the elastic limit, stress (σ) is proportional to strain (ε). This is also called the principle of elasticity:

$$\sigma \propto \varepsilon$$

In the form of equality, we need a proportionality constant (K) known as the elastic constant of material or coefficient of elasticity or modulus of elasticity or elastic modulus of the material.

$$\sigma = K\varepsilon$$

Depending upon different types of elasticity tests, the proportionality constant also named accordingly. For linear elongation of materials, the proportionality constant is known as Young's modulus, for shear resistance test, it is called shear coefficient of the materials, etc. Modulus of elasticity plays an important role in its behavior under different stress and loading conditions. A standard stress-strain experiment on a ductile material produces a stress-strain curve as shown in Figure 3.1, from which we can study the material properties under various loading conditions. In this graph, the portion generated from origin and moves ahead in a straight liner path is called the elastic range of material or the proportional limit. Within this range, when the load is removed, the structure regains its original shape and size, and we say the material was within the elastic range. Different material has a diverse elastic range. After this proportional limit, nonlinearity arrives in the stress-strain curve associated with stress-induced plastic flow in the material. By studying this graph, we can make an excellent analytical comparison on the limit up to which the materials can retain their original shape and size. By adequately checking these test curves, we can select the right kind of material for our intended work.

DOI: 10.1201/9781003081227-5

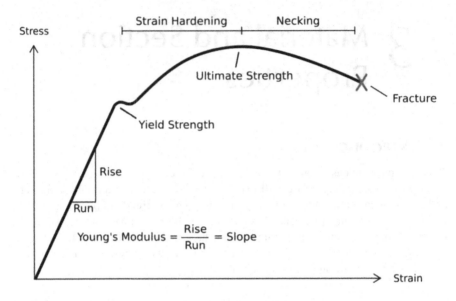

FIGURE 3.1 Typical stress-strain curve for ductile material.

The engineering measures of stress and strain denoted as σ_e and ε_e, respectively, are determined from the measured load and deflection using the original cross-sectional area of the specimen A_0 and original length L_0 as:

$$\sigma_e = \frac{P}{A_0}, \ \varepsilon_e = \frac{\delta}{L_0}$$

If the stress σ_e is plotted against ε_e, an engineering stress-strain curve is obtained. The engineering stress-strain curve must be interpreted with caution beyond the elastic limit because the specimen's dimension experiences substantial change from its original values. Using actual or true stress instead of engineering stress can provide more direct measures of the material's response in this plastic flow range.

3.3 YOUNG'S MODULUS OR MODULUS OF ELASTICITY

As stated in the previous section, Young's modulus (E) is the proportionality constant of the material under linear deformation. It is the slope of the stress-strain curve in the elastic limit. So it is defined as linear stress versus linear stain, and it can be written as:

$$E = \frac{P/A}{\Delta l/l}$$

where, P is the tension or compression force applied on the specimen within the elastic limit, A is the cross-sectional area of the specimen, Δl is the change in length due to linear elongation or contraction, and l is the actual or original length prior to start of test.

3.4   SECANT MODULUS

Secant modulus is the slope of a line drawn from the origin of the stress-strain diagram and intersecting the curve at the point of interest. It is denoted by E_s. This is also known as the static modulus of elasticity. In Figure 3.2, the secant modulus has been shown in different stress levels. Secant modulus can also be shown as a percentage of Young's modulus (e.g., 0.6E or 0.8E). It is used to describe the stiffness (i.e., the load required to produce unit deformation) of the material in its inelastic zone of the stress-strain curve.

3.5   TANGENT MODULUS

The slope of the tangent at any point in the stress-strain curve is known as the tangent modulus of elasticity (E_T). Based on the tangent locations, it can have different values. For example, a tangent modulus becomes equal to Young's modulus of elasticity if drawn within the stress-strain curve's linear range. Outside the linear region, it is always less than Young's modulus. In Figure 3.2, the tangent modulus is drawn at half of the ultimate stress level. It is mainly used to describe the stiffness of the material in the nonlinear range. Structural materials that do not have any well-defined yield point like concrete, tangent modulus provide a better opportunity to study the stress-strain characteristics and their elasticity range.

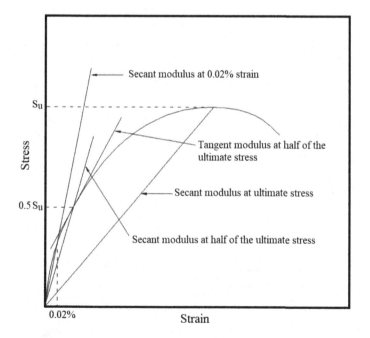

FIGURE 3.2   Different modulus of elasticity.

3.6 SHEAR MODULUS OR MODULUS OF RIGIDITY

Within elastic limit, the ratio of shear stress to shear strain is known as modulus of rigidity or shear modulus G. It is used to explain how a material resists a transverse deformation in its elastic limit.

3.7 YIELD STRENGTH

At the yield point in the stress-strain graph, materials begin to have permanent (unre-coverable) deformation. Some materials have well-defined yield points, whereas some do not have. In the absence of a distinct yield point, a 0.2% offset is considered to get an approximate yield point as shown in Figure 3.3 (a). The stress and strain corresponding to the yield point are called yield stress and yield strain, respectively.

The yield point is crucial in steel structural design and analysis. At this stress, the material specimen continues to elongate without an increase in stress. Sometimes, at this point, a sudden decrease in stress is also observed. Thus, we get two yield points, the upper yield point and the lower yield point. The corresponding stresses are called upper yield stress and lower yield stress, respectively. The lower yield point is gener-ally considered as the true characteristic yield point of a material.

The upper and lower yield points are shown for low carbon steel in Figure 3.3 (b). The stress corresponding to the lower yield point is the accurate yield stress of the material. The yield strength is defined in this case as the average stress at the lower yield point.

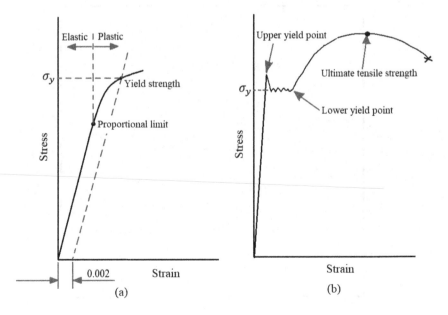

FIGURE 3.3   (a) Yield strength using offset method and (b) upper and lower yield points for low carbon steel.

3.8   ULTIMATE STRENGTH

The maximum tensile load, compressive load, or shear load a material capable of carrying divided by the original cross-sectional area is called ultimate tensile strength, ultimate compressive strength, or ultimate shear strength of the specimen.

In Figure 3.3 (b), the ultimate tensile strength of the material is pointed out. Beyond the point of ultimate tensile strength, the material appears to strain soften in the engineering stress-strain graph so that each increment of additional strain requires smaller stress and necking also starts after this point.

3.9   MODULUS OF RUPTURE IN BENDING

A specimen will fail (rupture) at the point of maximum stress where it exceeds the ultimate strength of the material. The modulus of rupture in bending, also known as flexural strength or transverse rupture strength, or bend strength, gives the maximum load-carrying capacity right before the material breaks or yields. The modulus of rupture is force per unit area, which is generally calculated by dividing the ultimate bending moment a section can carry just before failure by the section modulus of interest.

3.10   MODULUS OF RUPTURE IN TORSION

The modulus of rupture in torsion is the same as that for the modulus of rupture in bending. In this case, the only difference is that, instead of bending moment, torsional resistance needs to be considered, and it needs to be divided with torsional section modulus.

3.11   POISSON'S RATIO

Poisson's ratio is the ratio between two strains. It is defined as the ratio of transverse strain to the corresponding axial strain on a material stressed along one axis. When a bar is elongated, its cross-sectional area gets reduced. The change in diameter to the original diameter is called the transverse or lateral strain. On the other hand, the linear change in length to its original length is called the axial or longitudinal strain.

3.12   COEFFICIENT OF THERMAL EXPANSION

The coefficient of thermal extension is defined as the change in the original length of the specimen due to a per degree Celsius rise in temperature. Mathematically it can be expressed as:

$$\alpha = \frac{\Delta L}{L_0 \times \Delta T}$$

where ΔL is the change in length, α is the thermal expansion coefficient, L_0 is the original length, and ΔT is the temperature change. For different materials, the coefficient of thermal expansion assumes different values. Specialist literature needs to be studied to determine the coefficient of thermal expansion for the chosen material. However, the abovementioned coefficient is also called the coefficient of linear thermal expansion. There are other two thermal expansion coefficients, surface expansion (β) and volume expansion (γ). The former is defined as the change in area due to a per degree change in temperature of the specimen, and the latter is defined as the change in volume due to a per degree change in temperature. There exists a relationship between these three coefficients for the same material specimen that is given next:

$$\alpha = \frac{\beta}{2} = \frac{\gamma}{3}$$

3.13   ELASTIC ASSUMPTIONS

The deformation is said to be elastic if the external force produces a deformation that disappears with the removal of the load. Thus, the elastic behavior implies the absence of any permanent deformation.

The assumptions of linear elasticity are as follows:

1. The body is continuous, which means the whole-body volume is considered to be filled with continuous matter without any void.
2. The body is perfectly elastic, which means the body should wholly obey Hooke's law of elasticity.
3. The body is homogeneous, which means the elastic properties are the same throughout the body.
4. The body is isotropic, which means that the elastic properties of the body remain the same in all directions.
5. Displacements and strains components are small.

3.14   STURCTURAL NONLINEARITY

Although our preliminary discussion based on elasticity grounds seems perfect, due to some structural failure mechanism, we came to know that this phenomenon arises due to the nonlinear properties inherent to the parent material used as structural elements. Also, these failure modes cannot be analyzed or quantifiable based on the elastic failure modes of the structures that we are accustomed to. Some important engineering phenomena as listed next can only be accessed based on nonlinear analysis:

1. Collapse or buckling of structures due to sudden overloads.
2. Progressive damage behavior due to long-lasting severe loads.
3. For certain structures (e.g., cables), nonlinear phenomena need to be included in the analysis, even for service load calculations.

Real-world behaviors are primarily nonlinear that cannot be obtained through simplified linear models. In recent years, thus, the need for nonlinear analysis is mainly due to the following reasons:

1. To use optimized structures.
2. To use new materials.
3. To address safety-related issues of structures more rigorously.

The sources of nonlinearity are due to multiple system properties, for example, materials, geometry, nonlinear loading, and constraints. Here are some examples:

1. Geometric nonlinearity: In this case, a structural component experiences large deformation, which causes it to experience nonlinear behavior. A typical example is a fishing rod.
2. Material nonlinearity: Here, the component goes beyond the yield limit. As a result, the stress-strain relationship becomes nonlinear, and the material deforms permanently.
3. Contact: When two components come into contact, they can experience an abrupt change in stiffness, resulting in localized material deformation at the region.

To perform nonlinear analysis, we essentially need to adopt the following philosophy:

1. Stay with relatively small and reliable analytical models.
2. Perform a linear analysis first.
3. Refine the model by introducing nonlinearities as desired at various stages.

Interested students are encouraged to refer to various textbooks and references on this subject to gain knowledge. This textbook is based on elastic analysis of the structure, and all stress and strain levels of the structure will be such that the material will always be within the elastic limit.

3.15   CROSS-SECTIONAL AREA

A cross-sectional area represents an intersection of an object by a plane along or across its axis. We usually take transverse cross sections while determining the stress calculations. Cross-sectional areas depend on the geometry of the structural element. A rectangular beam will have a rectangle cross section, whereas a circular column will have a cross section like a circle in transverse direction.

3.16   CENTER OF GRAVITY AND CENTROID

The center of gravity (CG) is the point on an object through which its entire weight acts in the direction of gravity, while the centroid is the geometric center of the object. For isotropic and homogeneous objects with a regular geometric shape, the

CG of weight and geometric center (i.e., centroid) coincide. For anisotropic objects, the geometric center and weight CG do not coincide.

3.17 ELASTIC NEUTRAL AXIS

Let us first visualize the bending of a beam under any flexural loading (a load is called flexural load when it is acting in a perpendicular direction to the longitudinal axis of the beam, i.e., if the loading causes bending of a structure). So, under this loading, the beam will deflect in the downward direction and become slightly curved, as shown in Figure 3.4. The top fiber of the beam element will be in a state of compression, whereas the bottom fiber of the beam will be in a state of tension. Thus, between these two extreme fibers, there will be one location where the fibers are unaffected due to this bending. That particular beam level where the net stress concentration is zero is called the neutral plane, and the imaginary line drawn along this plane is called the neutral axis of the beam. As we are dealing with structural bending within the elastic limit, this neutral axis is sometimes also called the elastic neutral axis of the member under study. For more detail on this, refer to Chapter 5 of this book.

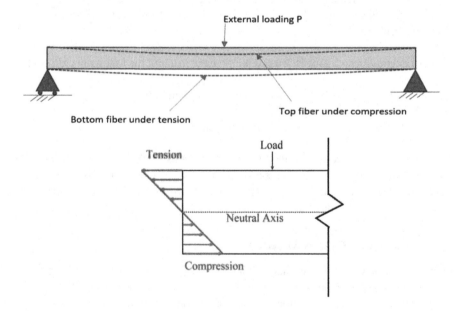

FIGURE 3.4 Elastic deformation of beam under flexural loading and neutral axis.

3.18 SECOND MOMENT OF AREA AND RADIUS OF GYRATION

The second moment of area or moment of inertia is a term used to describe the capacity of a cross section to resist rotation. It is always considered with respect to a reference axis such as X–X or Y–Y, as shown in Figure 3.5. The reference axis is usually the centroidal axis. It is a mathematical property of a section concerned

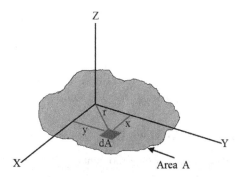

FIGURE 3.5 Second moment of area or moment of inertia.

with a surface area and its distribution about the reference axis (axis of interest). Mathematically, it is written as:

$$I_{x-x} = \int_A y^2 dA$$

$$I_{y-y} = \int_A x^2 dA$$

where y is the distance from the X-axis to area dA and x is the distance from the Y-axis to area dA.

For regular geometric shaped members, like a rectangle with width b and depth d, the moment of inertia of the section is:

$$I_{x-x} = \frac{bd^3}{12}$$

The radius of Gyration or Gyradius is the several related measures of the size of an object, a surface, or an ensemble of points. It is calculated as the root mean square distance of the objects' parts from either its CG or an axis.

In structural engineering, the two-dimensional radius of gyration is used to describe the distribution of cross-sectional area in a beam around its centroidal axis. It is the distance at which the entire area must be assumed to be concentrated so that the product of the area and the square of this distance will equal the moment of inertia of the actual area about the given axis. Here the radius of gyration is given by the following formula:

$$k_x = \sqrt{\frac{I_x}{A}}$$

where I_x is the moment of inertia about the X-axis, and A is the area. If no axis is specified, centroid axis is assumed. Here the radius of gyration is used to compare how various structural shapes will behave under compression along an axis. It is used to predict buckling in a compression member or beam. It is used to describe the cross-sectional area distribution in a column around its centroidal axis. If more area is distributed further from the axis, it will offer greater resistance to buckling. The most efficient column section to resist buckling is a circular pipe because it has its area distributed as far away as possible from the centroid.

3.19   ELASTIC SECTION MODULUS

Elastic section modulus (S) is the ratio of moment of inertia of the cross section (I) to the distance from the neutral axis to the most extreme fiber (y). So, it can be written as $S = I/y$. For the rectangular section, the elastic section modulus of the section is:

$$Z = \frac{I}{b/2} = \frac{bd^2}{6}$$

Section modulus is a geometric property of a given cross section and is used to design beams or flexural members.

4 Basic Concepts of Generalized Coordinates, Lagrangian, and Hamiltonian Mechanics

4.1 INTRODUCTION AND CONCEPT OF GENERALIZED COORDINATES

This chapter is independent and mostly has no direct connection with all other chapters in this book. Variational principle as a subject is very interesting and has a wide range of applicability in almost all fields of science and engineering. Hence, a short introduction to this subject with few examples and some important theorems related to this beautiful analytical technique has been provided here. Interested students may find this chapter as a steppingstone toward more advanced mechanics and structural analysis books. Moreover, this technique has a wide range of applicability in structural dynamics.

In this chapter, we will gain a brief insight into the vast area of mathematical field called generalized coordinates and its subsequent development and application in formulating Lagrangian and Hamiltonian mechanics. We are more familiar with the Cartesian coordinates where three orthogonal axes (namely, x, y, z) play the role of reference axes, and all other geometrical properties of particle or objects are expressed with respect to these axes. Although Cartesian coordinates or more specifically coordinate geometry helps us to formulate algebraic equations related to the orthogonal coordinates, sometimes it is found that all problems of mechanics are not so simple or appropriate enough to express the geometrical properties via only Cartesian coordinates. Also, it is found that a particular mechanical problem, which has very complex structure in Cartesian coordinates to express its equation of motion, becomes much easier if we use some other parameters that act as coordinates related to certain mechanical properties (like position and momenta of the particle), the analysis of the system to understand its behavior becomes much easier. There is also another advantage of using generalized coordinates. Equations expressed in terms of generalized coordinates are applicable to any coordinate system; one needs only to put appropriate values at the appropriate place in the expression that is made up of generalized coordinates. In the following section, we will develop and discuss some examples to see this powerful method in action.

DOI: 10.1201/9781003081227-6

4.2   CONCEPT OF CONFIGURATION SPACE AND PHASE SPACE

Configuration space is a set of geometric coordinates that define the configuration of a system. If the system consists of a single-point particle, then configuration space for that particle would be the three coordinates (x,y,z) that define its location in three-dimensional space. Time is also taken as a parameter whenever we are dealing with a dynamical system. So, for a static system of a single point, three coordinates form the configuration space, whereas for dynamical system of a single point, three Cartesian coordinates and a time parameter form the configuration space of the system. So, in mathematical language, we can write:

For static system of single particle configuration space $=(x,y,z)$

For dynamical system of single particle configuration space $=(x,y,z,t)$

For a rigid body or a system consisting of many particles, the configuration space consists of n-tuples forming the configuration space under study. So, for rigid body or many particle systems, the configuration space can be expressed mathematically as follows:

For static system of many particles $=(x_n,y_n,z_n)$

For dynamic system with many particle $=(x_n,y_n,z_n,t)$

So basically, configuration space is what we are most familiar with from our basic introduction to coordinate geometry in secondary standard mathematics. However, as it is evident from the mathematical forms, configuration space does not provide enough information about the nature of motion, deflection, and rotation of the objects under study. To understand the nature of the object under application of force or moment, we need to extend the configuration space to include more information about the behavior of the system under application of forces and moments. When we include some parameter in the configuration space to express the nature of motion of the particle/particles under study, then we form phase space. A phase space of particles or particle is the space consisting of Cartesian coordinate as well as momentum of the particle. So, the form of phase space coordinate is something like this:

Phase space for any dynamic system $=(x_n,y_n,z_n,p_{xn},p_{yn},p_{zn},t)$

For static system, there will be no momentum. Hence, phase space and configuration space become same for static systems. Although we are writing x, y, z, it needs to be understood that any set of parameters that define configuration of the system are equally valid to be included as coordinates in phase or configuration space.

In the context of generalized coordinates, the corresponding momentum in such cases is called generalized momentum. Generalized momentum might have different form than mass times velocity, which we are most familiar with from our basic concept of mechanics. The actual form and equation expressing the configuration of a system is dependent upon the coordinate system that we are using to express it. An example will be most welcome here to establish the concept of generalized

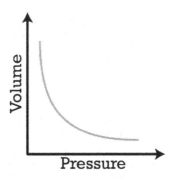

FIGURE 4.1 Boyle's law of gas – volume vs. pressure curve.

coordinates. From secondary school days, we have studied Boyle's law that states that when the temperature remains constant, the pressure and volume of a gas are inversely proportional. Therefore, pressure times volume is constant, and what we get from here is the rectangular hyperbola with the mutual orthogonal axes as pressure and volume parameters of gas. P versus V diagram for gas obeying Boyle's law is shown in Figure 4.1 for better understanding.

In this plane, the coordinate of any point does not represent its physical position in two-dimensional Cartesian plane, whereas in this plane, every point represents the state of gas having this much pressure for this much volume and vice versa. So, distance between two points in this plane has got no meaning in contrast to usual distance between two points in Cartesian plane. However, this plane is most appropriate for gases to express their state and nature of the change of state due to the change in the pressure or volume parameters. So, we see that, it is not suitable to express the configuration of a system with usual Cartesian coordinates all time. However, from this example, we can state that for the problem at hand, our generalized coordinates for a gaseous system is the pressure and volume in P–V plane. So, as stated earlier, generalized coordinates may not be coordinates at all that we are familiar with from the concepts of coordinate geometry. The power of generalized coordinates helps us to use parameter as coordinate of a system, which is suitable to express its configuration under phase space or configuration space. Phase space plays an important role toward formation of two most powerful tools to analyze mechanical systems under different situation. So, in the next section, we will study the formation of Lagrangian and Hamiltonian mechanics by the help of phase space of the system.

4.3 INTRODUCTION TO LAGRANGIAN AND HAMILTONIAN FORMULATION OF MECHANICS

From our earlier discussion, we can take the generalized coordinates for a particle in phase space as $(q, \dot{q}, t)$ and try to form the equation of motion with this. The Lagrangian of any system is defined as the difference between the kinetic and potential energy of the system under observation. Now, in this section, we will express the generalized coordinate by letter q and generalized velocity by $\dot{q}$. Time parameter is kept in its usual form t. So, Lagrangian of a system is the function:

Lagrangian of a system $= L(q,\dot{q},t)$

And the Euler-Lagrange equation of motion is defined as:

$$\frac{d}{dt}\left(\frac{\partial L}{\partial \dot{q}}\right) - \frac{\partial L}{\partial q} = 0$$

Typically, this equation is derived from the laws of calculus of variation. Euler's theorem on calculation of minima for a function of several variables has this form. Interested readers may refer to any book mentioned in the reference for the same. Above equation is not only applicable for mechanical systems but for all class of problems where we are dealing with functions of several variables. More specifically, Lagrangian defined above is a functional whose arguments are functions. To see how powerful the above equation is, let us consider the following problem of basic Euclidian geometry and prove the assertion that the shortest distance between any two points in plane is the straight line joining the two points.

Example 4.1: Prove that the shortest distance between two points in a plane (two-dimensional for sake of ease of calculation) is the straight line joining the two points.

SOLUTION: We know from basic coordinate geometry that distance between any two points (x_1, y_2) and (x_1, y_2) in two-dimensional Cartesian plane is given by:

$$s^2 = (x_1 - x_2)^2 + (y_1 - y_2)^2$$

When the two points are sufficiently close to each other, we can express the above equation in differential form as follows:

$$ds^2 = (dx)^2 + (dy)^2$$

or,

$$ds = dx\sqrt{1 + \left(\frac{dy}{dx}\right)^2}$$

Now, from the above Euler-Lagrange equation, we can choose the parameters of Lagrangian (functional) as follows:

$$L = ds; \; q = y, \; \dot{q} = \frac{dy}{dx} = \dot{y}; \; \frac{d}{dt} = \frac{d}{dx}$$

In this coordinate system, since the Lagrangian is not a direct function of y, we have,

$$\frac{\partial L}{\partial y} = 0$$

and,

$$\frac{\partial y}{\partial \dot{y}} = \frac{\dot{y}}{\sqrt{1+\dot{y}^2}}$$

Inserting all the above terms in the Euler-Lagrange equation, we get,

$$\frac{d}{dx}\left(\frac{\dot{y}}{\sqrt{1+\dot{y}^2}}\right) = 0$$

which implies that,

$$\left(\frac{\dot{y}}{\sqrt{1+\dot{y}^2}}\right) = \text{constant} = c$$

So, this solution is only valid if and only if, $\dot{y} = a$
    where 'a' is another constant related to earlier constant by the following:

$$a = \frac{c}{\sqrt{1-c^2}}$$

So, from $\dot{y} = a$, we get $dy = adx$, which upon integration yields,

$$\int dy = a \int dx$$

or,

$$y = ax + b$$

which is the equation of straight line.

    So, from the above example, we can understand how powerful the Euler-Lagrange equation is, and it is applicable for any problem involved finding out the extremum of functionals irrespective of the dynamical systems.

Example 4.2: Using Euler-Lagrange equation solve the equation of elastic line and provide physical explanation of the solution.

SOLUTION: Here we want to apply the Euler-Lagrange equation to a well-known equation of an elastic line. From the concept of strength of materials, we know that the elastic line equation for beams is in the following form:

$$\frac{d^2y}{dx^2} = -\frac{M}{EI}$$

where y is the transverse direction in which deflection takes place, x is the axis of beam, and M is the bending moment acting on the beam due to any external transverse loading. EI is the product of elastic modulus of the beam and moment of inertia of the beam cross section, respectively. We want to investigate this equation and our goal is to find the condition for maximum bending moment. So, let us define the Lagrangian of the system under study as:

$$L = \frac{d^2y}{dx^2} + \frac{M}{EI}$$

If the beam is subjected to a uniformly distributed force of intensity w, then bending moment from any section from distance x from its left support is given by:

$$M = \frac{wl}{2}x - \frac{wx^2}{2}$$

So, with this expression inserted in the above equation, we get,

$$L = \frac{d^2y}{dx^2} + \frac{1}{EI}\left(\frac{wl}{2}x - \frac{wx^2}{2}\right)$$

So, Lagrangian is function of both y and x. Hence, we need to solve this in two ways. First, we will set:

$$L = \frac{d^2y}{dx^2} - \frac{M}{EI}$$

with,

$$\frac{\partial L}{\partial y} = 0$$

And, Euler-Lagrange equation yields,

$$\frac{d}{dx}\left(\frac{d^2y}{dx^2} - \frac{M}{EI}\right) = 0$$

or, expanding the terms,

$$\frac{d^3y}{dx^3} - \frac{d}{dx}\left(\frac{M}{EI}\right) = 0$$

or,

$$\frac{d^3y}{dx^3} = \frac{d}{dx}\left(\frac{M}{EI}\right) = \left(\frac{V}{EI}\right) \tag{4.1}$$

where V is the shear force acting in the transverse direction of the beam. In second case with the substitution of the moment expression, we will get,

$$\frac{d}{dx}\left\{\frac{d^2y}{dx^2} - \frac{1}{EI}\left(\frac{wl}{2}x - \frac{wx^2}{2}\right)\right\} = 0$$

or, completing the differentiation we get,

$$\frac{d^3y}{dx^3} - \frac{1}{EI}\left(\frac{wl}{2} - \frac{wx}{2}\right) = 0$$

or,

$$\frac{d^3y}{dx^3} = \frac{1}{EI}\left(\frac{wl}{2} - \frac{wx}{2}\right) \tag{4.2}$$

Comparing equations (4.1) and (4.2), we get,

$$V = \left(\frac{wl}{2} - \frac{wx}{2}\right)$$

which is precisely the expression for shear force induced in the beam due to external uniformly applied load w on the beam in the transverse direction. Here is the beauty of this method. We have neither analyzed the beam by drawing cross section at a distance x from left support, nor have we drawn any induced force acting at that section. But the equation helps us to derive the exact expression of the shear force at a distance x from the support. Same is applicable for moment also. We can start by assuming that the induced shear force at a distance x is V, and from that, we will get the expression of moment in this section. So, in summary, we can analyze the internal forces of beams without diving into tidier details of shear force and bending moment diagrams, by applying this method.

4.3.1   HAMILTON'S EQUATION OF MOTION

From the above discussion on Lagrangian system, it is found that Lagrange's equation of motion is second-order differential equation that imposes a tedious solving process upon integration. So, to avoid difficulty, we can try to formulate mechanical laws in first-order differential equations. That is what precisely we get using Hamilton's equation of motion. Hamilton's equation of motion in generalized coordinate is expressed as follows:

$$H = \sum_{i=1}^{N} p_i \dot{q}_i - L(q, p, t)$$

$$\dot{p}_i = -\frac{\partial H}{\partial q_i}$$

$$\dot{q}_i = \frac{\partial H}{\partial p_i}$$

In elementary terms, Hamiltonian function H is the sum of kinetic and potential energy of a system or particle under study. So, the form of equation as written above clearly indicates that we are having equations of motion involving first-order derivatives only. Although nothing new has been added in the overall mechanical laws, Hamiltonian provides a more excellent way to solve equations of motion and a profound understanding of the system's dynamics that help build a more elegant way of solving the problem.

We will not dive into much finer details with Hamilton's equation of motion since they are more relevant for understanding the dynamical system. Still, familiarity with these laws helps one study a different advanced topic in mechanics. Interested readers may consult the textbooks mentioned in the reference for a deeper understanding of these elegant laws.

4.4   CONCEPT OF SYMMETRY AND CONSERVATION LAWS

To this end, we will conclude the basic concept of conservation laws and their relation to symmetry of the system under study. As we have seen from Lagrange's equation of motion, when Lagrangian is independent of q, then the equation reduces to,

$$\frac{d}{dt}\left(\frac{\partial L}{\partial \dot{q}}\right) = 0$$

or,

$$\frac{d}{dt}(p) = 0$$

where the momentum of the particle is related to Lagrangian by the following equation:

$$p = \left(\frac{\partial L}{\partial \dot{q}}\right)$$

So, the equation simply implies that when the Lagrangian is not function of one of the coordinates then the corresponding momentum related to that coordinate axis becomes conserved or constant of motion. This coordinate q is called ignorable or cyclic coordinates. To this point, we quote few lines related to this phenomenon from Goldstein's book of classical mechanics:

> If the generalized coordinate corresponding to a displacement is cyclic, it means that translation of the system, as if rigid, has no effect on the problem. In other words, if the system is invariant under translation along a given direction, the corresponding linear momentum is conserved.

Thus, momentum conservation is directly related to the translational symmetry of the system. The same is applicable for angular momentum conservation also. Interested readers may refer to Goldstein's book on classical mechanics Chapter 2 to have a detailed understanding of the same.

5 Equilibrium and Support Reactions

5.1 INTRODUCTION

In this chapter, we will study various conditions regarding the equilibrium of structures. To establish the equations for the equilibrium of structures, we will also investigate various types of supports and end conditions based on which the overall structural stability can be ensured. Equilibrium of various internal and external forces and moments acting on the structures will also be explored.

5.2 EQUILIBRIUM OF STRUCTURES

Equilibrium of a structure or a particle is a state in which there is no net resultant force acting on the structure. Hence, under the condition of static equilibrium, there will be no motion due to the application of external forces and moments. Several forces from various directions as well as moments can act on a structure. In equilibrium there will be no net resultant force acting on the structure, effectively imposing structural static condition. Although we are mostly dealing with statics here, hence, topics like dynamic equilibrium or D'Alembert's force law will not be referred. So, our subject matter here is the static equilibrium of structure and its various conditions. Following Newton's law, when there is no resultant force acting on the structure, then the acceleration of the structure will be a null vector or zero. Following the same logic, the equilibrium condition in terms of vector equation can be written as:

$$\vec{F} = m\vec{a} = \vec{0}$$

Expanding the vector equation along three mutually perpendicular axes (x, y, z), we can write:

$$\vec{F}_x + \vec{F}_y + \vec{F}_z = \vec{0}$$

Here all the above vectors are mutually independent; hence, we can write:

$$\vec{F}_x = \vec{0}$$

$$\vec{F}_y = \vec{0}$$

$$\vec{F}_z = \vec{0}$$

DOI: 10.1201/9781003081227-7

So, we get the first condition of equilibrium. It simply says when the total force acting on a system or a structure along three mutually perpendicular axes is individually equal to the null vector or zero, the structure or system is said to be in the state of equilibrium. Each component of force along three axes individually needs to be zero; otherwise, along with the nonzero component, there will be acceleration following Newton's second law of motion and thereby disturbing the equilibrium of the structure or the system. The above equation is also true for moment vectors. So, under equilibrium condition, the net moment acting on a structure at any fixed point of reference will be zero or null vector. Mathematically we can write this as:

$$\vec{M}_x + \vec{M}_y + \vec{M}_z = \vec{0}$$

So, the equilibrium condition tells us the following for moments as well:

$$\vec{M}_x = \vec{0}$$

$$\vec{M}_y = \vec{0}$$

$$\vec{M}_z = \vec{0}$$

In summary, we can reinstate the equilibrium condition in a much more compact form:

$$\sum \vec{F}_i = \vec{0}; \ \sum \vec{M}_i = \vec{0}$$

where i varies from 1 to 3, representing three mutually perpendicular axes under consideration and it is very important to note that moment summation has been taken with respect to a particularly chosen point. Moment due to all acting forces about this chosen point need to be considered in the above equation and it should not be such that moment about various points can make a body remain under equilibrium condition. In the next section, we will investigate two important types of equilibrium conditions, global and local equilibrium of structure and its elements. Expanding the vector equation in terms of components, we can write the equilibrium conditions as:

$$F_x\hat{i} + F_y\hat{j} + F_z\hat{k} = \vec{0} \quad \text{and,} \quad M_x\hat{i} + M_y\hat{j} + M_z\hat{k} = \vec{0}$$

where $\hat{i}$, $\hat{j}$, and $\hat{k}$ are the unit vectors along three mutually orthogonal axes (x, y, z), respectively. Refer to Figure 5.1 for force vectors along three mutually orthogonal axes for the ease of understanding.

5.2.1 GLOBAL AND LOCAL EQUILIBRIUM OF STRUCTURES

We can apply the above equations for equilibrium on a structure as a whole or its elements individually. The former case is called global equilibrium and the latter case is known as the local equilibrium. This can be better understood with the following example. Let us consider a simple two-dimensional (2D) truss system with an

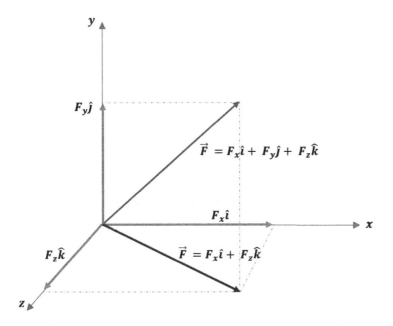

FIGURE 5.1   Force vector in a three-dimensional (x, y, z) plane.

external load P acting on the structure as shown in Figure 5.2; though, we will learn about truss in Chapter 7 in detail.

Now to calculate the support reactions at A and B due to the externally applied load, we need to consider the equilibrium of the structure with the external load along with the support reactions. The force diagram for the truss is shown in Figure 5.3 (for a list of support reactions depending on its different types refer to Section 5.6).

From the force diagram shown below, we can calculate the support reactions by the application of equilibrium equations along two axes (since this is a 2D truss). To determine the support reactions, we need to consider the total structure and its geometry without paying attention to its internal members and their orientations. So, to get support reactions we have to apply the concept of global equilibrium condition

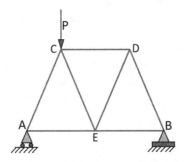

FIGURE 5.2   Two-dimensional truss with external loading.

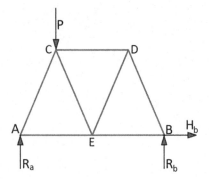

FIGURE 5.3   Two-dimensional truss with external loading and support reactions – global equilibrium.

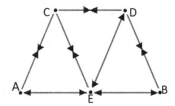

FIGURE 5.4   Two-dimensional truss members with axial force – local equilibrium.

of the complete structure as a whole. After finding the support reactions, we need to calculate the forces acting on each member of the truss. Now equilibrium conditions need to be applied to each member to find out the forces acting on them. This is the essence of the local equilibrium conditions. Thus, from the above discussion, it is found that the global equilibrium condition is applied to the whole structure, whereas local equilibrium condition is applied to each member comprising the structure separately to complete analyses. Individual truss members with the axial force acting on them are shown in Figure 5.4 according to the present loading and support condition for ease of understanding. After getting the values of support reactions using global equilibrium conditions and thereafter by the application of equilibrium equations for each member, we can easily calculate these axial forces. We will apply these concepts in Chapter 7 while analyzing 2D and 3D trusses.

To understand the forces acting on a system and its resultant and subsequently equilibrium condition, we need to carefully draw the free body diagrams, indicating all acting forces on the system or structure. In the next section, we will investigate the free body diagram of structures and will analyze the equilibrium conditions in more detail.

5.3   FREE BODY DIAGRAMS

Free body diagrams are the most essential part of structural analysis. As stated earlier, proper free body diagram with all acting forces at the proper location allows us to analyze the structure effectively and correctly. As stated in the previous section, while analyzing the 2D truss, we have applied the concept of global and local equilibrium conditions. Now, before we apply the equilibrium equations, we must draw a force diagram or free body diagram of the truss by removing all physical supports and showing support reactions at proper locations where they are acting in the actual structure. So, the free body diagram enables us to draw the force diagram as depicted in Figure 5.3 for a typical 2D truss. Without proper free body diagram, we will not be able to analyze and determine the unknown forces (like support reactions and member forces) acting on the structures. Also, improper schematic of the free body diagram will lead to erroneous structural analysis with faulty data.

5.4 SIGN CONVENTION

We will direct our attention toward the sign conventions that need to be considered while analyzing structures under equilibrium conditions. For support reactions under global equilibrium, we usually consider the vertical reaction force as positive, when it is directed toward the positive direction of y axis. Similarly, horizontal support reaction is taken as positive when it is directed toward the positive direction of x axis or along positive direction of z axis. Moments, on the other hand, are taken positive when they are directed in an anticlockwise direction. So, at the initial stage of analysis, we consider all unknown forces and moments positive and continue our analysis in this fashion. At the end, the algebraic signs of the calculated values indicate the proper directions of the support reactions and moments. A simple example is appreciated here to understand the concept. Let us consider a simple beam element under an axial pull P in the positive direction of x axis as shown in Figure 5.5 (a).

Under this loading, the free body diagram under global equilibrium condition may be drawn as shown in Figure 5.5 (b).

So as per the equilibrium condition along x axis, we can write:

$$R_x + P = 0$$

or,

$$R_x = -P$$

Since there is no external force acting on the beam in y direction, hence, there will be no support reaction in this direction. Applying equilibrium equation in y direction, we get simply $R_y = 0$.

From the above solution, we get the magnitude of horizontal support reaction as P and the algebraic sign in front of it indicates that the direction of the reaction force is contrary to what we assumed prior to the analysis. So, with proper sign convention and correct free body diagram, we will get the actual direction of reaction forces after carrying out the analysis using principles of global equilibrium conditions.

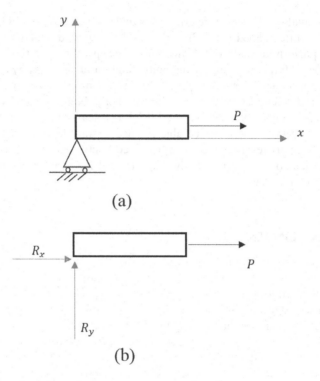

FIGURE 5.5 (a) Example problem of simple beam with horizontal load and (b) free body diagram of simple beam with axial load.

But one point is needed to take care of very seriously. We cannot change or mix our sign conventions while solving a problem. Whatever positive direction we have assumed initially, it should remain the same while solving the problem until its completion.

Now we will direct our attention toward the sign conventions for internal forces of a structure under local equilibrium conditions. So, our aim is to understand the origin of shear force and bending moments due to external loading acting on the structure. In the case of a beam under transverse loading, if we take any section at a certain distance from support then there will be internal forces and moments acting on that section, which try to balance the effects produced by transverse external loading. The resisting force induced at such arbitrary section is called shear force and the resisting moment generated at the same section is called the bending moment. The most commonly adopted positive directions of bending moment and shear force are shown in Figure 5.6 for ease of understanding.

While analyzing any structure to determine the shear force and bending moments, we will always use the diagram below or choosing the positive directions for shear force and moment. In the following section, we will use these sign conventions all along and we will get hands-on experience on its application.

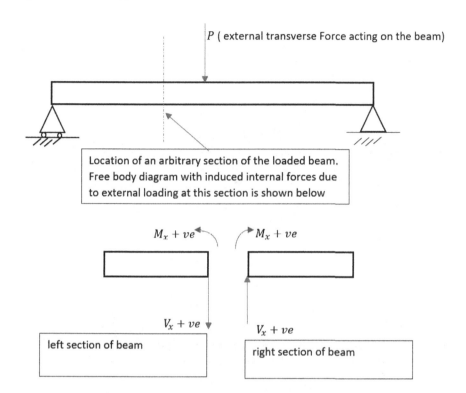

FIGURE 5.6 Simply supported beam under transverse point load.

5.5 EXTERNAL AND INTERNAL FORCES

In this section, we will study different types of external and internal forces acting on structural elements. By external force, we generally indicate the externally
applied loads acting on the structure as a whole or on a member. For example, in
the previous case of a 2D truss, we have seen that a point load is acting at node C
of the truss as a whole, and due to that load, we can calculate the support reactions
and member forces. So, the point load in this example is the external force and
forces generated on individual members due to this external force can be taken as
the internal forces.

But in any case, a structure needs to be in a state of equilibrium globally or
locally, as explained in the previous section. Normally for civil engineering structures, the external forces arises due to wind load, seismic load, impact load, self-
weight of the members comprising the structures, etc. Due to these loads, the internal
forces will be induced inside the structural members to maintain the equilibrium
condition without jeopardizing structural integrity. These induced forces are called
as internal reactions or member forces of the structures. To better understand the
effect of internal and external forces acting on structural elements, let us consider an
example. Let a simply supported beam be subjected to a uniformly distributed load

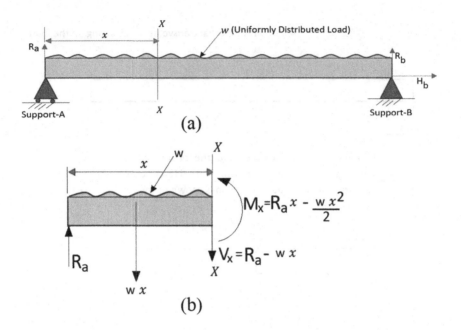

FIGURE 5.7 (a) Simply supported beam under uniformly distributed load throughout its length and (b) free body diagram of the beam up to section x–x from support A under flexural loading.

w as shown in Figure 5.7 (a). Under the application of this load, let us consider a section at a distance x from the left support 'A'.

Now, as the beam is subjected to a uniformly distributed load, the free body diagram of the beam up to section $X - X$ will be as depicted in Figure 5.7 (b).

So, we have drawn one vertical force V_x and moment M_x in the section shown in Figure 5.7 (b). These two quantities are new, and where do they come from? To understand this, let us first visualize the bending of a beam under any kind of flexural loading (a load is called flexural load when it is acting in a perpendicular direction to the longitudinal axis of a member, i.e., if the loading causes bending of a structure). So, under this loading, the beam will be deflected in the downward direction and will become slightly curved as shown in Figure 5.8 (a). The top fiber of the beam element will be in a state of compression, whereas the bottom fiber of the beam will be in a state of tension. If we consider a small element of beam under the externally applied load P as shown in Figure 5.8 (b), then that particular section tries to move the element toward its direction compared to the other parts of the beam, which are connected with supports. Let us consider a section at a distance x from the left support with separated elements to understand the concept.

However, the actual movement of beam element under section $p - q$ and $m - n$ is resisted by the other elements of the beam at its immediate vicinity. Now, in any case, the section should remain under equilibrium condition. Hence, to ensure that, there must be some forces that are needed to be developed at sections $p - q$ and $m - n$, which will counteract the external loading P applied on the element. This force that

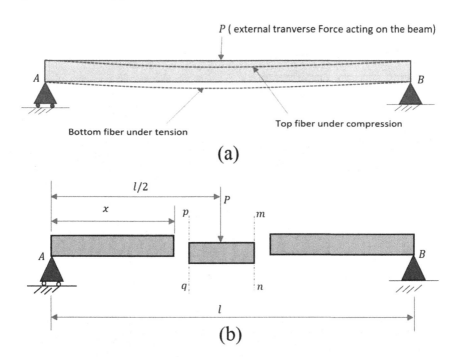

FIGURE 5.8 (a) Deflection diagram of beam under flexural loading and (b) tendency of a small element to move toward the applied loading on the beam.

is developed at these sections to counteract external load is called shear force. Now let us draw the force diagram to justify the concept and form the local equilibrium conditions, relating external load and shear force.

The positive direction of shear forces is maintained as per the sign convention discussed in the previous section.

Applying local equilibrium condition for section x for $0 \leq x \leq l/2$, we can write:

$$V_x - R_A = 0$$

or,

$$V_x = R_A$$

The free body diagram is shown in Figure 5.9 (a).

For the range $l/2 \leq x \leq l$, we need to consider Figure 5.9 (b) for the free body diagram to form the force equilibrium equation:

$$V_x - R_A + P = 0$$

or,

$$V_x = R_A - P$$

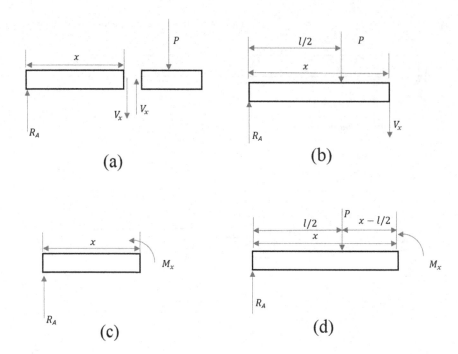

FIGURE 5.9 (a) Free body diagram of beam section with induced shear force at an arbitrary section $0 \le x \le l/2$, (b) free body diagram of beam section with induced shear force for $l/2 \le x \le l$, (c) free body diagram of beam for bending moment in $0 \le x \le l/2$, and (d) free body diagram of beam for bending moment in $l/2 \le x \le l$.

Since all the terms in the right-hand side of the above equation is known, hence, from this expression, we can calculate the value of shear force within the range $l/2 \le x \le l$. The free body diagram is shown in Figure 5.9 (b).

An interesting point is that shear force changes its sign at the point $x = l/2$, which is left as an exercise for the readers.

Now as discussed above for shear force, we can understand the formation of bending moments at the arbitrary section x from left support. To understand this concept, let us consider the Figure 5.9 (c) for the free body diagram with bending moments drawn at its positive orientation.

Now from local equilibrium condition, we can write for the range $0 \le x \le l/2$:

$$M_x - R_A x = 0$$

or,

$$M_x = R_A x$$

Since support reactions are already calculated from global equilibrium conditions at the beginning of the analysis, hence from the above expression, we can calculate the bending moment by varying x from 0 to $l/2$.

To get the bending moment for the range $l/2 \leq x \leq l$, we have to draw another section by increasing the range of x. We can do that by taking the section beyond the point $l/2$ and drawing the free body diagram as shown in Figure 5.9 (d). So, from this figure, we get after applying the local equilibrium condition, for the range $l/2 \leq x \leq l$:

$$M_x - R_A x + P\left(x - \frac{l}{2}\right) = 0$$

or,

$$M_x = R_A x - P\left(x - \frac{l}{2}\right)$$

So, from the above expression, we get the bending moments for rest of the portion of the beam under the applied loading. From these two equations, it can be seen that bending moment value is the same at a distance $l/2$ from the support, which is left as an exercise for the readers. It is to be noted that, though the shear force and bending moment are shown separately in Fig. 5.9 for the sake of understanding, they get induced simultaneously at a particular section.

The above concept of analysis can be applied to any type of lodging for beams by carefully drawing free body diagrams with forces and proper sign conventions. We will always get the above sets of algebraic equations, and solving the same, we not only get the magnitudes but also the direction of the shear and bending moments. A thorough understanding of the above method of analysis is the main building block to cover more advanced topics on this subject.

Now we have understood how the internal forces/internal reactions/member forces generate due to the application of external loads. The internal reactions or member forces can be of various types like axial force, bending moment, shear force, and torsional moment depending upon the type of members and nature of external loads in the structures. For example, if it is a member of a pin jointed structure or truss, the member force or the internal reaction will be generated solely as an axial force as these types of structures are loaded only at the joints. The member forces generated in a truss member is shown in Figure 5.10. In this figure, F_x and F_y are the external

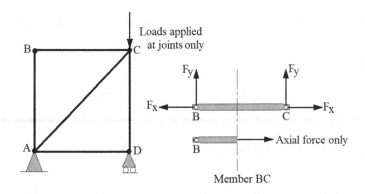

FIGURE 5.10 Member forces in pin jointed structures.

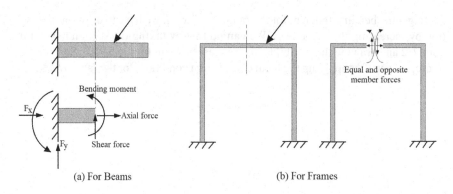

(a) For Beams   (b) For Frames

FIGURE 5.11 Member forces in rigid jointed structures for two dimensions: (a) for beams (b) for frames.

reaction forces at the pin joints. In case of rigid jointed structures, the generated member forces at a particular section (F_x, F_y, M_z) are shown in Figure 5.11 (a) and (b) for two dimensions. As the whole if the structure is in equilibrium, equal and opposite internal reactions or member forces generates at each cut section, as shown in Figure 5.11 (b).

5.6   TYPES OF SUPPORTS FOR STRUCTURES

Depending on the situation, a structure might have different support conditions. For example, beams and columns in a building frame may be fixed at the support or maybe simply supported. For civil engineering structures, the most common support types are listed in Table 5.1 with the respective support reaction forces. These support reaction forces are basically external reactions of the structures. A thorough understanding of these supports and support reactions is essential for analyzing different types of structures and their behavior under the application of external loading. Students are advised to go through and understand this table for all future analysis of structures that will be followed.

5.7   RELEASE OF INTERNAL REACTIONS OR MEMBER FORCES

Depending on the situation, it becomes necessary to remove or release any internal reaction forces or member forces such as axial force (AF), shear force (SF), bending moment (BM), or torsional moment (TM) from a specified location of a structural member. Some special types of joints are introduced in those particular locations of a structural member to release these internal reactions, as discussed below:

5.7.1   RELEASING BENDING MOMENT

To release the bending moment in a specified location, internal hinge is provided. The mechanism of introducing an internal hinge is shown in Figure 5.12 (a). By introducing an internal hinge, one can reduce bending moment at a particular section, or

TABLE 5.1

Different Types of Supports and Support Reactions

Type of Support	Symbolic Representation	Reactions in 2D	External Support Reactions in 2D and 3D
Fixed		R_x M_z R_y	In 2D Three (R_x, R_y, M_z) In 3D Six $(R_x, R_y, R_z, M_x, M_y, M_z)$
Hinged	or	R_x θ R_y or R_x R_y	In 2D Two (R_x, R_y) In 3D Three (R_x, R_y, R_z)
Roller	or θ	R or θ R	In 2D One $(R_y$ or $R)$ In 3D One $(R_y$ or $R)$
Rocker		R	In 2D One $(R_y$ or $R)$ In 3D One $(R_y$ or $R)$
Link	θ	R θ	In 2D One (R) In 3D One (R)
Horizontal guided roller		M_z R_y	In 2D Two (R_y, M_z) In 3D Two (R_y, M)
Vertical guided roller		R_x M_z	In 2D Two (R_y, M_z) In 3D Two (R, M)

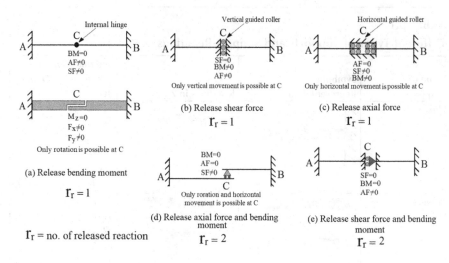

FIGURE 5.12 Release of internal reactions introducing special type of joints.

one can break the length of span of a very long beam so that the overall deflection and bending moment can be reduced. Suppose r_r is the number of released reaction for a particular special type of joint. So, for the first case of joint, i.e., the case shown in Figure 5.12 (a), only bending moment is released. Hence, $r_r = 1$. Now for 3D case, if internal hinge is provided, the number of released reaction would be 3, i.e., $R_x \neq 0$, $R_y \neq 0$, $R_z \neq 0$, $M_x = 0$, $M_y = 0$, $M_z = 0$. So if we can formulate the number of released reactions (r_r) in two and three dimensions, it will look like the following two equations, respectively.

$$r_r = \sum (m^* - 1); \text{ in 2D}$$

$$r_r = 3 \sum (m^* - 1); \text{ in 3D}$$

where $m^* =$ number of members connected to the internal hinge.

Example 5.1: Find the number of released reactions for the 2D framed structure as shown in Figure 5.13.

SOLUTION: Internal hinge is provided in joint D, H, and J in the above frame. At joint D, three members (*DA, DE, and DG*) are connected, at joint H four members (HE, HG, HI, and HK) are connected, and at joint J two members (JG and JK) are connected.

No. of released reactions are, $r_r = \sum (m^* - 1) = (3 - 1) + (4 - 1) + (2 - 1) = 6$.

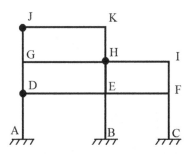

FIGURE 5.13   Example problem on released reaction.

5.7.2   RELEASING SHEAR FORCE

Vertical guided roller is introduced to release the internal shear force in a beam as shown in Figure 5.12 (b). Here bending moment and axial force remain as they are. As only the shear force becomes zero, here the number of released reaction, r_r is one.

5.7.3   RELEASING AXIAL FORCE

Horizontal guided roller is introduced to release the internal axial force in a beam as shown in Figure 5.12 (c). Here bending moment and shear force remain as they are. As only the axial force gets released, here the number of released reaction, r_r is one.

5.7.4   RELEASING AXIAL FORCE AND BENDING MOMENT

Horizontal roller is introduced to release the internal axial force and bending moment in a beam as shown in Figure 5.12 (d). Here the shear force remains as they are as the beam cannot move in vertical direction along position C. As the axial force and bending moment get released, here the number of released reaction, r_r is two.

5.7.5   RELEASING SHEAR FORCE AND BENDING MOMENT

Vertical roller is introduced to release the internal shear force and bending moment in a beam as shown in Figure 5.12 (e). Here axial force remains as they are as the beam cannot move in horizontal direction at location C. As the shear force and bending moment get released, here the number of released reaction, r_r is two.

6 Indeterminacy and Stability of Structure

6.1 INTRODUCTION

Indeterminacy and structural stability are discussed in detail with lots of solved example problems. Indeterminacy has been divided into two parts: static and kinematic indeterminacy. The principle of superposition has been discussed at the end. As can be seen by going through this book, structural analysis procedures vary completely depending on the degree of indeterminacy. This chapter is the cornerstone of structural analysis and a must-read for all whoever wants to master the theory and analysis of this subject.

6.2 STRUCTURAL INDETERMINACY

First, let us understand the indeterminacy or degree of indeterminacy. For that, let us consider two sets of algebraic equations as follow:

$$
\boxed{\begin{array}{c} Set-1 \\ 7x+5y=10 \\ 10x-y=20 \end{array}} \quad \text{and} \quad \boxed{\begin{array}{c} Set-2 \\ 7x+5y+8z=10 \\ 10x-y-4z=20 \end{array}}
$$

The first set of equations is solvable because the number of unknowns and equations are the same. From this set, the value of x and y comes out to be 1.93 and -0.702, respectively. As we can determine the values of unknowns from this set, this is a determinate set. But for the second set, the number of unknowns and equations is not the same. For this reason, the values of the unknowns cannot be determined from this set. This set is thus called an indeterminate set. The same concept is applicable in the structural analysis also. Here the unknown components are the forces in a given structure or member and available independent degrees of freedom of a particular structure. So, depending on the type of unknowns, the degree of structural indeterminacy can be classified into two types: (a) static indeterminacy (D_s) and (b) kinematic indeterminacy (D_k). The static indeterminacy is related to the unknown forces in a given structure, and kinematic indeterminacy is associated with the available independent unknown degrees of freedom of a particular structure.

6.3 STATIC INDETERMINACY AND STABILITY

When we observe most of the engineering structures around us, we may notice that multiple vertical members support one horizontal member. These different structural parts form a complex structural system separate from the simply supported beams and

DOI: 10.1201/9781003081227-8

structural elements. Statically indeterminate structures or redundant structures are those structures that cannot be analyzed with the help of equilibrium equations of statics only. Some additional compatibility equations are required to solve them entirely. There are a total of three static equations of equilibrium in two dimensions, i.e., $\sum F_x = 0$, $\sum F_y = 0$, and $\sum M_z = 0$ and in three dimensions, there are six, i.e., $\sum F_x = 0$, $\sum F_y = 0$, $\sum F_z = 0$, $\sum M_x = 0$, $\sum M_y = 0$, and $\sum M_z = 0$. Based on the type of reaction forces of a structure, static indeterminacy can be divided into two types: (a) external static indeterminacy (D_{se}) and (b) internal static indeterminacy (D_{si}). The external static indeterminacy is related to the unknown external support reactions, and the internal static indeterminacy is about unknown internal reactions or member forces. We can write:

$$D_s = D_{se} + D_{si}$$

or,

$$D_s = (r_e - \text{no. of equilibrium solutions})$$
$$+ \{3m - (3j + r_r - \text{no. of equilibrium solutions})\}$$

or,

$$D_s = (r_e - 3) + \{3m - (3j + r_r - 3)\}; \text{ in two-dimensional (2D)}$$

Similarly

$$D_s = (r_e - 6) + \{6m - (6j + r_r - 6)\}; \text{ in three-dimensional (3D)}$$

6.3.1 STATIC INDETERMINACY OF RIGID STRUCTURES

The structures, the members of which are connected to rigid or moment-resisting joints, are called rigid structures. For these types of structures, external loads can be applied both on the members and joints of the structure. In each member, there will be three internal reaction forces or member forces (F_x, F_y, and M_z) for a 2D case, but for three dimensions, in each member, there will be six internal member forces (F_x, F_y, F_z, M_x, M_y and M_z). These internal member forces are considered as unknowns. The other set of unknowns arrives from the external support reactions. The known quantities are the equations of equilibrium considered at each joint and the number of released reactions, as discussed in Chapter 5. Now the static indeterminacy in two dimensions can be written as:

$$D_s = No. \ of \ unknown - No. \ of \ known$$

or,

$$D_s = (r_e + 3m) - (3j + r_r); \text{ in two dimensions.}$$

and,

$$D_s = (r_e + 6m) - (6j + r_r); \text{ in three dimensions}$$

where r_e is the no. of external support reactions, m is the no. of members, j is the no. of joints, and r_r is the no. of released reactions in the structure.

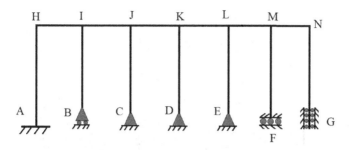

FIGURE 6.1 Example problem on the degree of static indeterminacy.

Example 6.1: Find the static indeterminacy (D_s) of the frame as shown in Figure 6.1 for the given support conditions.

SOLUTION: D_s = No. of unknowns – No. of knowns

$$D_s = (r_e + 3m) - (3j + r_r)$$

Finding no. of external reactions (r_e):

At support A = 3 (fixed support)

At support B = 1 (roller support)

At support C = 2 (hinged support)

At support D = 2 (hinged support)

At support E = 2 (hinged support)

At support F = 2 (horizontally guided roller)

At support G = 2 (vertically guided roller)

Total no. of external reactions (r_e) = $3+1+2+2+2+2+2 = 14$

No. of members (m) in the rigid frame = 13

Total no. of unknowns, $(r_e + 3m) = (14 + 3 \times 13) = 53$

Total no. of knowns, $(3j + r_r) = (3 \times 14 + 0) = 42$

Therefore, the degree of static indeterminacy for this 2D rigid frame, $D_s = 53 - 42 = 11$. That means, to analyze this frame fully, extra 11 equations are required.

Example 6.2: Find the static indeterminacy (D_s) of the 3D frame as shown in Figure 6.2 for the given support conditions.

SOLUTION: Total no. of unknowns, $(r_e + 6m) = (6 + 1 + 3 + 1 + 3 + 1) + (6 \times 13) = 93$

Total no. of knowns, $(6j + r_r) = (6 \times 12) + 0 = 72$

Therefore, the degree of static indeterminacy for this 3D rigid frame, $D_s = 93 - 72 = 21$. That means, to analyze this frame fully, extra 21 equations are required.

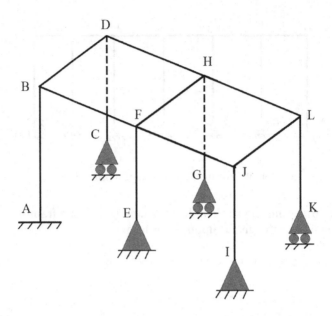

FIGURE 6.2   Example problem on the degree of static indeterminacy.

Example 6.3: Find the static indeterminacy (D_s) of the 2D frame as shown in Figure 6.3 for the given support conditions and internal hinges.

SOLUTION: Total no. of unknowns, $(r_e + 3m) = (3 + 2 + 2 + 2) + (3 \times 14) = 51$
Internal hinge is provided at joint I, F, and K.
No. of released reactions (r_r) due to the internal hinges $= \Sigma(m* - 1) = (2 - 1) + (4 - 1) + (3 - 1) = 6$; where $m* =$ number of members connected to the internal hinge.
Total no. of knowns, $(3j + r_r) = (3 \times 12) + 6 = 42$
Therefore, the degree of static indeterminacy for this 2D rigid frame, $D_s = 51 - 42 = 9$. That means, to analyze this frame fully, extra 9 equations are required.

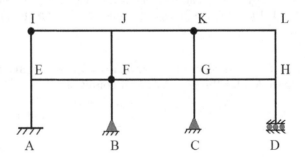

FIGURE 6.3   Example problem on the degree of static indeterminacy.

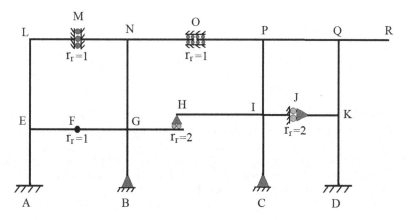

FIGURE 6.4 Example problem on the degree of static indeterminacy.

Example 6.4: Find the static indeterminacy (D_s) of the 2D frame as shown in Figure 6.4 for the given support conditions.

SOLUTION: Total no. of unknowns, $(r_e + 3m) = (3 + 2 + 2 + 3) + (3 \times 20) = 70$
Total no. of knowns, $(3j + r_r) = (3 \times 18) + 7 = 61$
Therefore, the degree of static indeterminacy for this 2D rigid frame, $D_s = 70 - 61 = 9$. That means, to analyze this frame fully, extra 9 equations are required.

Example 6.5: Find the static indeterminacy (D_s) of the 2D frame as shown in Figure 6.5 for the given support conditions.

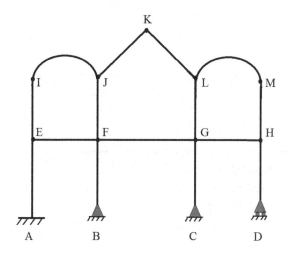

FIGURE 6.5 Example problem on the degree of static indeterminacy.

SOLUTION: Total no. of unknowns, $(r_e + 3m) = (3 + 2 + 2 + 1) + (3 \times 15) = 53$
Total no. of knowns, $(3j + r_r) = (3 \times 13) + 0 = 39$
Therefore, the degree of static indeterminacy for this 2D rigid frame, $D_s = 53 - 39 = 14$. That means, to analyze this frame fully, extra 14 equations are required.

Example 6.6: Find the static indeterminacy (D_s) of the frame as shown in Figure 6.6 for the given support conditions and internal hinges.

SOLUTION: Total no. of unknowns, $(r_e + 6m) = (6 + 3 + 3 + 1 + 1 + 1) + (6 \times 13) = 93$
Internal hinge is provided at joint B, J, and in-between member DH.
No. of released reactions (r_r) due to the internal hinges $= 3\Sigma(m* -1) = 3 \times \{(3 - 1) + (3 - 1) + (2 - 1)\} = 15$; where $m*$ is the number of members connected to the internal hinge.
Total no. of knowns, $(6j + r_r) = (6 \times 12) + 15 = 87$
Therefore, the degree of static indeterminacy for this 2D rigid frame, $D_s = 93 - 87 = 6$. That means, to analyze this frame fully, extra 6 equations are required.

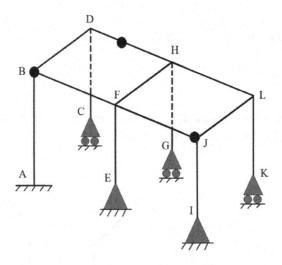

FIGURE 6.6   Example problem on the degree of static indeterminacy.

6.3.1.1   Shortcut Method for Determining Internal Static Indeterminacy of Rigid Structures

We have already found the values of external and internal static indeterminacies 2D and 3D rigid frames as follows:

$$D_{se} = (r_e - 3); \text{ in 2D}$$
$$D_{se} = (r_e - 6); \text{ in 3D}$$

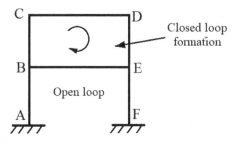

FIGURE 6.7   Closed-loop formations in rigid frames.

and,

$$D_{si} = \left\{ 3m - \left(3j + r_r - 3 \right) \right\}; \text{ in 2D}$$

$$D_{si} = \left\{ 6m - \left(6j + r_r - 6 \right) \right\}; \text{ in 3D}$$

From the abovementioned equations, we can see that finding external static indeterminacy is quite simple and straightforward. But finding the internal static indeterminacy is quite hectic. Chances of committing mistakes are quite high if sufficient attention is not paid. For this reason, an alternative shortcut method is developed to find the internal static indeterminacy of rigid frames. By this method, one can determine the internal static indeterminacy simply looking at the given frame. The internal static indeterminacy can be obtained quite easily by the following equations:

$$D_{si} = 3c - r_r; \text{ for 2D frames}$$

$$D_{si} = 6c - r_r; \text{ for 3D frames}$$

where c is the number of closed loops formed in the given frame. To understand the concept clearly, let us consider the rigid frame as given in Figure 6.7. Here one closed loop is formed by the members BC, CD, DE, and EB. So, internal static indeterminacy according to the abovementioned equation, $D_{si} = 3 \times 1 - 0 = 3$. Readers are encouraged to cross check the solution of the example problems 6.2–6.6 by this shortcut method again. Closed loop exists only in frames, not in beams or trusses.

Example 6.7: Find the static indeterminacy (D_s) of the frame as shown in Figure 6.1 for the given support conditions and internal hinges by the technique of closed-loop formation.

SOLUTION: $D_{se} = 14 - 3 = 11$; $D_{si} = 3c - r_r = (3 \times 0) - 0 = 0$; as there is no closed-loop formation ($c = 0$) in this problem. Therefore, $D_s = 11 + 0 = 11$. As we can see, it matches the earlier solution.

Example 6.8: Find the static indeterminacy (D_s) of the frames as shown in Figure 6.8 for the given support conditions and internal hinges by the technique of closed-loop formation.

SOLUTION: Case: a

$$D_{se} = (r_e - 3) = 3 - 3 = 0; \; D_{si} = 3c - r_r = 3 \times 0 - 1 = -1$$

$$D_s = D_{se} + D_{si} = 0 - 1 = -1$$

The frame is statically determinate but unstable. When D_s is negative. The given structure becomes unstable. We will learn about unstable structures in this chapter shortly.

Case: b

$$D_{se} = (r_e - 3) = 4 - 3 = 1; D_{si} = 3c - r_r = 3 \times 2 - 1$$
$$= 5 \, (\text{here two closed} - \text{loop forms}).$$

$$D_s = D_{se} + D_{si} = 1 + 5 = 6$$

This frame is statically indeterminate to the 6th degree.

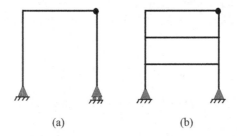

(a)   (b)

FIGURE 6.8 Example problem on the degree of static indeterminacy by closed-loop formations technique.

6.3.1.2 Degree of Statical Indeterminacy When Load is Applied to the Structure

External loads are imposed on structures. Depending on the loading magnitude and direction, the reaction forces are generated accordingly. In Figure 6.9, the nature of support reactions is shown depending on the external load for a fixed beam. In the first case, i.e., Figure 6.9 (a), we need to consider three equilibrium conditions ($\Sigma F_x = 0$, $\Sigma F_y = 0$, and $\Sigma M_z = 0$) as the inclined load is applied that produces two components during load application. But for the second case, i.e., Figure 6.9 (b), as only a vertical load is applied, there is no need to consider $\Sigma F_x = 0$. As a result, for

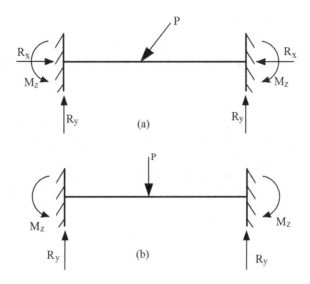

FIGURE 6.9 Nature of support reaction depending on external loads.

the second case, the effective number of equilibrium equations reduces down to two $(\Sigma F_y = 0 \text{ and } \Sigma M_z = 0)$.

Example 6.9: Find the static indeterminacy (D_s) of the fixed beam as shown in Figure 6.10 for the given load conditions.

SOLUTION:

$$D_{se} = (r_e - 2) = (5 - 2) = 3$$

$$D_{si} = \{3m - (3j + r_r - 3)\} = \{3 \times 2 - (3 \times 3 + 0 - 3)\} = 0$$

$$D_s = 3 + 0 = 3$$

In this problem, two support reactions were considered at each fixed end, and two equations of equilibrium were considered.

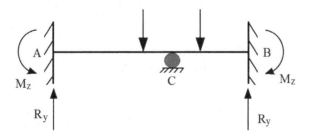

FIGURE 6.10 Example problem on finding static indeterminacy when external loads are applied.

6.3.2 STATIC INDETERMINACY OF PIN-JOINTED STRUCTURES

Pin-jointed structures or trusses are constructed by joining multiple members by frictionless pin joints. Here the structural loads are only allowed to apply to the joints of the trusses. As the members are connected through pin joints, no reactive moment is allowed to accumulate in these structures, i.e., the bending moments at joints turn out to be zero. That is why for pin-jointed structures, the moment equilibrium becomes unnecessary, and thus we need to consider the remaining two equilibrium conditions in each joint for the 2D case. For the 3D case, we need to consider three equilibrium conditions per joint.

Moreover, as the external loads are only allowed to apply to the joints, only axial force generates as the member force. We will study trusses in more detail in Chapter 7. However, in this chapter, we will focus on its degree of static indeterminacy only.

So, for the pin-jointed structures, the member forces (m) and external support reactions (r_e) are considered as unknowns, and the number of equilibrium equations at each joint (j) is the knowns.

Therefore,

$$D_s = No.\ of\ unknown - No.\ of\ known$$

$$D_s = (m + r_e) - 2j;\ \text{for 2D}$$

$$D_s = (m + r_e) - 3j;\ \text{for 3D}$$

Again, we can write, $D_s = D_{se} + D_{si}$; where external static indeterminacy,

$$D_{se} = (r_e - 3);\ \text{for 2D}$$

$$D_{se} = (r_e - 6);\ \text{for 3D}$$

and internal static indeterminacy,

$$D_{si} = m - (2j - 3);\ \text{for 2D}$$

$$D_{si} = m - (3j - 6);\ \text{for 3D}$$

If we add these D_{se} and D_{si}, we will obtain the value of the D_s given above for 2D and 3D cases.

Example 6.10: Find the degree of static indeterminacy for the following pin-jointed structure as shown in Figure 6.11.

SOLUTION:

$$D_s = (m + r_e) - 2j = (9 + 3) - 2 \times 6 = 0$$

The degree of static indeterminacy of this pin-jointed structure is 0 or the structure is statically determinate.

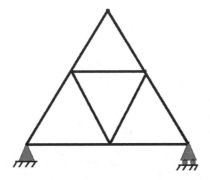

FIGURE 6.11 Example problem on the degree of static indeterminacy of pin-jointed structures.

6.3.2.1 Shortcut Method for Determining Internal Static Indeterminacy of Pin-Jointed Structures

Like the shortcut method to determine the internal static indeterminacy of the rigid structures, we can find the same here also. Only visual inspection is required to implement this method. The internal static indeterminacy of a pin-jointed truss depends on the triangle formation in its geometry. For example, consider Figure 6.12, where the degree of internal indeterminacy is found based on its geometric shape, i.e., triangle formations. Consider Figure 6.12 (a); here, the entire pin-jointed truss consisting of two nonoverlapping triangles. We cannot make any more nonoverlapping triangles from this geometry. So, in this case, the internal static indeterminacy is zero. In Figure 6.12 (b), we can observe one overlapping triangle is there. We can eliminate this triangle just by removing one member from this truss. So, this

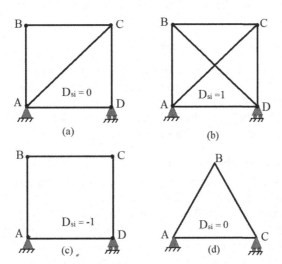

FIGURE 6.12 Finding internal static indeterminacy of pin-jointed structures by shortcut method.

structure is internally indeterminate to degree one. In Figure 6.12 (c), there is no triangle formed. One additional member is required to make two nonoverlapping triangles from this geometry. So this structure has internal static indeterminacy of degree −1. That means this geometry is internally unstable, or it will form a mechanism. The truss shown in Figure 6.12 (d) is also internally statically determinate.

Example 6.11: Find the degree of static indeterminacy for the following pin-jointed structure as shown in Figure 6.13.

SOLUTION: $D_{se} = (r_e - 3) = 3 - 3 = 0$; $D_{si} = -2$; here there is a deficiency of two members as shown in dotted lines in Figure 6.13.
$D_s = 0 - 2 = -2$; this structure is statically unstable. If we observe the pin-jointed structure carefully, we can also find that it will form mechanism.

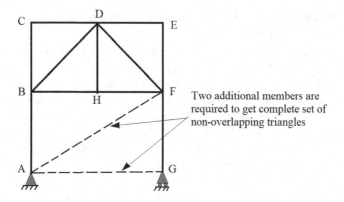

Two additional members are required to get complete set of non-overlapping triangles

FIGURE 6.13   Example problem on degree of static indeterminacy for pin-jointed structure.

6.3.3   EXTERNAL AND INTERNAL STABILITY OF STRUCTURES

A structure should be stable enough to serve its intended purpose i.e., it should not undergo rigid body motion under the influence of external loads, or it should not form any mechanism. Stability can be of two types: (a) external stability and (b) internal stability. External stability is related to the external support conditions and internal stability is related to the shape of the structure.

6.3.3.1   External Stability of Structures

Let us take the example of a simply supported beam. For this beam, the total unknown support reactions or force of constraints are three. Since this is a 2D structure, we have three equations of equilibrium at our disposal. Namely, $\Sigma F_x = 0$, $\Sigma F_y = 0$, $\Sigma M_z = 0$. So, the difference between unknown reactions and available equations is $3 - 3 = 0$. Hence, the structure is statically determinate.

   In the case of a propped cantilever, it has three unknown support reactions at the fixed and one unknown support reaction at the pinned roller end. Hence, the total unknown support reaction is $3 + 1 = 4$ and available statical equilibrium equation is 3. Hence, the difference between them is $4 - 3 = 1$. This means that the structure is statically indeterminate and there will be an additional equation that needs to be formed to completely analyze the structure. Indeterminacy related to support reactions only is called external indeterminacy. The examples we have provided above are all related to external indeterminacy. If external indeterminacy is positive, the structure is externally stable and if it is negative, the structure is externally unstable. Let us consider a beam supported at both ends by rollers. In this case, the only constraint force or unknown support reactions are vertical reactions at support. Now the available statical equilibrium, equations are three for 2D structures. Hence, the degree of indeterminacy in this case is $2 - 3 = -1 < 0$. Since the structure is supported on two rollers, hence it will slide along horizontal direction freely and is unstable. Hence, as we have stated, in the negative degree of indeterminacy structure becomes unstable. From a practical point of view, externally unstable structures need to be avoided. Hence, knowledge of external determinacy and its relation to the stability of the structure is very much important for all. Structural engineers and others too should have a clear idea about these salient features of structural analysis prior to start a career in structural engineering. For external stability of trusses, all support reaction components should not be parallel to each other, and all support reactions should not be concurrent, i.e., passing through the same point. See Figure 6.14 for stable and unstable trusses as an example of external instability.

In Figure 6.14 (a), the truss is supported on the roller at both ends; hence, only unknown support reactions are vertical ones as shown. As both these reactions are parallel to each other, hence, the structure is unstable. For the truss in Figure 6.14 (b), the support reactions are shown with arrows and not all of them are parallel to each other. Hence, the truss is a stable one.

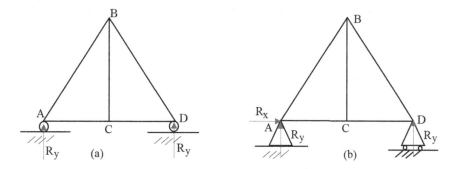

FIGURE 6.14   Example of (a) externally unstable truss (b) externally stable truss.

6.3.3.2   Internal Stability of Structures

A structure is internally stable, if it maintains its shape and remains a rigid body even after detaching from its supports. Conversely, a structure is internally unstable or nonrigid, if it undergoes large displacements under small disturbances when not

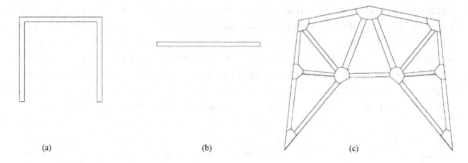

FIGURE 6.15 Examples of internally stable structures.

supported externally. In the second case, the structure forms mechanism under the influence of small disturbance. Some examples of internally stable and unstable structures are shown in Figures 6.15 and 6.16, respectively.

Note that in Figure 6.15, each of the structures maintains its original shape even after removing from their supports and moves as a rigid body. But the structures in Figure 6.16 are consisting of two rigid parts '*AB*' and '*BC*' that are connected by a hinge at location '*B*'. If removed from the supports, these two rigid parts will start rotating relative to each other about the hinge '*B*' under the influence of a small disturbance and thus forms mechanism. The second set of structures are internally unstable or nonrigid in nature. In reality, every engineering structure undergoes deformation when external loads are applied. But for rigid structures, these deformations are so small that those can be neglected. Rigid structures offer greater resistance against their shape changes whereas the nonrigid structures offer negligible resistance and deform easily when the external supports are removed. They can even collapse under their own weight when in unsupported condition.

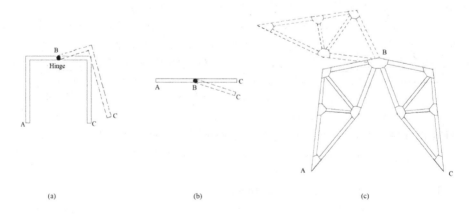

FIGURE 6.16 Examples of internally unstable structures.

6.4 KINEMATIC INDETERMINACY OF STRUCTURES

When loads are applied, a structure undergoes deformation. The kinematic indeterminacy or degrees of freedom of a structure, in general, are defined as the independent joint displacements (translations or rotations) necessary to specify the deformed shape of the structure when subjected to an arbitrary loading. In another way, we can say kinematic indeterminacy is the total number of unrestrained displacement (translations or rotations) components at the joint of that structure. If the displacement component of the joint cannot be determined by the compatibility (related to the shape of the structure) equation, then the structure is called kinematically indeterminate.

In Figure 6.17, the deformed and undeformed shapes of a pin-jointed structure are shown in two dimensions under the influence of external loads. The joint A is hinged supported and joint B is on roller. Joint A cannot move in any direction and joint B can only slide along X-direction. The other joints (C and D) of the truss are free and can move in both directions. After deformation took place, suppose joint B moves d_1 distance along X-direction. Likewise, joint C moves d_2, and d_3 distance along X- and Y-directions, respectively, and joint D moves d_4, and d_5 distance along X- and

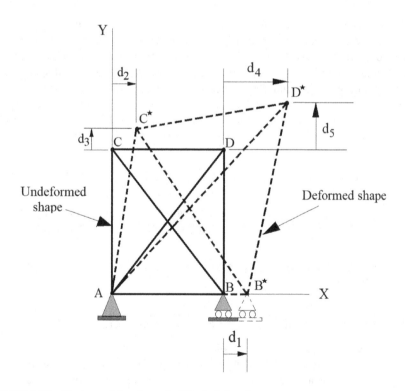

FIGURE 6.17 Degrees of freedom of a 2D pin-jointed structure.

TABLE 6.1

Types of Degrees of Freedom as Per Joint Condition

Types of Joint	No. of Degrees of Freedom	Types of Degrees of Freedom
2D truss joint	2	$\Delta x, \Delta y$
3D truss joint	3	$\Delta x, \Delta y, \Delta z$
2D beam/frame joint	3	$\Delta x, \Delta y, \theta_z$
3D beam/frame joint	6	$\Delta x, \Delta y, \Delta z, \theta_x, \theta_y, \theta_z$

Y-directions, respectively. So, as per the definition above, the degrees of freedom of this pin-jointed structure can be written as follows:

$$d = \begin{Bmatrix} d_1 \\ d_2 \\ d_3 \\ d_4 \\ d_5 \end{Bmatrix}$$

In the abovementioned example, we can observe that this pin-jointed structure has a total of five unrestrained displacement components at the joints. The two supports provided the restraints at A and B. If no supports were provided, then the number of joint displacements for this unsupported structure would be eight (two displacements per joint). But we have found the number of degrees of freedom was five. So, we can understand, a total of three displacement components were stopped by the supports provided. Now, we can quickly formulate the number of degrees of freedom of a structure as next:

$$\begin{Bmatrix} Number\ of\ degrees \\ of\ freedom\ or\ degrees\ of \\ kinematic\ indeterminacy \end{Bmatrix} = \begin{Bmatrix} Number\ of\ joint \\ displacements\ by \\ the\ unsupported \\ structure \end{Bmatrix} - \begin{Bmatrix} Number\ of\ joint \\ displacements \\ restrained\ by \\ supports \end{Bmatrix}$$

Depending on the joint types, the number of degrees of freedom per joint can be shown in Table 6.1.

6.4.1 Kinematic Indeterminacy of Pin-Jointed Structures or Truss

We have seen from Table 6.1 that a joint of a 2D truss has two degrees of freedom. So, in unsupported conditions, the total degrees of freedom of a 2D truss, as shown in Figure 6.18 (a), would be $2 \times 3 = 6$. But this truss is not externally stable as it will undergo a rigid body motion under the influence of external loads. To prevent this, we need to restrain the truss in some joints with supports, as shown in Figure 6.18 (b).

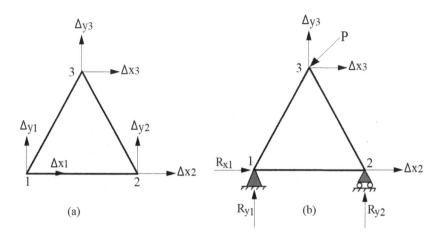

FIGURE 6.18 (a) Unsupported degrees of freedom and (b) supported degrees of freedom and support reactions of a 2D truss.

These supports will stop the movement of those joints in some specified directions and thereby generate the reaction forces.

So as per the earlier discussion, we can also find the number of degrees of freedom or degree of kinematic indeterminacy (D_k) as:

$$D_k = 2j - r_e; \text{ for 2D}$$

$$D_k = 3j - r_e; \text{ for 3D}$$

where j is the no. of joints of the pin-jointed structure, r_e is the no. of restrained/support reactions.

6.4.2 KINEMATIC INDETERMINACY OF RIGID-JOINTED STRUCTURES

Beams and frames are so-called rigid-jointed structures. In beam and frame members, the external load can be applied anywhere, and the joint rigidity can be of various types. Members can undergo bending due to the presence of transverse loads. That is why, additionally, reacting moments can be generated at the restrained joints depending on the support conditions and applied loads.

We have seen from Table 6.1 that a joint of a 2D beam/frame has three degrees of freedom. So, in unsupported conditions, the total degrees of freedom of a 2D beam/frame, as shown in Figure 6.19 (a), would be $3 \times 2 = 6$. But this beam/frame is not externally stable as it will undergo a rigid body motion under the influence of external loads. To prevent this, we need to restrain the member with joint supports, as shown in Figure 6.19 (b). These supports will stop the movement of those joints in some specified directions and thereby generate the reaction forces in those directions.

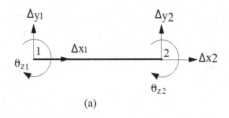

(a)

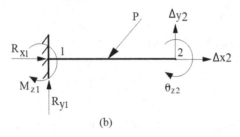

(b)

FIGURE 6.19 (a) Unsupported degrees of freedom and (b) supported degrees of freedom and support reactions of a 2D rigid-jointed member.

So as per the earlier discussion, we can also find the number of degrees of freedom or degree of kinematic indeterminacy (D_k) for a rigid-jointed structure as:

$$D_k = 3j - r_e; \text{ for 2D}$$

$$D_k = 6j - r_e; \text{ for 3D}$$

where j is the no. of joints of the pin-jointed structure, r_e is the no. of restrained/ support reactions.

The abovementioned two equations are true when the members are extensible or compressible axially. But if the members are inextensible or axially rigid, the equations can be rewritten as:

$$D_k = 3j - r_e - m; \text{ for 2D}$$

$$D_k = 6j - r_e - m; \text{ for 3D}$$

where m is the no. of inextensible members.

This can be understood by looking at Figure 6.20. In Figure 6.20 (a), we can see an extensible member whereas in Figure 6.20 (b), we can see an axially rigid member under the external load P. In the first case, the beam will undergo a horizontal deformation of $\Delta x2$ at joint 2, whereas in the second case it will not.

When the members of a frame are inextensible or rigid, then rigid body displacement should be considered combinedly for multiple members. Consider the following example problem to clarify this.

If there are released reactions, then the number of degrees of freedom will be increased as per the release arrangements. For example, if internal hinge is provided,

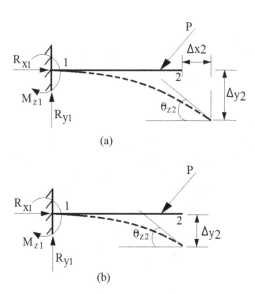

FIGURE 6.20   Degrees of freedom of (a) an extensible (b) inextensible rigid-jointed member in 2D.

the bending moment gets released over there. So, in this case, the number of released reaction (r_r) would be 1. Now, finally, we can add the influence of the released reactions in the abovementioned formulation and get the final form:

$$D_k = 3j - r_e - m + r_r; \text{ for 2D}$$

$$D_k = 6j - r_e - m + r_r; \text{ for 3D}$$

Note: If you want to calculate the degrees of kinematic indeterminacy (D_k) using the above formula, you must consider the location of the released reactions as a separate joint. Then only the correct answer will come.

Example 6.12: Find the degree of kinematic indeterminacy for the following frame as shown in Figure 6.21(a). Take all the members to be axially rigid.

SOLUTION: If the frame members were extensible/compressible, then the degrees of freedom would be as per Table 6.2, i.e., total 6 degrees of freedom is possible if the members are extensible/compressible.

As per the equation given previously, the degree of kinematic indeterminacy of this frame (for members are axially rigid).

$$D_k = 3j - r_e - m$$

$$D_k = 3 \times 3 - 3 - 2 = 4$$

TABLE 6.2

Degrees of Freedom If the Members Are Extensible/Compressible

Joint-B	Joint-C
∇x_B	∇x_C
∇y_B	∇y_C
θ_{ZB}	θ_{ZC}

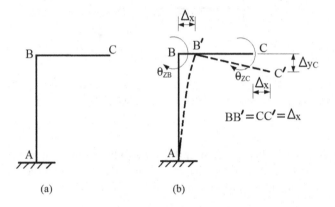

(a)   (b)

FIGURE 6.21 Example problem on degree of kinematic indeterminacy for axially rigid frame.

Here we can see the degrees of freedom gets reduced by 2 when the members are axially rigid. This is because, ∇y_B will not be there anymore as the column AB cannot get shortened or elongated and for beam BC instead of two axial deformations (i.e., ∇x_B and ∇x_C) at its ends, the whole assembly will move horizontally to the ∇x amount as shown in Figure 6.21 (b).

Example 6.13: Find the degree of kinematic indeterminacy for the frame shown in Figure 6.1.

SOLUTION:

$$D_k = 3j - r_e - m + r_r$$

$$D_k = 3 \times 14 - 14 - 0 + 0 = 28$$

As nothing has been mentioned in the problem regarding the axial stiffness of the members, we will consider the members as extensible/compressible.

Therefore, the degree of kinematic indeterminacy = 28.

Example 6.14: Find the degree of kinematic indeterminacy for the frame shown in Figure 6.3.

SOLUTION:

$$D_k = 3j - r_e - m + r_r$$

$$r_r = \sum (m^* - 1) = (4-1) + (3-1) + (2-1) = 6$$

Therefore, $D_k = 3 \times 12 - 9 - 0 + 6 = 33$

As nothing has been mentioned in the problem regarding the axial stiffness of the members, we will consider the members as extensible/compressible.

Therefore, the degree of kinematic indeterminacy = 33.

Example 6.15: Find the degree of kinematic indeterminacy for the frame shown in Figure 6.4.

SOLUTION:

$$D_k = 3j - r_e - m + r_r = 3 \times 18 - 10 - 0 + 7 = 51$$

As nothing has been mentioned in the problem regarding the axial stiffness of the members, we will consider the members as extensible/compressible.

Therefore, the degree of kinematic indeterminacy = 33.

Note that, the location of the released reactions was considered as separate joints.

6.4.3 SUMMARY OF ALL FORMULATIONS FOR STATIC AND KINEMATIC INDETERMINACY

Finally, we are putting all the formulations for static and kinematic indeterminacy shown in this chapter in a tabular form (Table 6.3) as shown next. All the notations mentioned in the equations are already explained in this chapter.

TABLE 6.3
Summary of All Formulations for Static and Kinematic Indeterminacy

Type of Structure		Static Indeterminacy (D_s)		Kinematic Indeterminacy (D_k)
		D_{se}	D_{si}	
Truss	2D	$r_e - 3$	$m - (2j - 3)$	$(2j - r_e)$
	3D	$r_e - 6$	$m - (3j - 6)$	$(3j - r_e)$
Beams/Frames	2D	$r_e - 3$	$(3c - r_r)$ or $\left\{3m - (3j + r_r - 3)\right\}$ for beams, $c = 0$	$3j - r_e - m + r_r$
	3D	$r_e - 6$	$(6c - r_r)$ or $\left\{6m - (6j + r_r - 6)\right\}$ for beams, $c = 0$	$6j - r_e - m + r_r$

6.5 PRINCIPLE OF SUPERPOSITION

The principle of superposition is the most important axiom or method or statement/concept for theory of structural analysis. It may be introduced as follows: the total

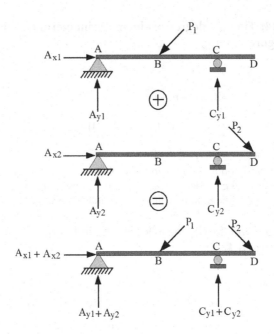

FIGURE 6.22   Principle of superposition.

displacement or internal stress, at a point/location in a structure acted upon by several external loadings, can be calculated by adding together the displacements or internal stress generated by each of the external loads acting separately at the same location as that of original structure. For this statement or axiom to be valid, it is important to understand that a linear relationship exists among the loads, stresses, and displacements. Thus, to apply the principle of superposition, two points need to be satisfied by the structure as described next:

1. Material, which the structural member is made up of, should behave in a linear-elastic manner, so that Hooke's law is valid, and therefore the load will be proportional to displacement. In other words, material should be elastic in nature as per the stress-strain curve explained at the beginning of this section.
2. The geometry and shape of the structural elements must not undergo significant change when the loads are applied, i.e., plane section should remain plane before and after loading. Large displacements will completely change the position and orientation of the loads.

Throughout this book, these two requirements will be found to be satisfied in all cases. Hence, we may not declare it at all places wherever we will apply this principle. Only at the end of this book, in the 'Plastic Analysis' section, we will introduce another concept that will not follow this principle and a separate plastic analysis principle will be introduced to that section under that particular different concept. In Figure 6.22, the principle of superposition has been explained.

7 Plane Trusses and Space Trusses

7.1 INTRODUCTION

In this chapter, we will study the detailed analysis procedures for plane and space trusses. Truss is an important type of structural element that has multiple applications for constructing large load-carrying structures like bridges, roofs, and buildings. A sound understanding of the truss system and its analysis procedure is a must-have for all working engineers and researchers. So, students are advised to go through this chapter in detail to gain a good understanding of various analysis procedures for the plane as well as space trusses. We will explore different methods for analyzing the trusses, and by analysis, we actually mean to be able to calculate all external and internal unknown forces acting on the truss. Analysis is the first step before carrying out detailed design of structural members. So, proper analysis of truss with all forces acting internally and externally will help us to do the design work efficiently and correctly.

7.2 COMMON TYPES OF TRUSSES

A framework consisting of members joined at their end by flexible connections to form a rigid structure is called a truss. Structural elements, such as I-beams, angles, channels, and square hollow sections connected at their ends by welding, rivets, or bolts, form a complete truss for a particular application. When the truss members are all lying in the same plane, it is called a plane truss. The combination of two or more plane trusses in different planes is said to form space trusses. There are many types of plane trusses used in bridges, roofs, and highway over bridges. The most commonly used types of plane trusses are shown in Figure 7.1 for reference.

7.3 CLASSIFICATION OF COPLANAR TRUSSES

The basic element of any truss is the triangular arrangement of elements as shown in Figure 7.2 (a). Three bars connected at the ends with pins, rivets, bolts, or welding, form a rigid or internally stable triangular element. The triangular element is important to notice since by connecting in this fashion, the element becomes rigid, and there will be no relative movement among the members. This is not applicable for four or more members connected at ends to form a quadrilateral or polygonal element as shown in Figure 7.2 (b). Such an element is free to move at the joints, and it will form a mechanism. To make it stable, we can connect any one of its diagonals and, thus, forming two triangular elements. So, for trusses, triangular elements always play a key role in structural stability.

DOI: 10.1201/9781003081227-9

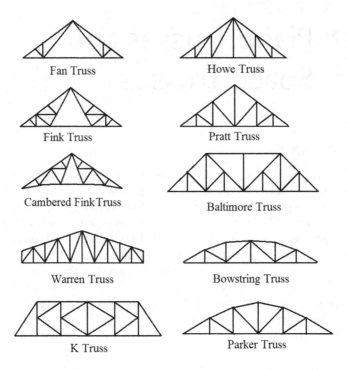

FIGURE 7.1 Different types of trusses.

Toward this end, we can say that trusses can be of three types such as simple truss, compound truss, and complex truss. A simple truss is the one made up of a single triangle formed by three connected members and, thereafter, enlarging its basic truss element depending on the span required as shown in Figure 7.3. However, the final form and overall dimension of the truss will be calculated based on the actual site requirements. The basic truss unit contains three members. Now if we want a simple truss for a longer span, every time we need to add two new members with

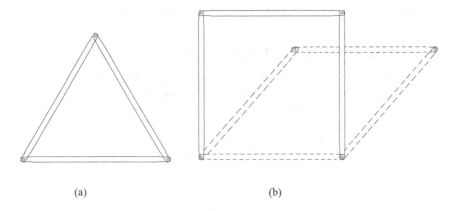

  (a)   (b)

FIGURE 7.2 Fundamental triangular element for plane truss.

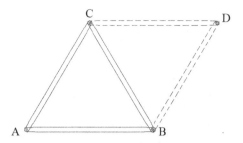

FIGURE 7.3 Every time addition of two new members and a new joint in simple truss extension.

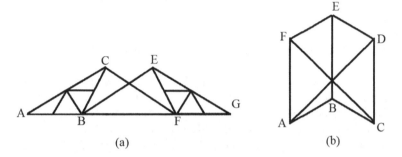

FIGURE 7.4   (a) compound truss and (b) complex truss.

a new joint from the basic truss unit. If we formulate the number of members (m) present for a simple truss, it will take this form: $m = 3 + 2 \times (j - 3) = 2j - 3$; j being the no. of joints.

A compound truss can be formed by combining two or more triangular simple trusses together to form a single rigid body. See Figure 7.4 (a) for better understanding of compound truss. Here two simple trusses *ABC* and *EFG* are connected by three members BE, CF, and BF. Note that these three members are nonparallel and nonconcurrent to each other, which makes the whole truss system internally stable or rigid.

A complex truss is the one that cannot be classified as simple or compound. A typical complex truss is shown in Figure 7.4 (b) for reference.

7.4   ASSUMPTIONS ON ANALYSIS OF TRUSSES

Elements of trusses are all axially loaded members. By this, what we mean, for trusses, there will be no induced moment due to external loading on the members forming the truss. So, the members are said to be in a state of tension or compression only. Even under certain situations, there may be some members in which no force is being induced due to external loading. Calculation of forces of various members of the truss is carried out by applying the local equilibrium conditions to those members. And the support reactions acting on the overall truss will be calculated based on global equilibrium conditions. There are several methods for analyzing plane truss that will be developed in the subsequent sections of this chapter. The important assumptions that help us to analyze trusses are as follows:

1. External loads act on joints only. If there is any load acting in between nodes, then we need to distribute it equally among two adjacent joints.
2. All members are subjected to axial loads only. There will be no moment acting on the members due to any kind of loading on the truss.
3. The connections between different members are perfectly hinged/pinned through frictionless connections.
4. Self-weight of the truss is either ignored or assumed to be equally distributed among its various members.
5. Even if members are connected by welding, bolt, etc., a nominal moment that can be generated due to imperfect fixity (other than frictionless hinge) is ignored.
6. No matter how many members are connected by gusset plates at a node, if axis of all members is passing through the same point at the joint, then the members are assumed to be hinged/pin connected. Hence, no moment will come into play at this joint. For better understanding of this particular point, please see Figure 7.5. As can be seen, the axis of all members is passing through the same point at the support, and hence, it is a pin/hinged connection without any moment induced at the members.
7. The vector representing a force acting on a joint or a section is drawn on the same side of the joint or section. Since truss members are a two-force system (tension or compression-only), before analysis, we assume the member is in tension and draw a force vector along the axis of the member at the same joint where the member is located. After carrying out the detailed analysis, we get the algebraic sign and value for the acting force on that particular member, and then we get the perfect knowledge about the tension/compression state of the member.
8. Force on the member acting towards the joint or section is taken as compression, whereas force away from the joint or section is taken as tension. See the Figure 7.6 for better understanding.

We have presented the assumptions of truss analysis in the above paragraph without any example. As we will go through the examples, we will get a clear idea how these

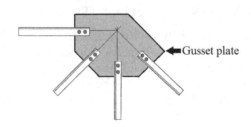

FIGURE 7.5   A typical truss member connection using gusset plates.

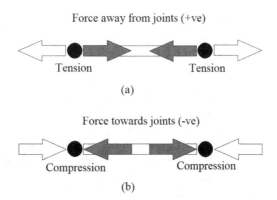

FIGURE 7.6   Tension/compression convention for truss members.

assumptions help us analyze any particular truss correctly and efficiently. Provision for expansion and contraction arising from temperature changes and for deformation/ deflections caused by the applied loads is usually made at one of the supports for big trusses. Roller, rocker supports, or some kind of slip joint are provided to manage such effects.

7.5   ARRANGEMENT OF MEMBERS OF COPLANAR TRUSSES – INTERNAL STABILITY

We have already explained the different types of trusses used to model our actual site requirement. Coplanar trusses are the simplest models that provide us easy analysis and design procedures. Moreover, noncoplanar three-dimensional (3D) trusses are made by combining several coplanar trusses with tie members at the top and bottom of the structure. Various members of coplanar trusses are arranged in a way so that from any side, each is seen to be composed of triangular shapes. The three-member triangular shapes provide the best to form a rigid structure to construct the overall truss. Thus, for stability, we need to connect various triangular elements of the truss in a fashion so that there are no nonconnected nodes available in the truss. If there is any nonconnected node, the force distribution among various members will be uneven, and ultimately it will cause the structure to collapse. However, it may be understood that we may mistakenly provide more internal members than that are required for the overall structural stability of the truss. In such a case, when there are more members than needed for structural stability, they become statically indeterminate truss. We cannot analyze such type of truss completely by applying only equilibrium equations. In the following sections, we will discuss the conditions and equation to determine whether any given truss is statically determinate or indeterminate. The term rigid is applied in the sense of being non-collapsible and in the sense that deformation/ deflections of the members due to induced internal forces and strains are minuscule to be neglected.

7.6 STATIC DETERMINACY, INDETERMINACY, AND INSTABILITY OF COPLANAR TRUSSES AND THEIR SOLUTION METHODS

Continuing from the previous discussion, let us concentrate on determining whether a truss is statically determinate or indeterminate. We should remember that this is the most crucial step before diving into analyzing the truss completely. For truss, we have two kinds of determinacy – internal and external. External determinacy is related to the nature and number of supports provided to the truss as a whole. Say, for example, if we provide two pin/hinge supports in a truss, it will be statically indeterminate because we can not calculate the horizontal support reactions that will be acting on the two pin supports from the equation of global equilibrium. And the degree of indeterminacy will be one, which is the difference between the total number of unknown support reactions and the number of equilibrium equations available. From this example, we can set the condition or criteria to check the external indeterminacy of any given truss, which is as follows:

$$D_{se} = (r_e - 3); \text{ for two-dimensional (2D) truss}$$

$$D_{se} = (r_e - 6); \text{ for 3D truss}$$

where D_{se} is the degree of external indeterminacy, r_e is the number of unknown support reactions in a particular direction, number of global equilibrium equations available = 3 or 6 for 2D and 3D, trusses, respectively, in general.

For one-pin and one-roller supports for a truss in 2D, we have total unknown support reactions equal to 3 (two at pin and one at roller) depending on the direction of the external loading. If only vertical loading is applied on the truss, no horizontal support reaction will be generated, and, in that case, horizontal equilibrium equation is also not required. If inclined load is applied, the unknown support reactions will be 3, and the available equations for the global equilibrium is also equal to 3, namely,

$$\sum F_x = 0, \ \sum F_y = 0, \text{ and } \sum M_z = 0; \text{ (for 2-D trusses)}$$

And, hence, the degree of indeterminacy is = 3 − 3 = 0, i.e., the truss is statically determinate externally, and we can determine the unknown support reactions completely by applying equations of global equilibrium only.

$$\sum F_y = 0 \text{ and } \sum M_z = 0$$

At this stage, if someone feels to brush up equilibrium conditions and equations for structures, it is highly recommended to go through Chapter 6 of global and local equilibrium conditions as we will use those concepts all through this book for analyzing structures.

Having gained sufficient knowledge about the external degree of indeterminacy, we will now discuss the internal degree of indeterminacy for a 2D truss. Let us think of a truss made up of m number of members connected by j number of joints and suppose the total number of support reactions are r (mostly this will be three for plane trusses supported at one end by pin and at the other end by roller). We already know that for m number of members and r number of support reactions, total unknown forces are $= m + r$. Since there are j number of joints, total number of equilibrium equations available will be $2j$. Hence, for a statically determinate stable truss, the following necessary condition should be met:

$$m + r = 2j$$

For instance, if,

$$m + r > 2j$$

And if, for simplicity, we take $r = 3$, then we have more members than that are required for the overall stability of the truss. Hence, at this situation, it is called a truss having redundant members inside it. And hence, equations of statics are not sufficient enough to calculate all the unknown member forces. Thus, the truss becomes statically indeterminate internally.

Another instance is the following,

$$m + r < 2j$$

And as per previous discussion, let us take $r = 3$. So, this indicates that the truss is having fewer members than are required for the overall stability. Thus, the truss is not a stable one, and it will collapse eventually or form a mechanism. So, this condition refers to unstable trusses. Hence, following this logic, we can determine at the very beginning, before doing any analysis, whether it is statically determinate or not. If it is statically indeterminate, we will need additional equations other than equilibrium conditions to determine all unknown member forces.

For space trusses or 3D trusses, we can develop the same type of equations or conditions for checking their degree of indeterminacy. For space trusses or 3D trusses having m number of members, j number of joints, and r number of support reactions, we have the following relationship:

$$m + r = 3j; \text{ for determinate truss.}$$

For trusses having redundant members, the above relationship will become:

$$m + r > 3j$$

And for unstable trusses, the equation will be:

$$m + r < 3j$$

The degree of static indeterminacy as discussed in Chapter 6 is defined by the following expression:

$$D_s = m + r_e - 2j \ \text{(for plane truss)}$$

$$D_s = m + r_e - 3j \ \text{(for } 3-D \, / \, \text{space truss)}$$

For statically determinate trusses, $D_s = 0$, which is self-evident from the above discussions.

So, before we proceed to analyze a truss, we should first check whether it is statically determinate or not. That being said, we have arrived in a position to learn various methods of truss analysis. We will study the following methods for truss analysis:

1. Method of joints
2. Method of sections
3. Method of tension coefficients
4. Graphical method of truss analysis
5. Henneberg's method of solution for complex trusses

7.6.1 METHOD OF JOINTS

Among the various methods listed above, the method of joint is the simplest and most straightforward method for truss analysis. To apply this method, we need to pay attention only to the equations of global and local equilibrium, and no special consideration needs to be taken as such. In this method, we need to work joint by joint of the truss. We will always start from a joint with no more than two unknown forces (which must not be colinear). So, the appropriate choice is to start from the nodes connected with the supports, because at the start before doing anything, we calculate the support reaction acting on the supports due external loading on the truss. Let us proceed with this method with a simple example to better understand its underlying concepts.

In Figure 7.7, a simple truss with an external point load P acting on node C is shown. All the members of the truss are of same length l. Hence, the angle between members will be 60°. Now before doing any force calculations, first, let us check whether this truss is statically determinate or not. For this plane truss we have:

$$m = 7$$

$$r = 3$$

$$j = 5$$

So, the degree of static indeterminacy, $D_s = m + r_e - 2j = 10 - 10 = 0$

Thus, the given truss is statically determinate.

Next, we need to calculate the support reactions acting on the supports due to external loading. As explained in the earlier chapter of equilibrium conditions, we need to consider the truss with external loading and support reactions and apply global equilibrium conditions to calculate the support reactions as per Figure 7.8.

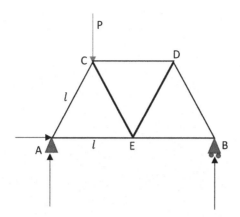

FIGURE 7.7 Plane truss for analysis example.

Now applying global equilibrium condition in x direction, namely, $\sum F_x = 0$ yields,

$$H_A = 0$$

And taking moment of all forces about support B, we get,

$$R_A \times 2l - P \times (2l - l\cos 60°) = 0$$

or,

$$R_A = \frac{3P}{8}$$

Now, for global equilibrium in y direction we have, $\sum F_y = 0$, which means,

$$R_A + R_B = P$$

or,

$$R_B = P - R_A = \frac{5P}{8}$$

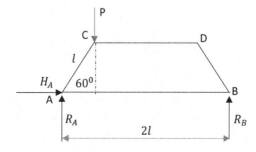

FIGURE 7.8 Global equilibrium condition for determining unknown support reactions.

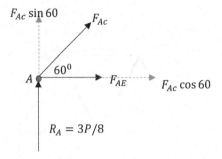

FIGURE 7.9 Method of joint and free body diagram.

So, we get the unknown support reactions with their proper direction from the above expressions. Now we have arrived at the point from where we can start analyzing the truss internally to calculate its member forces by the *method of joints*.

Let us begin from joint A as shown in Figure 7.9, since already we have a known support reaction i.e., the number of unknown forces is two at this joint, and draw the free body diagram with all members replaced by tension forces (assumed) as shown below:

Now we are dealing with local equilibrium conditions to calculate member forces induced on members due to external loading. So, from the above free body diagram, applying equilibrium condition in y direction, we get,

$$\sum F_x = 0$$

or,

$$F_{AC} \; \text{Sin} \; 60 + R_A = 0$$

or,

$$F_{AC} = -\frac{R_A}{\sin 60} = -\frac{3P}{2\sqrt{3}}$$

So, we get the member force F_{AC} with its appropriate direction. The minus sign indicates that the member is under compression. Now applying local equilibrium condition in x direction, we get,

$$\sum F_x = 0$$

or,

$$F_{AE} + F_{AC} \cos 60 = 0$$

or,

$$F_{AE} = -F_{AC} \cos 60 = \frac{3P}{4\sqrt{3}}$$

So, we get the member force on the member AE, and the positive sign indicates that the said member is under tension as was assumed.

We may now proceed sequentially to other joints and draw appropriate free body diagrams to calculate the member forces in those members. Complete analysis of this truss applying this concept is left as an exercise to the reader.

Thus, we get a complete idea about the method of joints and how this method helps us to determine the unknown member forces in various members of a given truss due to external loading. This method is very simple and elegant. However, for large trusses, this method becomes very tedious and time consuming. For a large truss, if we want to calculate member force of any particular member, then this method doesn't give immediate result. To be able to do that, we need a more sophisticated method that we are about to learn in the next section.

7.6.2 Method of Sections

This method is elegant in a sense that we can calculate the member force on any member directly from this without carrying out detailed step-by-step analysis. We can immediately calculate the member force induced on any member due to external loading by taking a suitable section passing through that member. Sections drawn through truss should be such that it passes from bottom to top or top to bottom of the entire truss. Also, it should be borne in mind that no chosen section should contain more than three members. With this simple logic, let us calculate the member force on the member DE of the previous example directly after calculating the degree of indeterminacy and support reactions. So, the section is as shown in Figure 7.10.

Once section is drawn, we need to replace the members by member forces through which the section passes. Hence, the free body diagram of the truss after taking section can be as shown in Figure 7.11.

Now, for local equilibrium condition in y–direction, we have $\sum F_y = 0$
or,

$$F_{DE} \sin 60 - R_B = 0$$

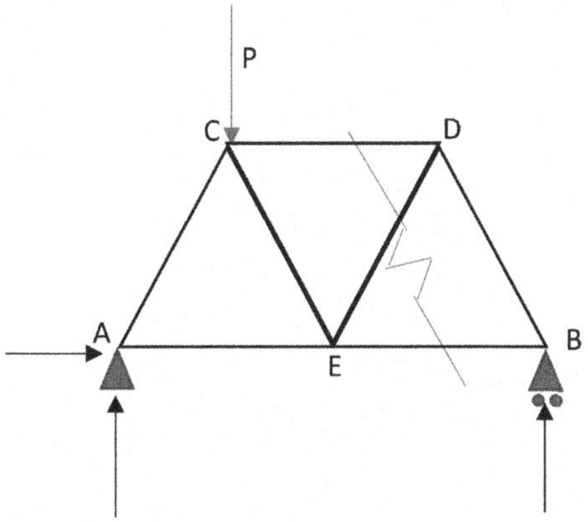

FIGURE 7.10   Method of section and placement of imaginary section line through truss.

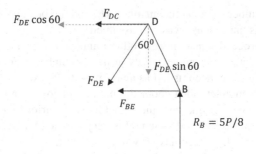

FIGURE 7.11 Free body diagram with member forces through as per the section line.

or,

$$F_{DE} = \frac{5P}{4\sqrt{3}}$$

So, member *DE* is under tension, and the member force acting on it is $5P/4\sqrt{3}$, and our requirement is satisfied. We are able to calculate the member force directly by taking the appropriate section through that member and then applying local equilibrium conditions. Also, from the above free body diagram, if we take moment about joint *B* of all forces, then we should get the member force at *D*. A complete truss analysis by this method is left as an exercise for the readers. Also, it is also interesting to check whether analyses by these two different methods yield the same result.

Hint to calculate member force DC from above free body diagram:
Taking moment about joint *B* yields:

$$F_{DC} \times l \sin 60 + F_{DE} \sin 60 \times l \cos 60 + F_{DE} \cos 60 \times l \sin 60 = 0$$

Since we have already calculated F_{DE}, from the above expression, we can calculate the unknown member force F_{DC}.

In drawing the above free body diagram, we have not considered the left section of the truss. If we draw the left section of the truss, then the following free body diagram as shown in Figure 7.12, appears.

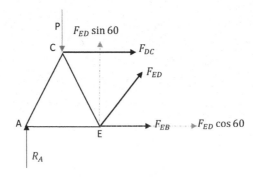

FIGURE 7.12 Left section free body diagram of the truss.

From the above diagram, by applying local equilibrium condition in y direction, we get:

$$\sum F_y = 0$$

or,

$$F_{DE} \sin 60 + R_A - P = 0$$

or,

$$F_{ED} = \frac{P}{\sin 60} - \frac{R_A}{\sin 60} = \frac{5P}{4\sqrt{3}}$$

Exactly same as that obtained from the right section, we have considered in the first analysis.

Hence, by taking left and right sections and sketching the appropriate free body diagrams, we can easily get the unknown member forces by simple application of local equilibrium equations. Applying this concept for the whole truss to determine all unknown member forces is left as an exercise for the readers.

7.6.3   METHOD OF TENSION COEFFICIENTS

The method of tension coefficient was developed by R.V. Southwell in 1920 and is applicable for both plane and space trusses. This method is very elegant in the sense that it can be directly converted into a computer program for quicker analysis. To understand this method, let us take a single member AB of a plane truss in (x,y) plane and also let us write down the coordinates of the end nodes for the same.

The coordinate of node A = (x_a, y_a), and coordinate of node B = (x_b, y_b) as per Figure 7.13. As per the concept, the member force (T_{ab}) is to be considered as tensile

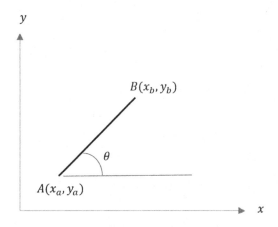

FIGURE 7.13   Member coordinates for tension coefficient method.

for this member AB. The components of the member force (T_{ab}) along the coordinate axis can be written as:

$T_x = T_{ab}\cos\theta$, and $T_x = T_{ab}\sin\theta$; where θ is the inclination of the member to the x-axis. Here $\cos\theta$, $\sin\theta$ are the direction cosines of the force vector and can be expressed as follows:

$$\cos\theta = \frac{x_b - x_a}{l_{ab}} = \frac{x_{ab}}{l_{ab}}, \text{ and } \sin\theta = \frac{y_b - y_a}{l_{ab}} = \frac{y_{ab}}{l_{ab}}; \text{ where } l_{ab} = \sqrt{x_{ab}^2 + y_{ab}^2}$$

is the length of the member AB.

Now, we can replace $\cos\theta$ and $\sin\theta$ in the expressions for T_x and T_y and rewrite the equations as:

$$T_x = T_{ab} \times \frac{x_{ab}}{l_{ab}} = t_{ab} \times x_{ab}, \text{ and } T_y = T_{ab} \times \frac{y_{ab}}{l_{ab}} = t_{ab} \times y_{ab}$$

Here the parameters, $\frac{T}{l}$ is the force per unit length of a member and is known as *tension coefficient* of the member.

So, if somebody wants to calculate the x component of tensile force acting on this member, then he/she needs to multiply the tension coefficient t_{ab} with the difference in x coordinate of the two nodes as explained above. That means, the x component of force on member $AB = t_{ab} \times (x_b - x_a)$, and similarly y component of force acting on this member will be as $t_{ab} \times (y_b - y_a)$. The tension coefficients are assumed to be positive initially. But if it turns out to be negative after calculation, the force in that member is taken as compressive. Now consider a truss joint j where three members are connected as shown in Figure 7.14.

In Figure 7.14, the joint i of a truss is connected to other truss joints l, m, and k with three members and a concentrated load P is acting on it. Now, applying the above concept of tension coefficient, we can formulate the equilibrium condition of joint j as below:

$$\sum F_x = 0 = t_{jk} \times x_{jk} + t_{jl} \times x_{jl} + t_{jm} \times x_{jm} + P_x = 0$$

$$\sum F_y = 0 = t_{jk} \times y_{jk} + t_{jl} \times y_{jl} + t_{jm} \times y_{jm} + P_y = 0$$

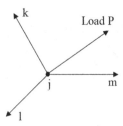

FIGURE 7.14 Joint j is connecting three members and load P is applied on it.

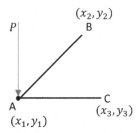

FIGURE 7.15   Example on tension coefficient method.

In compact form, the above two equations can be written as:

$$\sum t_{jk} \times x_{jk} + P_x = 0$$

$$\sum t_{jk} \times y_{jk} + P_y = 0$$

Example 7.1: Establish the equilibrium equations along both the coordinate axis for the truss joint A shown in Figure 7.15 by tension coefficient method.

SOLUTION: In this figure, all the coordinates of the two connected members are shown, and there is an external point load P acting on y direction at joint A. Following the above rules for tension coefficients, we can write the equilibrium equations as:

$$\sum F_x = 0$$

$t_{AB} \times (x_2 - x_1) + t_{AC} \times (x_3 - x_1) = 0$; since there is no external load acting on x direction.

$$\sum F_y = 0$$

$t_{AB} \times (y_2 - y_1) + t_{AC} \times (y_3 - y_1) - P = 0$; since P is acting toward the joint, it is compressive in nature.

All the coordinates and external loads are known to us. Hence, we can solve these simultaneous equations to get the unknown tension coefficients.

To summarize this method of analysis, we can follow the below procedure:

a. Mark the positive and negative directions of (x, y) for plane trusses and (x, y, z) for space trusses.
b. Start with the assumption that all members are in tension, i.e., member forces are away from the joint at which all are connected.
c. Then write down the equilibrium equations at each joint as per the procedure just shown for a typical truss with external loading.

d. Solve the equations for unknown tension coefficients.
e. Finally calculate the member forces by the relationship $T_{AB} = t_{AB} L_{AB}$

This method is to some extent linked with the method of joints since we need to move from joint to joint for calculating the unknown tension coefficients. Let us work out few member forces by this method for the same truss we have considered in our previous section. All the coordinates are presented as per the geometry of the truss. Origin has been assumed to be situated at joint A.

Let us begin our work from node A. At this node, there are two members AC and AE. And there is a support reaction acting on node A as found from our previous analysis. So, in terms of tension coefficient in x direction, we can write:

$$t_{AE} \times (l - 0) + t_{AC} \times \left(\frac{l}{2} - 0\right) = 0$$

or,

$$t_{AE} = -t_{AC}/2$$

And for the y direction, we can write,

$$t_{AE} \times (0 - 0) + t_{AC} \times \left(\sqrt{3}/2l - 0\right) - P + R_A = 0$$

or,

$$t_{AC} = \frac{5P}{4\sqrt{3}l}$$

So, from previous expression we get,

$$t_{AE} = -\frac{5P}{8\sqrt{3}l}$$

Hence, member force on AC, $T_{AC} = t_{AC} \times L_{AC}$

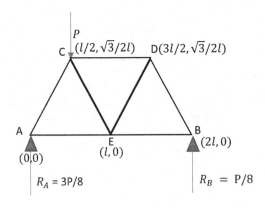

FIGURE 7.16 Complete truss with the coordinates for tension coefficient.

or,

$$T_{AC} = \frac{5P}{4\sqrt{3}l} \times l = \frac{5P}{4\sqrt{3}}$$

and,

$$T_{AE} = -\frac{5P}{8\sqrt{3}l} \times l = -\frac{5P}{8\sqrt{3}}$$

So, we get the magnitude as well as direction of forces of the member (compression/tension) from the above method. This can be checked with respect to the earlier methods whether the direction and magnitude of force of the members are same. As can be understood, this method mostly relies on the method of joints, since we need to move joint by joint to get the tension coefficients and corresponding forces acting on the members. Thus, for quick and elegant approach for calculating any arbitrary member forces, the method of section is the most effective. For complete truss analysis using computer programs, the method of tension coefficients is most suitable. We can build an algorithm to incorporate large trusses by demanding very few inputs to analyze it completely once the program analysis is complete. Complete analysis of this truss using tension coefficients for all members is left as an exercise for the reader.

7.6.4   GRAPHICAL METHOD OF TRUSS ANALYSIS

Now, we will learn how to analyze a truss by graphical method. In this method, no real calculation is being made. Only a suitable choice of scales for drawing member forces yields the desired result of unknown member forces. To be able to do that, we need to learn a few basic theorems and notations for preparing ourselves for graphical analysis. This method was largely developed by several analysts in the early 19th century (the most famous among them were Maxwell and Betti). Although no equations are required to be dealt with in this method, one must have a good understanding of the force vector, triangle rule of vector addition, parallelogram rule of vector addition, etc. To this end, we need to understand Bow's notation for describing the force vector by the space enclosed by them. It will be clear from the following diagrams.

7.6.4.1   BOW'S NOTATION

In this notation, all force vectors are represented by the space included by them. Let us draw some coplanar force vector with spaces marked as shown in Figure 7.17.

In the given diagram, we have drawn three concurrent forces with directions indicated by arrows. These force vectors are not named as per usual convention for vectors. Instead, we have marked the spaces surrounding the vectors by letters A, B, C, etc. So, starting from the left, the first force vector will be denoted as AB, second one will be BC, and the third will be denoted as CA. So, for triangle rule of vector addition, if we draw the triangle with small letters a, b, c, etc. to indicate the force vectors, then it will be something as shown in Figure 7.18.

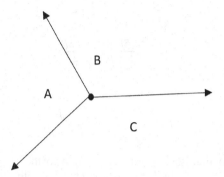

FIGURE 7.17 Bow's notation for graphical analysis.

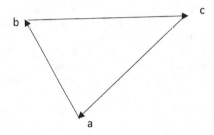

FIGURE 7.18 Force triangle following Bow's notation.

So, from the triangle of forces, it is closed under vector addition. It means that the forces are not having net resultant force. Hence, the concurrent force system is in a state of equilibrium. The force triangle is drawn with a suitable scale to represent the forces. Also, once we start with any one of the vectors *ab*, *bc,* or *ca*, the triangle of forces automatically falls in place following the rule for vector addition. Bow's notation helps us to neatly draw the force system replicating the actual forces acting on any structure or members like this way. For nonconcurrent forces, we indicate the forces in the same way of surrounding spaces as we did for concurrent forces. See Figure 7.19 for a system of nonconcurrent forces and their nomenclature for better understanding.

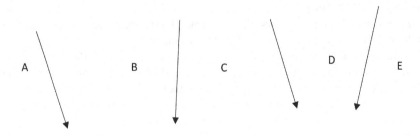

FIGURE 7.19 Several nonconcurrent forces.

So, for the above nonconcurrent forces, the force vectors are *AB*, *BC*, *CD*, and *DE* as per Bow's notation. These forces are not in equilibrium. So, the force polygon will be open-ended, and the final vector joining the first node to the last node of the force polygon will be the resultant force for all these forces. To establish this fact graphically, we will now draw some particular force diagrams with suitable scales for calculating the resultant force and its location/position. Let us take an arbitrary point *O* and draw the forces *AB*, *BC*, *CD*, and *DE* tail to tail as with proper direction as per Figure 7.19.

Above diagram is known as the polar diagram. The pole is the point '*o*'. All forces are drawn in a suitable scale say 1 mm = 1 kN from tail to tail as shown above. From the individual small triangles, we can see that all the internal forces except force *oa* and *oe* remain left (which are indicated by the dotted arrows). So, the resultant for all given nonconcurrent forces is the force *ae*, connecting vector from point *a* to point *e*. So, from this diagram, we can get the value of the resultant force by measuring the length of arrow *ae* and applying the chosen scale factor to calculate how much kN will be resultant force.

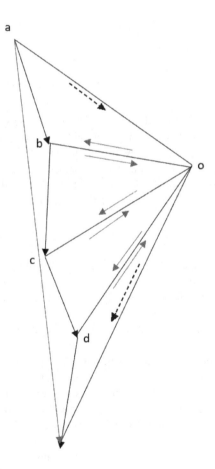

FIGURE 7.20   Polar diagram.

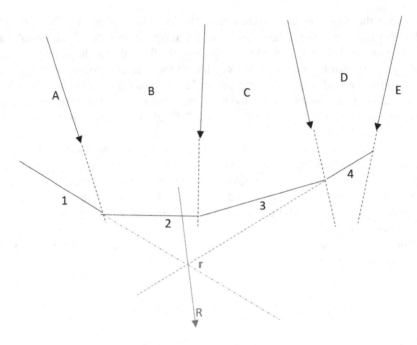

FIGURE 7.21   Funicular polygon for system of nonconcurrent forces.

To find the location of the resultant force, we need to draw the diagram as shown in Figure 7.21, where line 1 is drawn at *a* suitable distance from force *AB*, parallel to *oa* line in polar diagram. Similarly, all other lines are drawn parallel to sides of *ob*, *oc*, and *od* as per polar diagram. As can be seen from the polar diagram, the resultant force is represented by the line *ae*, and as per triangle rule for vector addition, we get *ae* = *ao* + *oe*. So, we extend the line 1 and line 4 in the above diagram to intersect at the point *r* as shown in Figure 7.21. Now we can draw the line through point *r* parallel to the line *ae* of the polar diagram. Hence, by this method, we get the location of the resultant force for nonconcurrent forces and it is near to the force *AB*. The above shape of the line 1, 2, 3, 4, etc. looks very similar to a loaded string or rope. If we hang several loads from a tied rope, then it will take a shape something like the above diagram lines 1–2–3–4. This diagram is known as funicular polygon. A funicular polygon may be thought of a possible configuration of equilibrium of a rope or string, suspended from its ends, and is loaded to the given system of coplanar forces. It needs to be understood that a string takes the profile of a straight-line segments when it is subjected to a series of concentrated forces. In case of distributed loads, string assumes a smooth curve profile. The same concept can be applied to general class of loading to determine location and magnitude of resultant force. However, we will focus on the methods of truss analysis using the graphical method just explained. Even this method can be applied to determine center of gravity for plane areas with irregular geometry. Interested reader may refer to some excellent textbook mentioned at the end of the chapter reference for the same.

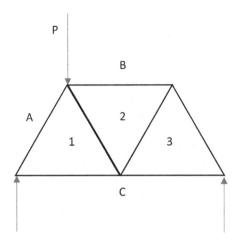

FIGURE 7.22   Truss for graphical analysis.

Let us consider the same simple truss with point load at its top left joint as shown below. But in case of graphical analysis, we need to apply Bow's notation for forces acting on the system. For that, let us define the spaces surrounding the truss as shown in Figure 7.22.

So, as per Bow's notation, the support reactions will be *AC* and *CB*, whereas the unknown member forces will be *A1, 12, 23, 2B, 1C, 3C, 3B*, etc.

As we already know, to analyze any truss, we need to first find out support reactions at the supports. For determining support reactions, we also know that whole truss is visualized as a single rigid beam like structure with loads at the distance from supports as shown in the drawing. Hence, in graphical analysis, we need to draw a funicular polygon as shown in Figure 7.23, representing the truss as a rigid

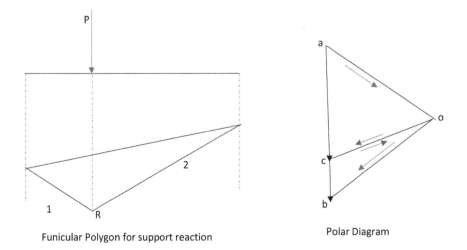

Funicular Polygon for support reaction   Polar Diagram

FIGURE 7.23   Funicular polygon and polar diagram for truss example.

beam with point load on top of it. So, the funicular polygon and polar diagram will be something like as presented below. Polar diagram has been drawn by choosing a suitable scale say 1 mm = 1 kN and then the lines *oa* and *ob* are drawn. It is to be understood that first line of funicular polygon is to be drawn from any point as shown below, and it is parallel to the line *oa* and meeting the projected line of external applied force at some point *R* as shown in Figure 7.23.

Now from *R*, a second line is drawn parallel to line *ob* of polar diagram up to the left end projection line of beam. This line is named as line *2* in the funicular polygon. Now the line joining the two ends of line *1* and line *2* is known as the closing line of the funicular polygon. This line is the one that helps us to determine the support reactions from polar diagram. Once the closing line is drawn, a parallel line from *o* to the closing line is drawn which intersect the *ab* line at some point *c*. Hence, *ac* segment will be the left support reaction, and *cb* will be the right support reaction that can be directly measured from the polar diagram. It is also to be noted that this method of determination of support reaction is applicable for any determinate beams with point loadings. Once the support reactions are known then we can proceed to analyze the truss by simply force triangle and force polygon method remembering the fact that we should start from joint where at least one force is known to us. Also, it is to be noted that scale chosen during drawing the polar diagram should remain same for the entire problem we are dealing with. Scale factor once chosen cannot be changed in the middle of calculations. Hence, for the joint *AC1A*, we have the following force triangle as shown in Figure 7.24.

Since all forces are in equilibrium, force triangle will be closed under vector addition. From the above force triangle, *ac*, is already known and has been drawn with the chosen scale factor. Hence, all other forces fall neatly following the vector rule of addition. To draw the forces *c1* and *a1*, we can draw two lines parallel to the truss members from point *c* and point *a*, and the intersection of these two lies will be the point *1*. Once the triangle is drawn, we can measure the length and multiply it with the chosen scale factor to obtain the force values in the member. Also, from the force triangle, we immediately get the state of the member. For example, from force triangle, we get that member *C1* is under tension while member *A1* is under compression.

Now we can proceed to joint *AB21* where three members are connected. Of these three member forces, we already calculated the force on the member *A1* in

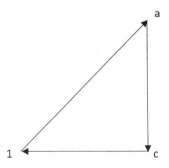

FIGURE 7.24   Joint AC1A force triangle.

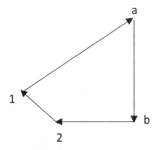

FIGURE 7.25   Force polygon for joint AB21.

the previous analysis. So, the force polygon for this joint will be something like as shown in Figure 7.25.

First of all, we need to draw the force *ab* which is the known external point load acting on the truss as per the chosen scale. Then line *a1* is drawn as this force is already known from previous analysis, and this line is parallel to the member *A1*. Once these two lines are drawn next, we need to draw two lines, namely, from point *b*, *b2* parallel to member *B2* and from point *1*, *12* parallel to member *12*. These two lines will intersect at the point *2*. All forces will be directed as shown in the force polygon, since this should also be closed under vector addition rule under equilibrium conditions. As all known forces are drawn as per the chosen force scale, we can measure the length of lines *12* and *b2* to get the force values in the member. Also, from force polygon, it is immediately clear that member *B2* is under compression, member *12* in under compression.

All other joints and force polygons can be drawn to scale as per above method to complete the truss analysis as a whole. Once calculated, member forces from graphical analysis can be compared to that obtained from other earlier defined methods to compare the error in graphical analysis. Graphical analysis provides a nice tool to analyze trusses with excellent accuracy compared to the other theoretical analysis. Professor Maxwell made outstanding contributions toward the graphical analysis of structures. For detailed discussion and understanding on the same, students are encouraged to consult the books mentioned in the reference.

7.6.5   HENNEBERG'S METHOD OF SOLUTION FOR COMPLEX TRUSSES

The member force analysis for a complex truss can be determined using the method of joints; however, the analysis procedure will need to write two equilibrium equations for each of the *j* joints of the truss and then solving the complete set of $2j$ equations simultaneously. This approach may be tedious for hand calculations, especially in the case of large complex trusses. Therefore, a more direct qualitative method for analyzing a complex truss, referred to as the method of substitute members or Henneberg's method, will be presented here in a stepwise fashion.

a. ***Reduction to a stable simple truss***

At first, we need to determine the reactions at the supports and begin by imagining how to analyze the truss by the method of joints, i.e., progressing from

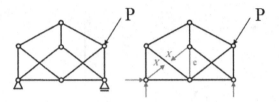

FIGURE 7.26 Henneberg's method of bar replacement by axial force.

joint to joint and solving for each member force. If a joint is reached where there are three unknown forces, we need to remove one of the members at the joint and replace it by an imaginary member elsewhere in the truss. By doing this, reconstruct the truss to be a stable simple truss.

b. ***External loading on simple truss***

Load the simple truss with the actual loading P, then determine the force on each member i. But first of all, we need to find out the reactions. Then one could start at any joint say A to determine the forces in connected member at that node, then progressing successively to other joints where not more than three members meet.

c. ***Remove External loading on simple truss.***

Just imagine a simple truss without the external load P. Now let us place equal but opposite collinear unit loads on the truss at the two joints from which the member was removed as shown in Figure 7.26. If these forces develop a force on the ith truss member, then by proportion an unknown force x in the removed member would exert a force on the ith member. We replace one bar by its axial force X. Because the structure becomes unstable, we add one extra bar (at another place) to ensure mechanical stability. The axial force on the extra bar should be zero. Using this condition, we can calculate the force on the removed bar.

Let us introduce two more terms to define force system in this method. First one is $N_e(P)$ that represents force due to external loading with proper support reactions. Second one is $N_e(X=1)$ is the force caused by self-equilibrium forces X with no support reactions. And we will use superposition principle as per the following scheme to obtain the result:

$$N_e = N_e(P) + N_e(X) = N_e(P) + XN_e(X=1) = 0$$

From which we get,

$$X = -\frac{N_e(P)}{N_e(X=1)}$$

where, $N_e(X=1)$ is the force due to unit load $X=1$ in equilibrium condition. To understand this method of analysis, refer to the following example.

Example 7.2: Determine the Axial Force on the Indicated Member of the Following Truss Using Henneberg's Method of Analysis.

SOLUTION: The given truss is complex in nature. We cannot analyze this truss by the method of joint or the method of sections without involving large number of equilibrium equations. Hence, Henneberg's method is most appropriate for this type of complex truss analysis.

First of all, we determine support reaction for tis truss by applying global equations of equilibrium.

$$V_A = \frac{40.3 + 70.3}{4} = 82.5 \text{ kN}, \ R_B = 40.1 - 70.3 \frac{40.1 - 70.3}{4} = -42.5 \text{ kN}, \ H_A = 70 \text{ kN}$$

Next, we replace the indicated bar with self-equilibrium force X as shown in Figure 7.27 (c). Since we have removed a bar from the original truss, we have

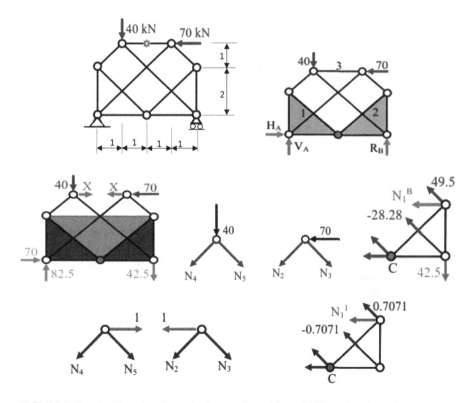

FIGURE 7.27 (a) Henneberg's method example problem. (b) Henneberg's method example problem. (c) Henneberg's method example problem. (d) Henneberg's method – Basic member forces calculation. (e) Henneberg's method – basic member forces calculation. (f) Henneberg's method – self equilibrium force calculation.

provided an additional new bar connecting the far ends of the truss as shown in figure to maintain stability and the overall integrity of the structure.

Now let us analyze the truss for basic load in the members connected at the nodes of the target member by the method of joints.

From Figure 7.27 (d), we get,

$$\sum F_y = 0$$

$$N_2 = -N_3$$

$$\sum F_x = 0$$

$$-\frac{\sqrt{2}}{2}N_2 + \frac{\sqrt{2}}{2}N_3 - 70 = 0$$

or,

$$N_3 = 49.5 \text{ kN}$$

From the left-hand side force diagram, we get,

$$\sum F_x = 0$$

$$N_4 = N_5$$

$$\sum F_y = 0$$

$$40 + \frac{\sqrt{2}}{2}N_4 + \frac{\sqrt{2}}{2}N_5 = 0$$

$$N_5 = -\frac{40}{\sqrt{2}} = -28.28 \text{ kN}$$

Let us denote the unknown member force on the indicated member as N_1. As per Henneberg's method, we need to carry out two separate analysis to determine N^B_1, which is normal member force coming from basic force analysis, and N'_1, which is generated from self-equilibrium force analysis with unit loads as will be indicated in the force diagram. At first, we will determine N^B_1 and then N'_1.

$$\sum M_c = 0$$

$$\frac{\sqrt{2}}{2}N_5 \times 2 + \frac{\sqrt{2}}{2}N_3 \times 4 + N_1^B \times 2 = 0$$

or,

$$N_1^B = -7.5 \text{ kN}$$

Now we will carry out analysis for self-equilibrium forces.

$$\sum F_y = 0$$

$$N_2 = -N_3$$

$$\sum F_x = 0$$

$$-\frac{\sqrt{2}}{2}N_2 + \frac{\sqrt{2}}{2}N_3 - 1 = 0$$

$$N_3 = 0.7071 \text{ kN}$$

$$\sum F_y = 0$$

Again, we get from left side force diagram,

$$N_4 = -N_5$$

$$\sum F_x = 0$$

$$1 - \frac{\sqrt{2}}{2}N_4 + \frac{\sqrt{2}}{2}N_5 = 0$$

or,

$$N_5 = -0.7071 \text{ kN}$$

$$\sum M_c = 0$$

$$\frac{\sqrt{2}}{2}N_5 \times 2 + \frac{\sqrt{2}}{2}N_3 \times 4 + N'_1 \times 2 = 0$$

$$N'_1 = -0.5 \text{ kN}$$

We have finally,

$$N_1 = -\frac{N^B_1}{N'_1} = \frac{-7.5}{-0.5} = 15 \text{ kN}$$

So, by applying Henneberg's analysis method, we can easily determine force on any member without dealing with a large number of equilibrium equations, which may result if we adopt the method of joints or the method of sections.

7.7   COMPOUND TRUSSES

In the beginning of this chapter, it was shown that compound trusses are formed by joining two or more simple trusses together either by bars or by joints. Occasionally this type of truss is best computed by applying both the method of joints and the method of sections. Hence, the same procedure for simple truss can be considered for compound truss analysis.

7.8 SPACE TRUSSES

A space truss is the 3D version of plane trusses. An ideal space truss is formed by joining of rigid members connected by ball and socket joints. As discussed earlier for a plane truss, a three-member structure is connected to form a triangle, which is the basic non-collapsible structure for truss. Similarly, for space truss, six rigid members are connected together by ball and socket joint in the form of tetrahedron, providing the basic non-collapsible space truss unit. Thus, refer to Figure 7.28 for the basic tetrahedron formed by six members as an example of a simple space truss unit.

In the above tetrahedron, members *AB* and *BC* are joined together by ball and socket joint at the node *B*. To make this triangular shape stable and rigid, we need to connect a third member *AC* at the base. Similarly, other faces of the tetrahedron are formed. Hence, after connecting all the members, we have something as shown in Figure 7.29.

From the above geometrical object, it can be clearly seen that the unit is quite stable even without any supports. However, we need to provide supports as per the site condition and functional requirement of the truss. Thus, by this logic, we can form the basic unit of simple 3D truss tetrahedrons. Similarly, we can form large space truss by combining several unit tetrahedral trusses and joining the members accordingly. Same was found applicable for the case of simple trusses where we can join desired number of triangular units to form large simple truss. For analysis of space trusses, it is much more convenient to use the vector form of equations of equilibrium than the scalar form. So, in 3D space, we can put the coordinates of each joint by choosing suitable (x, y, z) axes and denoting member forces in terms of component of forces and unit vectors along three axes. More clarity will be obtained once we analyze a simple space truss by vector method. The method of joints discussed earlier for plane truss can be made

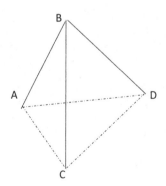

FIGURE 7.28   Fundamental building element (tetrahedron) for space truss.

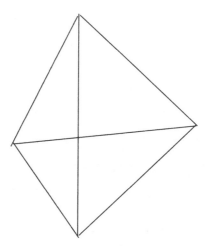

FIGURE 7.29   A typical space truss fundamental tetrahedron.

directly applicable for space trusses by adopting simple vector equilibrium equation as follows for each joint:

$$\sum \vec{F} = \vec{0}$$

where, $\vec{0}$ is the null vector. We always start from a joint where at least one force is known to us and not more than three unknown forces are present. Similarly, the method of section developed in earlier chapter for plane trusses may be extended to analyze space trusses also. For that method, we need to apply force vector and moment vector equilibrium equations simultaneously to get the unknown forces. The vector equilibrium equations for the method of section for space truss will be something as the following:

$$\sum \vec{F} = \vec{0}$$

and,

$$\sum \vec{M} = \vec{0}$$

If the unknown support reactions are six, then according to degree of indeterminacy formula, for statically determinate space trusses, we have the following equality in terms of member numbers and joint numbers.

$$m + 6 = 3j$$

Also, it needs to be noted that above moment and force vector equations, there will be total six number of equations. Hence, during drawing a section for space truss, we need to be careful that the section does not pass through more than six members.

In general, the statical degree determinacy and stability of space truss can be denoted as:

$$b + r < 3j \rightarrow \text{statically indeterminate} - \text{unsatble truss}$$

$$b + r = 3j \rightarrow \text{startically determinate} - \text{check stabilty}$$

$$b + r > 3j \rightarrow \text{statically indeterminate} - \text{check stability}$$

where:
b = number of members of the space truss;
r = number of reactions;
j = number of joints.

Example 7.3: Determine the reactions at the supports and the force in each member of the space truss as shown in Figure 7.30.

SOLUTION: Static Determinacy: The truss contains nine members and five joints and is supported by six reactions. As, $m + r = 3j$ and the reactions and the truss members are properly arranged, it is statically determinate.

Zero-force members: At joint D, three members, AD, CD, and DE, are connected. Of these members, AD and CD lie in the same (xz) plane, whereas DE does not. Since no external loads or reactions are applied at the joint, member DE is a zero-force member.

$$F_{DE} = 0$$

DE is a zero-force member. The remaining two members, DA, and DC are not colinear. So, they must be zero-force members.

$$F_{AD} = 0$$

$$F_{DC} = 0$$

Support reactions:

$$\swarrow + \sum F_Z = 0$$

$$B_Z + 70 = 0$$

$$B_Z = 70 \text{ kN} \nearrow$$

Taking moment about the Y-axis,

$$\circlearrowright + \sum M_Y = 0$$

$$B_X \times 3 + 70 \times 6 - 70 \times 3 = 0$$

$$B_X = 70 \text{ kN} \leftarrow$$

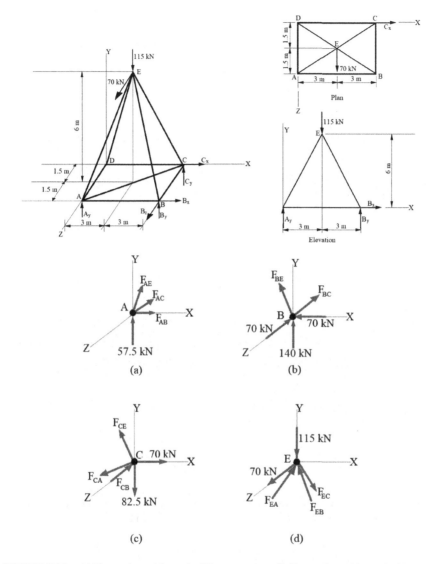

FIGURE 7.30 (a) Example problem of a 3D space truss. (b) Example problem of a 3D space truss: Joint equilibriums.

$$\rightarrow + \sum F_x = 0$$

$$-70 + C_X = 0$$

$$C_X = 70 \text{ kN} \rightarrow$$

Taking moment about the X-axis,

$$\circlearrowleft + \sum M_X = 0$$

$$-A_Y \times 3 - B_Y \times 3 + 115 \times 1.5 + 70 \times 6 = 0$$

$$A_Y + B_Y = 197.5$$

$$\uparrow + \sum F_y = 0$$

$$A_Y + B_Y + C_Y - 115 = 0$$

$$197.5 + C_Y - 115 = 0$$

$$C_Y = 82.5 \ kN \downarrow$$

Taking moment about the Z-axis,

$$\circlearrowright + \sum M_Z = 0$$

$$B_Y \times 6 - 82.5 \times 6 - 115 \times 3 = 0$$

$$B_Y = 140 \ kN \uparrow$$

$$A_Y = 57.5 \ kN \uparrow$$

The projections of the truss members in the X, Y, and Z directions, as obtained from Fig. 7.30 (b) and their lengths computed from these projections are tabulated in the following.

Joint A: See Fig. 7.30 (b) (i)

$$\uparrow + \sum F_y = 0$$

$$57.5 + \left(\frac{Y_{AE}}{L_{AE}} \right) F_{AE} = 0$$

in which the second term on the left-hand side represents the Y component of F_{AE}. Substituting the values of Y and L for member AE from the Table 7.1, we write,

$$57.5 + \left(\frac{6}{6.87} \right) F_{AE} = 0$$

$$F_{AE} = 65.84 \ kN \ (C)$$

$$\swarrow + \sum F_z = 0$$

$$-\left(\frac{3}{6.71} \right) F_{AC} + \left(\frac{1.5}{6.87} \right) \times 65.84 = 0$$

$$F_{AC} = 32.15 \ kN \ (T)$$

$$\rightarrow + \sum F_x = 0$$

TABLE 7.1

Member Lengths of the 3D Space Truss

| Member | Projection | | | Length (m) |
	X (m)	Y (m)	Z (m)	
AB	6	0	0	6
BC	0	0	3	3
CD	6	0	0	6
AD	0	0	3	3
AC	6	0	3	6.71
AE	3	6	1.5	6.87
BE	3	6	1.5	6.87
CE	3	6	1.5	6.87
DE	3	6	1.5	6.87

$$F_{AB} + \left(\frac{6}{6.71}\right) \times 32.15 - \left(\frac{3}{6.87}\right) \times 65.84 = 0$$

$$F_{AB} = 0$$

Joint B: See Fig. 7.30 (b) (ii)

$$\rightarrow + \sum F_x = 0$$

$$-\left(\frac{3}{6.87}\right) F_{BE} - 70 = 0$$

$$F_{BE} = 160.3 \; kN \; (C)$$

$$\swarrow + \sum F_z = 0$$

$$-70 - F_{BC} + \left(\frac{1.5}{6.87}\right) \times 160.3 = 0$$

$$F_{BC} = 35 \; kN \; (C)$$

As all the unknown forces at joint B have been determined, we will use the remaining equilibrium equation to check our computations:

$$\uparrow + \sum F_y = 0$$

$$140 - \left(\frac{6}{6.87}\right) \times 160.3 = 0$$

Hence, it is fine.

Joint C: See Fig. 7.30 (b) (iii)

$$\uparrow + \sum F_y = 0$$

$$-82.5 + \left(\frac{6}{6.87}\right) F_{CE} = 0$$

$$F_{CE} = 94.46 \ kN \ (T)$$

7.9   ZERO-FORCE MEMBERS OF TRUSSES

Some members do not carry any force in a truss system, or the force magnitude in those members is zero. These types of members are called zero-force members. These types of members are added to increase the stability or rigidity of the truss system and to provide support during different loading conditions. The method of joints will be greatly simplified, if one can identify zero-force members at the beginning.

By inspecting the truss joints properly, we can identify the zero-force members in the following ways:

1. If two noncollinear members are connected at a joint, and no external force or support reactions are acting at that joint, then those two members are the zero-force members. In Figure 7.31 (a) of the truss system, members *AB, AD, CE, EF* are the zero-force members. We can easily proof this by considering equations of equilibrium at joint *A* and *E*. In Figure 7.31 (b), we can see the zero-force members have been removed before beginning the analysis.
2. If three members form a truss joint, of which two are collinear, and there are no external loads, or support reactions acting on the joint, then the third noncollinear member is the zero-force member. In Figure 7.32 (b) of the truss system, members *BE* and *CE* are the zero-force members.

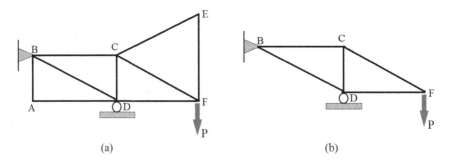

FIGURE 7.31   (Example I): Truss system (a) with zero-force members (b) without zero-force members.

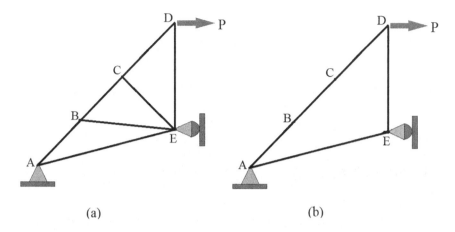

(a)   (b)

FIGURE 7.32   (Example II): Truss system (a) with zero-force members (b) without zero-force members.

In Figure 7.32 (b), we can see the zero-force members have been removed before beginning the analysis.

3. If all the members and external forces at a joint lie in the same plane but one member at that joint is out of plane, then force in the out of plane member is zero. In Figure 7.33, members *AB, AF, AD*, and force *P* are in the same plane but member *AC* is out of plane. This member *AC* is a zero-force member.

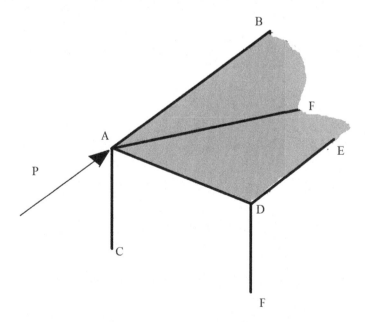

FIGURE 7.33   (Example I): Zero-force members for 3D space truss.

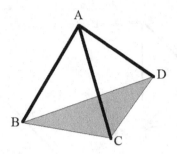

FIGURE 7.34 (**Example II**): Zero-force members for 3D space truss.

4. If three members at a joint do not lie in the same plane and there is no external load or support reaction at that joint, then the force in the three members are zero-force members. In Figure 7.34, members *AB, AC,* and *AD* are zero-force members.

8 Beams and Frames, Shear, and Bending Moments

8.1 INTRODUCTION

Apart from trusses discussed in the previous chapter, members of which are always subjected to axial forces only, the members of rigid frames and beams may be subjected to shear forces and bending moments (as discussed in Chapter 5) as well as axial forces under the action of external loads. The analysis and determination of these internal forces and moments are necessary for the design of such structures. The aim of this chapter is to gain a deeper understanding of the analysis of internal forces and bending moments that develop in beams, and the members of plane frames, under the action of coplanar systems of external forces and moments. We start by defining the three types of internally induced forces and moments – axial forces, shear forces, and bending moments – that will act on any arbitrary cross sections of beams and the members of plane frames. Then we will discuss the construction of shear and bending moment diagrams by the method of sections. We will also discuss qualitatively the deflected shapes of beams and the relationships between loads, shears, and bending moments. Also, we will develop the methods for constructing the shear and bending moment diagrams using these relationships and equations. Finally, we will introduce the concept of classification of plane frames as statically determinate, indeterminate, and unstable, and last but not least, analysis of statically determinate plane frames.

8.2 AXIAL FORCE, SHEAR, AND BENDING MOMENTS

Internal forces were defined in Chapter 5 as the forces and moments exerted on a portion or section of a structure by the rest of the structure and its loading conditions. Consider, for example, the simply supported beam shown in Figure 8.1 (a). Same kind of figure was used in Chapter 5 while discussing the bending moment and shear force.

As the applied force is inclined, hence, there will be horizontal force of the same, which will act along the axis of the beam. Hence, the horizontal support reaction at B will balance the same satisfying the equilibrium condition along x axis $\left(\sum F_x = 0 \right)$ of the beam. This implies $H_B = P_x$. Similarly, the other internal shear force and bending moment at the arbitrary section x from the left support can be calculated by applying the global equilibrium equations for force and moments.

It is to be noted that without these internal forces and moments (as drawn in the free-body diagram above), section x will not be in a state of equilibrium. Also, under a general coplanar system of external loads and reactions, two perpendicular force components and a moment or couple are necessary at any arbitrary section to

DOI: 10.1201/9781003081227-10

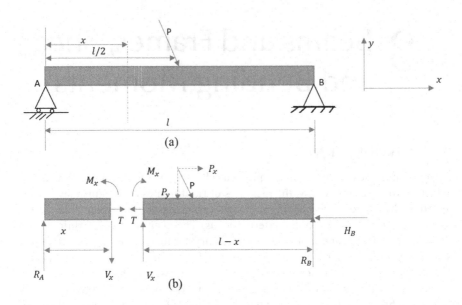

FIGURE 8.1 (a) External loading applied on a beam, (b) Its free body diagram showing internal resistive forces at a section along with the applied force and support reactions.

maintain that section of the beam in equilibrium. The two internal forces are in the direction of, and perpendicular to, the centroidal axis of the beam at section x, as shown in Figure 8.1 (b). The internal force T in the direction of the centroidal axis of the beam is called the axial force, and the internal force V_x in the direction perpendicular to the centroidal axis is referred to as the shear force. The moment M_x of the internal couple is termed the bending moment at the section x. To calculate the axial force acting along the axis of the beam, we need to focus on local equilibrium conditions for the sections of beams under consideration. From the local equilibrium condition, we can write from $\sum F_x = 0$, $T + H_B = 0$, or $T = -H_B$. Since H_B has already been calculated from the global equilibrium condition, we now have the complete information of the magnitude and direction of the axial force acting in the beam at the centroidal axis. Similarly, applying the local equilibrium equations for the left section of the free-body diagram, we can write, $R_A - V_x = 0$, or $V_x = R_A$, and $\circlearrowleft + \sum M_{x-x} = 0$ implies $M_x - R_A \times x = 0$, which gives $M_x = R_A \times x$ for the range $0 \leq x \leq l/2$ (same was derived in detail in Chapter 5 also). So, we have formed equations for calculating the shear force and bending moments induced in a beam under external loading. These equations help us to form shear and bending moment diagrams for the beams, which will be explained in detail in the next section.

8.3 SHEAR AND BENDING MOMENT DIAGRAMS FOR A BEAM

Shear and bending moment diagrams indicate the graphical variation of values and direction of these quantities along the length of the member. These diagrams can be constructed by taking sections indicated in the previous section and in Chapter 5. Multiple sections are chosen from one end of the member to the other (usually from left to right), considering a successive change in external loading along the length to determine the equations expressing the shear and bending moment in terms of the distance of the section from the starting point usually chosen as the origin. Shear and bending moment values determined from these equations are then drawn as ordinates against the position with respect to the chosen member end as abscissa to obtain the desired shear and bending moment diagrams. The said procedure is established by an example that follows next.

Example 8.1: Determine and draw the shear and bending moment diagram for the following beam with respect to the given external loading condition.

SOLUTION: From global equilibrium condition using moment equilibrium equation, $\Sigma M = 0$, we can write the same by taking moment of all forces about right support point (see the free-body diagram Figure 8.3 of entire beam with equivalent force).

So, from the above free-body diagram, we can write using the said moment equilibrium condition with respect to right support B:

$$R_A \times l - wl \times \frac{l}{2} = 0$$

or,

$$R_A = wl \times \frac{l}{2} \times \frac{1}{l} = \frac{wl}{2}$$

Now applying global force equilibrium condition along y direction, $\Sigma F_y = 0$

$$R_A + R_B = wl$$

or,

$$R_B = wl - R_A = \frac{wl}{2}$$

Hence, the two unknown support reactions are calculated first. Then we take any arbitrary section at a distance x from left support as shown in Figure 8.2 and we

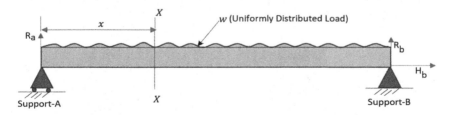

FIGURE 8.2 Simply supported beam under uniformly distributed load w throughout its length.

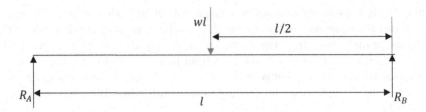

FIGURE 8.3 Free-body diagram of beam for global equilibrium condition with equivalent concentrated load wl at midspan.

can draw the free-body diagram indicating all acting forces in the beam as shown in Figure 8.4.

Now applying local equilibrium condition for force in y direction, we get,

$$R_A - wx - V_x = 0 \text{ for } 0 \le x \le l$$

$$\text{or, } V_x = R_A - wx$$

So, shear force is a function of x, when $x = 0$, we have $V_x = R_A - 0 = R_A = wl/2$, and for $x = l/2$, we have $V_x = 0$. This expression for shear force is valid for the entire span of the beam as the nature of loading is same throughout the span. Now, if $x = l$, from the abovementioned expression, we get $V_x = R_A - wl = wl/2 - wl = -\frac{wl}{2}$. Since shear force is linear function of x, hence, it will be a straight line and the values of shear force will vary from $wl/2$ to 0 from left support to midspan and from 0 to $-wl/2$ from the midspan up to the right support. The shear force diagram is shown in Figure 8.5.

Once the shear force diagram is drawn, we can now focus on drawing the bending moment diagram of the beam. To do that, let us use the free-body diagram of Figure 8.4 and take the moment of all forces to the left of point o as shown in the same. Hence, we can write following local moment equilibrium condition, $\circlearrowleft + \Sigma M = 0$,

$$M_x + wx \times \frac{x}{2} - R_A x = 0$$

for the range of $0 \le x \le l$, because nature of loading is same throughout the span.

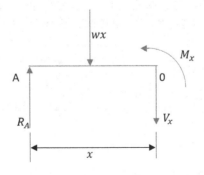

FIGURE 8.4 Free-body diagram of section x from left support of beam.

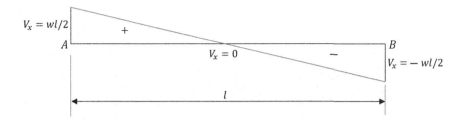

FIGURE 8.5   Shear force diagram for the entire span of the beam.

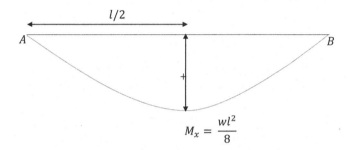

FIGURE 8.6   Bending moment diagram for the entire span of the beam.

On solving, we get,

$$M_x = -wx \times \frac{x}{2} + R_A x$$

So, from the abovementioned equation for $x = 0$, we get $M_x = -wx \times (x/2) + R_A x = 0$. So, bending moment is zero at the support. And for $x = l/2$, we get $M_x = -wl/2 \times (l/4) + R_A l/2 = wl^2/8$, which is positive. Now, if $x = l$, we get $M_x = -wl \times (l/2) + R_A l = 0$.

So, the algebraic equation for bending moment at any arbitrary section x is parabolic in nature with the maximum value at the apex of the parabola as drawn in Figure 8.6.

Example 8.2: Find the shear force and bending moment for the following cantilever beam shown in Figure 8.7.

SOLUTION: The equivalent loading diagram is shown in Figure 8.7 (b).
    Consider entire beam as free body,

$$\circlearrowleft + \sum M_B = 0$$

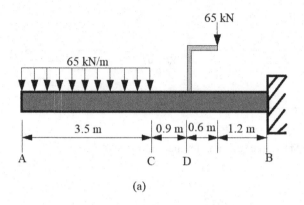

(a)

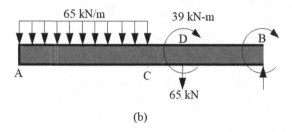

(b)

FIGURE 8.7   Problem on shear force and bending moment for a cantilever beam.

$$-M_B + 65 \times 1.8 - 39 + 227.5 \times 4.45 = 0$$

$$M_B = 1090.375 \text{ kN} - m$$

$$\uparrow + \sum F_y = 0$$

$$-65 \times 3.5 - 65 + V_B = 0$$

$$V_B = 292.5 \text{ kN}$$

1. From *A* to *C* (see Figure 8.8 (a))

$$\uparrow + \sum F_y = 0$$

$$-65x - V = 0$$

$$V = -65x \text{ kN}$$

$$\circlearrowleft + \sum M_{1-1} = 0$$

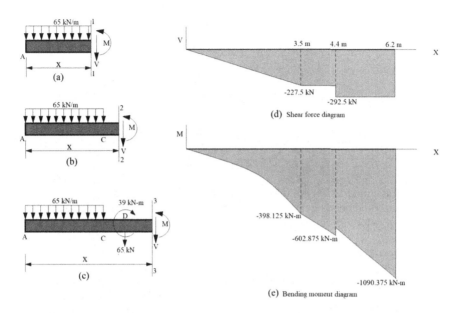

FIGURE 8.8 Problem on shear force and bending moment for a cantilever beam: SF and BM at different sections.

$$65x\frac{x}{2}+M=0$$

$$M=-32.5x^2 \text{ Kn}-m$$

This V and M is valid in the region $0 < x < 3.5$ m interval.

2. From C to D (see Figure 8.8 (b))

$$\uparrow+\sum F_y =0$$

$$-227.5-V=0$$

$$V=-227.5 \text{ kN}$$

$$\circlearrowleft+\sum M_{2-2}=0$$

$$227.5(x-1.75)+M=0$$

$$M=398.125-227.5x$$

This V and M is valid in the region $3.5 < x < 4.4$ m interval.

3. From D to B (see Figure 8.8 (c))

$$\uparrow+\sum F_y =0$$

$$-227.5 - 65 - V = 0$$

$$V = -292.5 \text{ kN}$$

$$\circlearrowleft + \sum M_{3-3} = 0$$

$$227.5(x - 1.75) + 65(x - 4.4) - 39 + M = 0$$

$$M = 723.125 - 292.5x \text{ kN} - m$$

This V and M is valid in the region $4.4 < x < 6.2$ m interval. The shear force and bending moment diagrams are shown in Figure 8.8 (d) and (e), for various sections of the beam.

So, following the abovementioned procedure shown in the example problems, and varying the range of the section x along the beam, bending moment and shear force for any general class of loading can be drawn. The values, including the sign in front of them, will dictate the positive and negative directions of shear forces and moments and should be drawn accordingly in the diagram.

8.4   QUALITATIVE DISCUSSION ON THE DEFLECTED SHAPE OF BEAMS

The qualitative deflected shape of a structure is simply a rough sketch of the neutral surface of the structure, in the deformed or bend position, under the action of applied external loading conditions. This shape is also called the elastic line for the structure. These diagrams, which can be drawn without any knowledge of the values of deflections/bending, provide valuable information about the behavior of structures and are very useful in computing the numerical values of actual deflections/bending. Detailed theoretical analysis or calculating the deflection of the structure under external loading condition will be developed in the next chapter.

According to the sign convention described in Section 5.1, a positive bending moment or sagging moment bends a beam concave downward (or toward the negative y direction as shown in Figure 8.6), whereas a negative bending moment bends a beam concave upward (or toward the positive y direction). Thus, the sign of the curvature at any point along the axis of a beam can be obtained from the bending moment diagram itself. Using the signs of bending moments, a qualitative deflected shape (elastic curve) of the beam, which is consistent with its support conditions, can be easily sketched (see Figure 8.9). For example, deflection line diagram of the beam, which we have incorporated in Example 8.1, is shown in Figure 8.8.

Consider another example of a simply supported beam with a cantilever overhung near support B as shown in Figure 8.10. Point load is acting on the beam as shown in Figure 8.10.

It is to be noted that a qualitative deflected shape is approximate, as it is drawn based on the signs of curvatures; the values of deflections along the axis of the beam

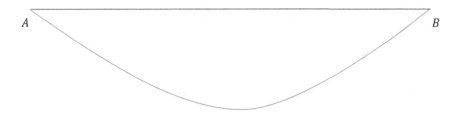

FIGURE 8.9 Deflection line diagram for the simply supported beam with uniformly distributed loading.

are not known. In any case as stated earlier, the deflected shape of beam always follows the nature of bending moment diagram. Also, in the case of bending moment diagrams, there may be points of discontinuity, but for the deflected shape, and there will not be any point of discontinuity along its entire range. Deflection line, or more specifically, elastic line, is always a continuous function and does not have any discontinuity.

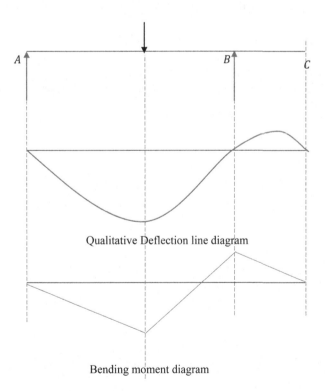

FIGURE 8.10 Deflection line diagram and bending moment diagram for the simply supported beam with a cantilever overhung.

8.5 RELATIONSHIPS BETWEEN LOADS, SHEAR, AND BENDING MOMENTS

The formation of shear and bending moment diagrams can be analyzed by using the basic differential relationships that relate the loads, the shears, and the bending moments. To derive these relationships, consider a beam subjected to an arbitrary loading, as shown in Figure 8.11. As discussed in the previous sections, we are considering a simply supported beam with uniformly distributed load all over its entire span. But it is to be noted that the relationships that will be formed ultimately are applicable to all types of beams with any general class of loading.

Let us take a small segment of beam of length dx. The free-body diagram of this beam segment with all internal and external forces acting on it will be something like as shown in Figure 8.12.

Now taking moment of all forces about the left portion of beam, we get from moment equilibrium equation:

$$(V_x + dV_x)dx + (M_x + dM_x) - M_x - wdx\frac{dx}{2} = 0$$

Since the last quantity is very small (square of differential quantities), hence, can be neglected. So, the modified differential equation takes the following form:

$$(V_x + dV_x)dx - (M_x + dM_x) + M_x = 0$$

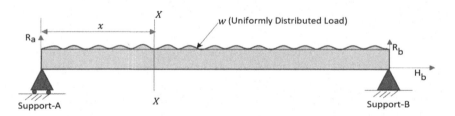

FIGURE 8.11   Simply supported beam under uniformly distributed load w throughout its length.

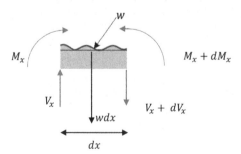

FIGURE 8.12   Free-body diagram of beam segment.

or,

$$V_x dx + dV_x dx - (M_x + dM_x) - M_x = 0$$

Since the second term from left is the product of two differential quantities, we can neglect that to get:

$$(V_x)dx - (dM_x) = 0$$

Implies,

$$V_x = \frac{dM_x}{dx}$$

Hence, we get the differential equation relating to the bending moment at any section to the shear force at the same section of the beam element. This differential equation also tells us that the slope of bending moment line is equal to the shear force at that point of the beam element. This equation also indicates that at the point of maximum bending moment, the shear force will be zero. So, at that point, the shear force line will cross the beam axis and will direct toward the other portion with respect to the beam axis. Refer to the shear force and bending moment diagram drawn in Figures 8.5 and 8.6, respectively. For that simply supported beam with uniformly distributed load, bending moment is maximum at the midspan of the beam, and hence, the shear force value at the same point is zero and the shear force line crosses the axis of the beam to progress in the upper half portion.

Integrating the abovementioned equation between two boundary points say x_A to x_B, we have:

$$M_x = \int_{x_A}^{x_B} V_x \, dx$$

which means that bending moment of two points x_A and x_B is equal to the area under the curve under the same boundary points of shear force diagram.

Now, considering the local force equilibrium equation along y direction, $\sum F_y = 0$, we get:

$$V_x - (V_x + dV_x) - w dx = 0$$

or,

$$\frac{dV_x}{dx} = w$$

So, we get the relationship between applied load intensity and the shear force at any arbitrary section x of the beam element. Mathematically, it indicates that the slope of the shear force line at any point of beam is the intensity of loading applied on the beam at that point.

In the case of concentrated loads, applying the local equilibrium condition in y direction, we can write:

$$P+(V_x+dV_x)-V_x = 0$$

or,

$$dV_x = -P$$

This equation indicates that the change in shear force at the point of application of point load is equal to the magnitude of the point load. Note that all quantities are related to each other through the derived differential equations and hence, an abrupt change in one value leads to the abrupt changes in other quantities also.

It is also to be mentioned here that the relationship between bending moment and shear force was earlier derived in Chapter 2 applying the concept of Euler Lagrange's equation. Same needs to be checked against the above-derived relationships for better in-depth understanding.

Example 8.3: Find the shear force and bending moment for the following simply supported beam shown in Figure 8.13.

SOLUTION: Finding reaction forces:

$$\circlearrowleft + \sum M_A = 0$$

or,

$$-50 \times 3 - 160 \times 6 - 80 \times 9 + R_B \times 12 = 0$$

$$\therefore R_B = 152.5 \text{ kN}$$

$$R_A = (50 + 160 + 80) - 152.5 = 137.5$$

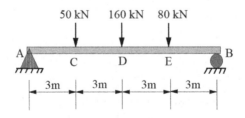

FIGURE 8.13 Problem on a shear force and bending moment diagram for a simply supported beam.

Shear force diagram:

Point A: Since a positive (upward) concentrated force of 137.5 kN magnitude acts at point A, the shear diagram increases abruptly from 0 to +137.5 kN at this point.

Point C:

The shear just to the left of point C,

$S_{B,L} = S_{A,R}$ + area under load diagram between just to the right of A to just to the left of C.

$$\therefore S_{B,L} = 137.5 + 0 = 137.5 \text{ kN}$$

and,

$$S_{B,R} = 137.5 \text{ kN} - 50 \text{ kN} = 87.5 \text{ kN}$$

Point D:

$S_{D,L} = S_{C,R}$ + area under the load diagram between just to the right of C to just to the left of D.

$$\therefore S_{D,L} = 87.5 + 0 = 87.5 \text{ kN}$$

and,

$$S_{D,R} = 87.5 \text{ kN} - 160 \text{ kN} = -72.5 \text{ kN}$$

Point E:

$S_{E,L} = S_{D,R}$ + area under the load diagram between just to the right of D to just to the left of E.

$$\therefore S_{D,L} = -72.5 + 0 = -72.5 \text{ kN}$$

and,

$$S_{E,R} = -72.5 \text{ kN} - 80 \text{ kN} = -152.5 \text{ kN}$$

Point B:

$$S_{B,L} = -152.5 + 0 = -152.5 \text{ kN}$$

$$S_{B,R} = -152.5 \text{ kN} + 152.5 \text{ kN} = 0$$

The shear force diagram is shown in Figure 8.14.

Bending moment diagram:

Point A: No couple is applied at end A. So, $M_A = 0$

Point C: $M_C = M_A$ + area under the shear force diagram between A and C. $= 0 + 412.5 = 412.5$ kN m.

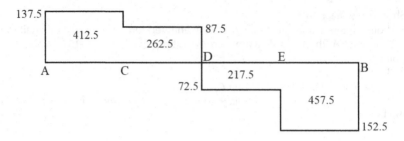

FIGURE 8.14 Shear force diagram for the example problem.

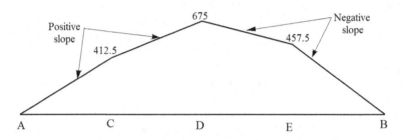

FIGURE 8.15 Bending moment diagram for the example problem.

Point D: $M_D = M_C +$ area under the shear force diagram between C and D. = 412.5 + 262.5 = 675 kN m.

Point E: $M_E = M_D +$ area under the shear force diagram between D and E. = 675 − 217.5 = 457.5 kN m.

Point B: $M_B = M_E +$ area under the shear force diagram between D and E. = 457.5 − 457.5 = 0 kN m.

The bending moment diagram is shown in Figure 8.15.

Example 8.4: Find the shear force and bending moment for the following simply supported beam shown in Figure 8.16.

SOLUTION: Finding reaction forces:

$$\rightarrow + \sum F_x = 0$$

$$A_x + 42 = 0$$

or,

$$A_x = -42 \ kN(\leftarrow)$$

$$\circlearrowright + \sum M_D = 0$$

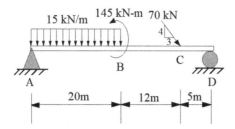

FIGURE 8.16 Problem on a shear force and bending moment diagram for a simply supported beam.

or,

$$-A_y \times 37 + (15 \times 20 \times 27) + 145 + 56 \times 5 = 0$$

$$\therefore A_y = 230.4 \text{ kN}(\uparrow)$$

$$\uparrow + \sum F_y = 0$$

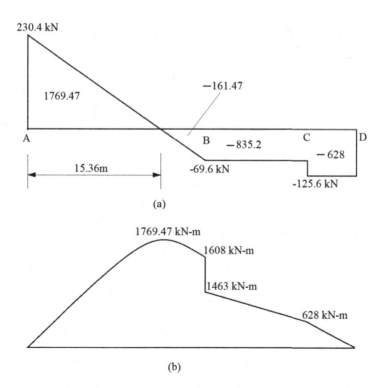

FIGURE 8.17 (a) Shear force and (b) bending moment diagram for a simply supported beam.

$$230.4 - (15 \times 20) - 56 + D_y = 0$$

$$\therefore D_y = 125.6 \text{ kN}(\uparrow)$$

Shear force diagram:
  Point A: $S_{A,R} = 230.4$ kN
  Point B: $S_{B,R} = 230.4 - (15 \times 20) = -69.6$ kN
  Point C: $S_{C,L} = -69.6 + 0 = -69.6$ kN
          $S_{C,R} = -69.6 - 56 = -125.6$ kN
  Point D: $S_{D,L} = -125.6 + 0 = -125.6$ kN
  $S_{D,R} = -125.6 + 125.6 = 0$ kN
  The shear force diagram is shown in Figure 8.17 (a).
  Bending moment diagram:
  Point A: No couple is applied at end A. So, $M_A = 0$
  Point E: $M_E = 0 + 1769.47 = 1769.47$ kN m (here slope of the bending moment
diagram is zero)
  Point B: $M_{B,L} = 1769.47 - 161.47 = 1608$ kN m
  $M_{B,R} = 1608 - 145 = 1463$ kN m
  Point C: $M_C = 1463 - 835.2 \approx 628$ kN m
  Point D: $M_D = 628 - 628 = 0$ kN m
  The bending moment diagram is shown in Figure 8.17 (b).

8.6   SHEAR AND BENDING MOMENT DIAGRAM OF FRAMES

In contrast to truss, frames are truss-like structures, which carry axial force as well as moments. Hence, for frames, both force equilibrium and moment equilibrium equations will be found useful to calculate the bending moments as well as axial and shear forces. Like beams, we need to draw the shear and the bending moment diagrams for each member as well. So, the connection among members of a frame is primarily rigid, which induces joint moments as well.

As a whole (external and internal), a frame is considered to be statically determinate if all forces, moments of each member of the frame can be determined by the application of force and moment equilibrium equations alone. No additional equations are required for a complete analysis of the statically determinate frames. In Chapter 6, the methods of determining the static indeterminacy of rigid frames were shown in a very easy and elegant way.

Shear and bending moment diagrams of a frame are formed based on the same principle as we have learned for beam elements. Since each frame element is capable of carrying bending moment and shear force, hence, for all members shear and bending moment diagrams need to be drawn as part of complete frame analysis. To be able to understand the process, let us analyze completely the following frame shown in Figure 8.18. Method of analysis presented here can be applied to any structurally determinate frame systems.

Let us determine the static determinacy of the above frame. For the given frame, $m = 3$, $j = 4$, $r_e = 3$, $r_r = 0$.

Thus, the degree of static indeterminacy, $D_s = (r_e + 3m) - (3j + r_r) = (3 + 3 \times 3) - (3 \times 4 + 0) = 0$. Hence, the frame is statically determinate. Once we get that we must

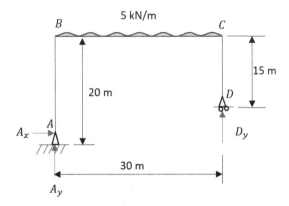

FIGURE 8.18 Example problem of a plane frame with external loading.

calculate the support reactions for the external loading applied on the structure. To be able to do that, we need to apply the global equilibrium condition, considering the frame as a whole.

Taking moment of all forces with respect to support A, we get:

$$5 \times 30 \times 30/2 - D_y \times 30 = 0$$

or,

$$D_y = 75 \text{ kN}$$

Now, applying global equilibrium equation in y direction, we can write:

$$D_y + A_y = 30 \times 5$$

or,

$$A_y = 150 - 75 = 75 \text{ kN}$$

Also applying global equilibrium condition in x direction, we get:

$$A_x = 0$$

Hence, we have calculated all support unknown support reactions of the frame by the application of global equilibrium conditions/equations.

Now, let us draw the BMD and SFD of each member by the same principle as that we have learned for beam elements. Let us begin the bending moment for member AB. Since there is no external load acting on the same, hence, there will be no bending moment or shear force at the span of the beam.

For member BC, proceeding from joint B, the moment at joint B should be zero as there is no moment coming from member AB at joint B. At midspan of member BC, the bending moment will be $w\ l^2/8$, which is equal to 562.5 kNm. And at joint C,

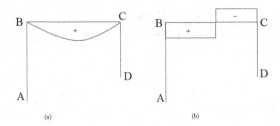

FIGURE 8.19 (a) Bending moment and (b) shear force diagrams of the example frame.

there will be no moment as for the span *CD*, there is no load acting on the frame and hence, there will be no moment. So, combining all the facts, we have the following bending moment diagram for the given frame.

Since there are no loads acting on the member *AB* and *CD*, the shear force diagram for the same has not been drawn. For any general class of loading, BMD and SFD can be drawn in the same way sketched earlier for the entire frame.

Example 8.5: Find the shear force and bending moment for the following frame shown in Figure 8.20.

SOLUTION: Let us determine the static determinacy of the above frame. For the given frame, $m = 3$, $j = 4$, $r_e = 3$, $r_r = 0$. Thus, the degree of static indeterminacy, $D_s = (r_e + 3m) - (3j + r_r) = (3 + 3 \times 3) - (3 \times 4 + 0) = 0$. Hence, the frame is statically determinate.

Finding reaction forces: $\circlearrowleft + \Sigma M_A = 0$

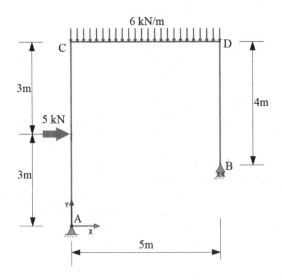

FIGURE 8.20 Example problem for shear force and bending moment of a 2D frame.

or,

$$-6 \times 5 \times 2.5 - 5 \times 3 + R_{By} \times 5 = 0$$

$$\therefore R_{By} = 18 \text{ kN}$$

$$\uparrow + \sum F_y = 0$$

or,

$$R_{Ay} + R_{By} = 6 \times 5$$

$$\therefore R_{Ay} = 12 \text{ kN}$$

$$\rightarrow + \sum F_x = 0$$

$$R_{Ax} = 5 \text{ kN}$$

The entire frame is in static equilibrium. So, its every part is in equilibrium also. First consider the column AC to be in equilibrium as shown in Figure 8.21.

$$\circlearrowleft + \sum M_A = 0$$

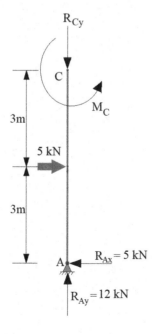

FIGURE 8.21 Equilibrium of column AC.

or,

$$-5 \times 3 + M_C = 0$$

$$\therefore M_C = 15 \text{ kN} - m$$

$$\uparrow + \sum F_y = 0$$

$$\therefore R_{Cy} = 12 \text{ kN}$$

Now consider the beam CD to be in equilibrium as shown in Figure 8.22.

$$\uparrow + \sum F_y = 0$$

or,

$$R_{Cy} + R_{Dy} = 6 \times 5$$

$$\therefore R_{Dy} = 30 - 12 = 18 \text{ kN}$$

$$\circlearrowleft + \sum M_C = 0$$

or,

$$-M_C - 6 \times 5 \times 2.5 + R_{Dy} \times 5 + M_D = 0$$

FIGURE 8.22 Equilibrium of column *BD*.

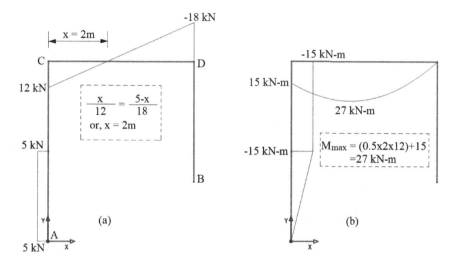

FIGURE 8.23   (a) Shear force diagram and (b) bending moment diagram.

or,

$$-15 - 75 + 90 + M_D = 0$$

$$\therefore M_D = 0$$

Consider the equilibrium of column BD as shown in Figure 8.22.

$$\uparrow + \sum F_y = 0$$

$$R_{Dy} = 18 \text{ kN}$$

The shear force and bending moment diagram is shown in Figure 8.23.

9 Deflections of Beams by Geometric Methods

9.1 INTRODUCTION

While acted upon by an external loading system, structures undergo deflection in the same direction as that of the applied loading. In case external loading acts on the structure in different directions, the resultant direction of the external loading system dictates the deflection of the structure in that direction. Deflection of structure is a major part of serviceability requirements related to the design of structure. Structures that may be safe under the loads may not be safe under the case of deflection. That is why all structures are being designed to satisfy both strength and serviceability requirements simultaneously and independently. Once a structure is found to be safe, satisfying both these conditions, it is regarded as a safe structure, and design analysis can be concluded at that point. In this chapter, we will study the deflection calculation for beams under different types of loading, and we will also learn geometric methods to determine the deflection due to the application of general class of loads acting on the structures externally. The methods of calculating structural deflection are classified into two parts, energy methods and geometric method. As the chapter name suggests, we will learn the geometric method of calculating structural deflection and energy method we will learn in a separate chapter. The deformations we will deal with in this chapter are the linear elastic deformations.

9.2 DEFLECTED SHAPES AND ELASTIC CURVE

Before we start developing the deflected shape geometric ideas, students are advised to go through Section 5.4 of the equilibrium condition chapter. In that section, we have already explained in detail that, under the application of transverse loading, beam members undergo deflection in the direction of the applied force. Due to such bending, the upper fibers of the beam element remain under compression, and the bottom fibers of the beam remain under tension. To understand the concept, see Figure 9.1 that is being reproduced here from Chapter 5.

Under these circumstances, we should be able to calculate the numerical value of deflection by solving certain equations. To develop the equation of deflection of a beam element under any general transverse loading, we need to look deeply into the bending feature and its relationship with the radius of curvature of bending. See Figure 9.2 for a typical beam element under any general transverse loading.

We have drawn the deflection line of the beam for any general class of transverse loading, and we have marked two close section lines at, m and m' which are ds apart

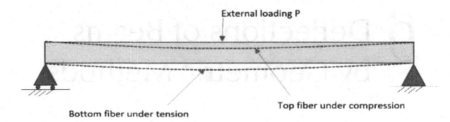

FIGURE 9.1   Deflection diagram of beam under flexural loading.

on the deflection line, to have a closer look into the deflected shape of the beam. Let us consider a small beam element ds between the two section lines. Connecting the radius of the curved line from common center point O, let us denote the small angle between the two radius lines as $d\theta$. In Figure 9.2, the tangent at m makes an angle θ with x axis. The geometric relationship between radius, angle of rotation of radius vector, and arc length assumes the following form:

$$ds = rd\theta$$

So,

$$\frac{1}{r} = \left|\frac{d\theta}{ds}\right|$$

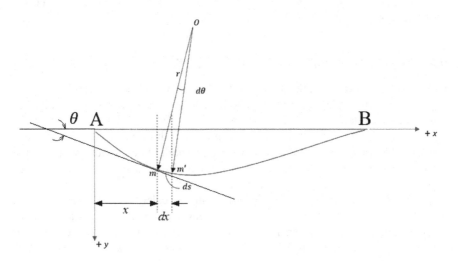

FIGURE 9.2   Differential beam element with radius of curvature for transversely loaded beams.

The modulus sign indicates we are only concerned about the numerical value of the radius of curvature. As the angle θ decreases if the point m move moves along the curve from A to B in Figure 9.2, hence, the actual sign will be as follows:

$$\frac{1}{r} = -\frac{d\theta}{ds}$$

In the case of pure bending, the relationship between moment and radius of curvature is as follows:

$$\frac{1}{r} = \frac{M}{EI_z}$$

where, I_z is the moment of inertia about the z axis of the beam element (transverse to the x direction of the beam). For all practical purposes, since the arc length will be very small, we may write the following approximations without causing any geometrical change on the curvature of the beam,

$$ds \approx dx$$

and,

$$\theta \approx \tan \theta = \frac{dy}{dx}$$

So, substituting the values of ds and θ, we can write,

$$\frac{1}{r} = -\frac{d^2 y}{dx^2}$$

And equating this with the moment relationship, we get,

$$\frac{d^2 y}{dx^2} = -\frac{M}{EI_z}$$

Thus, the above differential equation is the required relationship between deflection line and the bending moment of a beam element we sought for. It should be noted that the sign in this equation depends on the direction of the coordinate axis. Solving this differential equation with proper initial and boundary conditions, we can determine the deflection line of a beam element under any type of transverse loading system. Let us go through a simple example of a simply supported beam with point loading at the middle span of the beam. We want to calculate the deflection at middle point of the beam where the load is being applied. From the qualitative discussion, it is clear that deflection becomes maximum just under the load with maximum intensity. Hence, we can assume that for the simply supported beam, let us find the maximum deflection from the differential equation itself. This method of solving the differential equation twice to get the deflection is known as the double integration method, which will be introduced in the following section.

9.3 DOUBLE INTEGRATION METHOD

From Figure 9.3, and earlier discussion, the bending moment at any section x from the left support A is given by,

$$M = R_A \ x = \frac{P}{2} x$$

So, the value of bending moment at the midspan of the beam (i.e., at $x = l/2$) is:

$$M = R_A \ x = \frac{P}{2} x = \frac{Pl}{4}$$

Inserting the general expression for bending moment in the differential equation we have,

$$\frac{d^2 y}{dx^2} = -\frac{M}{EI_z} = -\frac{P}{2EI_z} x$$

or, integrating the above w.r.t x we have,

$$\frac{dy}{dx} = -\frac{P}{4EI_z} x^2 + c_1$$

where, c_1 is the integration constant. Integrating the above differential equation once again we get,

$$y = -\frac{P}{12EI_z} x^3 + c_1 x + c_2$$

Now, we know that for any beams, at support, deflection will be zero. So, at $x = 0$, $y = 0$. Putting this boundary condition in the above expression, we get,

$$c_2 = 0$$

So, we are left with the following expression for deflection line with one unknown parameter, c_1. To determine this parameter, we recall our earlier discussion on the location of maximum value of deflection. The maximum deflection should occur at the same point where the applied load has the maximum intensity. Hence, at $x = l/2$,

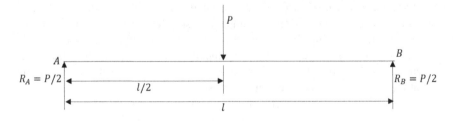

FIGURE 9.3 Simply supported beam with point load.

the deflection should be maximum. So, at this point, the slope of the deflection line should be zero, i.e., at this point $dy/dx = 0$.

Applying the mentioned boundary condition, we get,

$$0 = -\frac{P}{4EI_Z}\left(\frac{l}{2}\right)^2 + c_1$$

or,

$$c_1 = \frac{Pl^2}{16EI_Z}$$

With this, the final form of deflection line relationship becomes,

$$y = -\frac{P}{12EI_Z}x^3 + \frac{Pl^2}{16\ EI_Z}x$$

Now, putting $x = l/2$ to get the desired deflection value at the middle span of beam,

$$y = \frac{Pl^3}{48\ EI_Z}$$

Thus, by solving the differential equation with appropriate boundary conditions, we can determine the actual deflection at any point of a transversely loaded beam. Double integration method also provides the essential tool for solving the differential equation by numerical integration method to calculate the deflection with a good degree of accuracy.

9.4 MOMENT-AREA METHOD

Expanding the idea of double integration, we will learn a very simple yet powerful tool in calculating the deflection of beam. This method is known as moment area method. This method was developed by Charles E. Greene in 1873. To understand this method, let us once again write down the general differential equation for deflection of beams:

$$\frac{d^2y}{dx^2} = -\frac{M}{EI_Z}$$

Integrating the above general equation once between two arbitrary points A and B on the beam, we get:

$$\frac{dy}{dx} = \theta_{BA} = \theta_B - \theta_A = -\int_A^B \frac{M}{EI_Z}dx$$

This integral gives the change in slope between two arbitrary points A and B on the beam. Again, it should be noted that the sign in this equation depends on the direction of the coordinate axis. For example, if we take y-axis positive upwards,

$\theta \approx -\frac{dy}{dx}$. Now we will obtain plus instead of minus. θ_A and θ_B are the slopes of the elastic curve at point A and B, respectively, with respect to the undeformed axis of the beam. θ_{BA} denotes the angle between the tangents to the elastic curve at A and B and $\left| \int_A^B \frac{M}{EI_Z} dx \right|$ represents the area under the $\frac{M}{EI}$ diagram between points A and B. This concept is known as the first moment area theorem. Continuing further, if we integrate it once more, we have the following relationship:

$$y_{BA} = -\int_A^B \frac{M}{EI_Z} x \, dx$$

This relationship is the desired relationship that we are looking for. Here, y_{BA} represents the tangential deviation of point B from the tangent at A, which is the deflection of point B in the direction perpendicular to the undeformed axis of the beam from the tangent at point A, and $\left| \int_A^B \frac{M}{EI_Z} x \, dx \right|$ represents the moment of the area under the $\frac{M}{EI}$ diagram between points A and B about point B. This concept is known as the second moment area theorem. Unlike the double integration method, moment area theorem totally depends on the shape and type of bending moment diagram. So, to apply this theorem, one needs to be very careful about drawing the bending moment diagram accurately. These two concepts are explained through Figure 9.4.

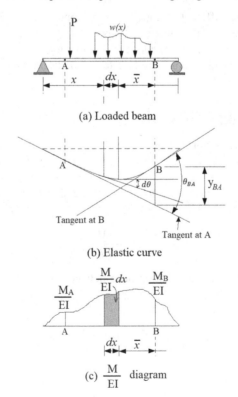

(a) Loaded beam

(b) Elastic curve

(c) $\dfrac{M}{EI}$ diagram

FIGURE 9.4 Deflection of a beam: moment area method.

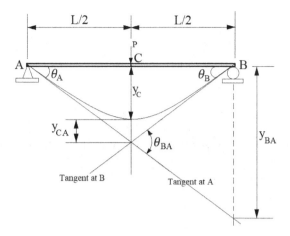

(a) Loaded beam and elastic curve

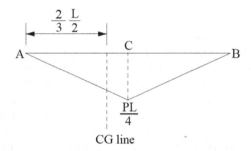

(b) Bending moment diagram and its CG location

FIGURE 9.5 Deflected shape of the beam and its bending moment diagram.

To understand this method, let us consider the same problem of simply supported beam with point load acting at its midpoint as shown in Figure 9.5.

The deflected shape and the bending moment diagram for this beam is shown as in Figure 9.5.

Now, to calculate deflection at midpoint of the beam, we need to draw two tangents at point A and B of the deflection diagram as shown in Figure 9.5 (a). Therefore, we can write $\theta_A = \frac{y_{BA}}{L}$ in which θ_A is assumed to be small, so that $tan\theta_A \approx \theta_A$. The tangential deviation y_{BA} can be expressed as:

$$y_{BA} = \text{moment of area of the } \frac{M}{EI} \text{ diagram between } A \text{ and } B \text{ about } B$$

i.e., $y_{BA} = \frac{1}{EI}\left[\left\{\frac{1}{2}\times\frac{L}{2}\times\frac{PL}{4}\times\left(\frac{1}{3}\times\frac{L}{2}\right)\right\}+\left\{\frac{1}{2}\times\frac{L}{2}\times\frac{PL}{4}\times\left(\frac{L}{2}+\frac{2}{3}\frac{L}{2}\right)\right\}\right]=\frac{PL^3}{16EI}$

Therefore, $\theta_A = \frac{y_{BA}}{L} = \frac{PL^3}{16EI}\times\frac{1}{L} = \frac{PL^2}{16EI}$

Again, from the second moment area theorem we can get the tangential deviation y_{CA}, as:

$$y_{CA} = \frac{1}{EI}\left[\frac{PL^2}{16} \times \frac{1}{3} \times \frac{L}{2}\right] = \frac{PL^3}{96EI}$$

If the deflection is small, we can write $(y_C + y_{CA}) = \frac{L}{2} \times \theta_A$,
where y_C is the deflection of the beam under the load, and y_{CA} represents the tangential deviation of point C from the tangent at A. Now, if we put the values of y_{CA} and, θ_A in the above expression, we get the value of y_C, which is the desired deflection amount at the midpoint of the beam under this loading condition.

Therefore, $y_C = \frac{L}{2} \times \theta_A - y_{CA} = \frac{L}{2} \times \frac{PL^2}{16EI} - \frac{PL^3}{96EI} = \frac{PL^3}{48EI}$

This is in exact agreement with the earlier found value of deflection calculated using double integration method.

Students are encouraged to apply this concept for several different kinds of beams with different kinds of loading. This method gives a very quick assessment of the deflection value and is purely geometrical in nature. However, this method has certain limitations. In the case of complex loading and irregular geometry of bending moment diagram, it is difficult to determine the distance of CG points from points of concern. Thus, without correct bending moment and CG distance, this method does not provide the correct result. However, for simple loading and symmetry, this method gives a very quick assessment of the deflection values. It is to be noted, for any general loading system, irrespective of symmetry, double integration method provided the most general way to get the desired result. Double integration method does not require any kind of diagram to complete the analysis. Only deflection line equation and few boundary conditions as per the nature of loading and support conditions are sufficient to determine the exact value of deflection at any point on the beam element. For ease of reference, few geometric shapes with area and distance of CG are shown in Figure 9.6, which may be found useful while applying the moment area method for different types of loads.

9.5 CONJUGATE BEAM METHOD

Continuing the discussion further on the deflection topic, let us learn another simple yet powerful geometric method known as conjugate beam method. We begin this from the same equation as that has been used for deflection line. This method was proposed by Prof. Otto Mohr in 1868 and later on developed by Prof. Muller-Breslau in 1885. In this method, sketching of elastic curve is not required. The name conjugate beam refers to an imaginary beam of same length as that of the original beam. The width of the beam is assumed to be of the length EI. And the beam is loaded with the bending moment diagram of the original beam. The slope and deflection are then calculated by the following two theorems known as Mohr's Theorem.

The shear force at any section of the conjugate beam is equal to the slope of the elastic curve at the corresponding section of the original beam.

The bending moment at any section of the conjugate beam is equal to the deflection of the elastic curve at the corresponding section of the actual beam.

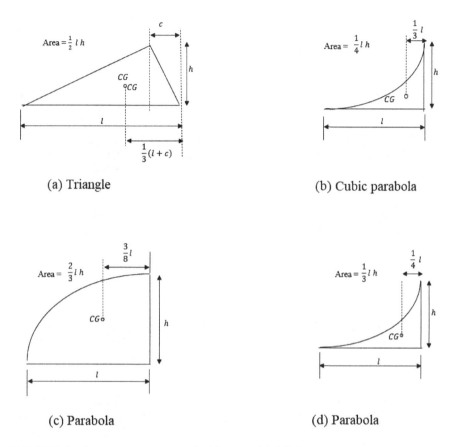

(a) Triangle

(b) Cubic parabola

(c) Parabola

(d) Parabola

FIGURE 9.6 Some common geometrical figures with CG distance and area.

To simplify the formation of conjugate beam, we simply load it with M/EI and then apply Mohr's theorems as stated above. Apart from the above theorems, there are a few key steps one needs to follow to apply the theorem correctly. See Table 9.1 for such points that need to be followed while constructing conjugate beams.

To better understand the process, we take the following example of a cantilever beam loaded with a point load at its free end as shown in Figure 9.7. We will apply the concept of conjugate beam to calculate the slope and deflection at the free end of the beam by conjugate beam method.

First, let us draw the bending moment diagram of the above cantilever beam. It is as shown in Figure 9.8.

The bending moment at the free end of the beam is zero and it varies linearly up to support. At the support, the bending moment is negative with value Pl. Now to draw the conjugate beam, we first draw the axis line of the beam as the same length as that of the original beam. Then, we draw the bending moment divided EI as applied loading on the same beam. Following the above table, the fixed end of cantilever beam will be replaced by free end and the free end of original beam will be replaced

TABLE 9.1

Supports for Conjugate Beam

Actual Beam	Conjugate Beam	Remarks
Fixed end supports	Free end supports	Slope and deflection at the fixed end of real beam is zero. So, the shear force and bending moment values of conjugate beam is zero at the two supports.
Free end	Fixed end	Slope and deflection at the free end of original beam is not zero. So, the shear force and bending moments at these locations will also not be zero.
Simply supported or roller supported end	Simply supported end	Slope at the supports for original beam exists but deflection is zero at both supports. So, in conjugate beam, shear force exists at the supports but bending moment should be zero at both supports.

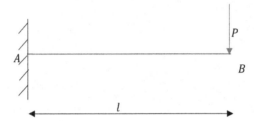

FIGURE 9.7   Cantilever beam with point load at its free end.

FIGURE 9.8   Bending moment diagram.

by fixed end in the conjugate beam. Thus, following these steps, the conjugate beam with all modified support conditions and loading will look something like as shown in Figure 9.9.

Now, we are left with a simple task. To calculate the deflection at free end of original beam, we simply need to calculate the bending moment at that point of the conjugate beam. From Figure 9.9, the bending moment at the support of the conjugate beam is given by:

$$y_B = \text{Bending moment at support } B = \frac{1}{2}\frac{Pl}{EI} \times l \times \frac{2l}{3} = \frac{Pl^3}{3EI}$$

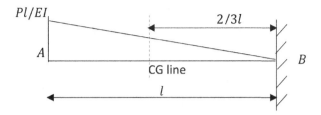

FIGURE 9.9 Conjugate beam of cantilever beam.

So, it is fairly simple process, and we get the correct expression for calculating deflection at the free end of the cantilever beam. Similarly, for slope of elastic line at this end of real beam, we need to calculate the shear force at the same point of the conjugate beam. Hence, slope will be equal to shear force at support B of conjugate beam, which is denoted as $\frac{1}{2} Pl/EI \times l = Pl^2/2EI$. These values can be crosschecked by the earlier discussed methods by calculating deflection and slope at the same point of the beam. Students are also encouraged to calculate the deflection of a simply supported beam with point load acting at the midspan of the beam. While applying conjugate beam method, students are encouraged to take care of the necessary changes that are required for supports.

We will conclude this section by giving another example of a cantilever beam applied by uniformly distributed load to its entire length as shown in Figure 9.10.

As we have done in our previous example, first we need to draw the actual bending moment diagram and then this diagram, divided by EI, will be applied as a load on the conjugate beam. Also, the free end of the real beam will be changed to fixed end in conjugate beam and the fixed end of the beam will be free end for conjugate beams. These diagrams are shown in Figure 9.11 for ease of understanding.

From the above diagram, we will now calculate the bending moment at support B of the conjugate beam. This will be $wl^2/2EI \times l/3 \times (l - 3l/8) = 5wl^4/48EI$. So, this is the desired deflection at the free end B due to this loading.

Again, students are encouraged to check this value by calculating deflection, applying other alternative methods introduced in the earlier sections of this chapter.

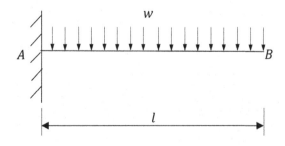

FIGURE 9.10 Cantilever beam with UDL.

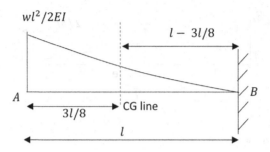

FIGURE 9.11   Conjugate beam with loading.

9.6   MACAULAY'S METHOD

As we have seen, the double integration method is very tedious and involves considering several correct boundary conditions depending on the geometry of the deflection line. Macaulay's method provides a simple method of integration in such a way that the constants of integration are valid for all sections of the beam; even though the law of bending moment varies from section to section. The following rules need to be followed while applying Macaulay's method of deflection calculation:

1. Origin is always to be considered at the left support of beam.
2. Positive and negative moments are considered anticlockwise and clockwise direction, respectively.

With the above mentioned points, let us calculate the deflection of the simply supported beam under the application of point load at its midspan. As stated, we consider that origin of the beam is at the left support of the beam. Now the bending moment at any section x from the left support is given by $M = Px/2$. In this case, the range of x varies from 0 to $l/2$; for a section more than $l/2$ from left support, the bending moment will be $M = Px/2 - P(x - l/2)$. Refer Figure 9.12 for ease of understanding.

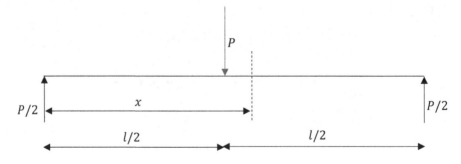

FIGURE 9.12   Simply supported beam with point load at midspan.

So, for the entire span of the beam, the bending moment equation can be written as follows:

$$M = Px/2 - P(x - l/2)$$

When we want to calculate the bending moment within the range $0 \leq x \leq l/2$, then, we will only use the first half of the equation. When we want to calculate bending moment within the range $l/2 \leq x \leq l$, we will use the complete equation as written above. Inserting the above expression of bending moment in elastic line equation we have:

$$EI \frac{d^2 y}{dx^2} = -\frac{Px}{2} + P(x - l/2)$$

Integrating the above equation, we get,

$$EI \frac{dy}{dx} = -\frac{P}{4} x^2 + C_1 + \frac{P}{2} \left(x - \frac{l}{2} \right)^2$$

It is to be noted that $(x - l/2)$ has been integrated as a whole and not part wise. Integrating once again we get:

$$EI \, y = -\frac{P}{12} x^3 + C_1 x + C_2 + \frac{P}{6} \left(x - \frac{l}{2} \right)^3$$

It may again be noted that the expression $\left(x - \frac{l}{2} \right)^2$ has been integrated as a whole and not part wise. We know that when $x = 0$, $y = 0$. Substituting these values in the above equation we get,

$$C_2 = 0$$

To get the above value of C_2, we need to keep in mind that for $0 \leq x \leq l/2$, we need to use the first half of the equation. We also know that at $x = l$, $y = 0$. So, for this range we need to use the entire expression as written earlier for deflection. Putting this value, we get,

$$0 = -\frac{P}{12} \times l^3 + C_1 \times l + \frac{P}{6} \left(0 - \frac{l}{2} \right)^3$$

or,

$$C_1 = \frac{Pl^2}{16}$$

Now, substituting this value in the differential equation, we get,

$$EI\, y = -\frac{P}{12}x^3 + \frac{Pl^2}{16}x + \frac{P}{6}\left(x - \frac{l}{2}\right)^3$$

Now, for $x = l/2$, we need to use the two parts of the expression, to the left of the doted section location line. So, deflection at midspan will be:

$$EI\, y = -\frac{P}{12}x^3 + \frac{Pl^2}{16}x = \frac{Pl^3}{48EI}$$

This is in exact deflection value that was derived earlier by other methods.

10 Energy Principles and Deflection of Beam

10.1 INTRODUCTION

In the previous chapter, we have learned several geometric methods to determine the deflection of beams under the application of different kinds of loading. All these methods primarily rely on the bending moment diagrams and their geometric shapes. Energy principles, on the other hand, are geometry independent. In energy methods, we need to construct the equation based on energy conservation principle or work energy theorems and on scalar relations. It means that we need not pay any attention to the nature of bending moment diagrams and several other details related to that bending moment diagrams. More will be understood in detail once we go through this chapter in detail and practice the necessary mechanisms to form energy equations.

10.2 STRAIN ENERGY AND PURE BENDING

Let us consider a straight prismatic simply supported beam AB of length l made of a homogeneous material with two transverse external loads of magnitude P acting on it, as shown in Figure 10.1.

From the Bending Moment Diagram(BMD) and Shear Force Diagram (SFD), we find that between C and D, there is no shear force acting in the beam and the moment at this portion of the beam is uniform with value $M = Pa$. This condition is known as the pure bending condition of a beam. Due to pure bending, the portion CD of the beam will take the form of a circular arc. In this deformed state, each cross-section of the beam, initially plane, is assumed to remain plane after bending. This beam types is called Euler beams, although this condition is not valid for large bending or impure bending state. Beams in which the plane sections before and after bending do not remain plane are called shear deformable beams or Timoshenko beams. Our principal focus will be on the Euler beams, and we will develop more details and deeper insight into these types of beams and their deflections.

As a result of the deformation, as shown in Figure 10.2, fibers on the convex side of the beam will be elongated slightly, whereas, on the concave side, the fibers will be shortened. Somewhere between the top and bottom fiber, there will be a layer of fiber that does not undergo any kind of change in length. This layer is known as the neutral layer or neutral axis of the beam.

From the above diagram, the plane of two cross sections mn and pq intersect at a common point O. Let us denote the angle between these two planes by $d\theta$, and noting that, $d\theta = dx/r$ where $1/r$ is the curvature of the neutral axis. Now, at this situation, we draw through point b on the neutral axis, a line $p'q'$ parallel to mn and indicating

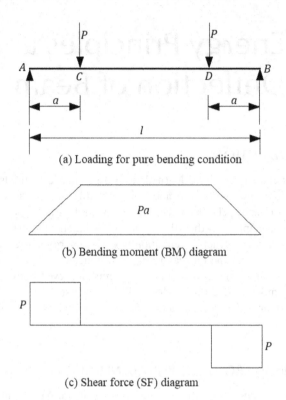

FIGURE 10.1   (a) Beam with transverse load, (b) BMD, and (c) SFD.

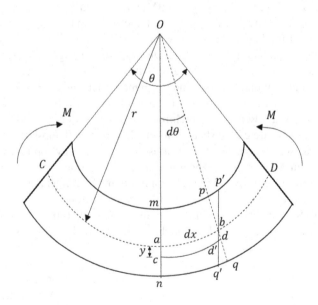

FIGURE 10.2   Deformed shape and neutral axis.

the original orientation of pq before bending. From this figure, we find that segment cd' of any fiber at the distance y from neutral axis elongates by an amount $d'd = yd\theta$. Since its original length is $cd' = dx$, the corresponding strain is:

$$\epsilon = \frac{yd\theta}{dx} = \frac{y}{r}$$

If a fiber on the concave side of the neutral axis is considered, the distance y will be negative, and the strain is also negative. Thus, all fibers on the concave side of neutral axis will be in a state of compression. And following similar logic, we can say, all fibers on the convex side of neutral axis will be under tension. Since we are working well within the elastic limit, stress will be proportional to strain. Thus,

$$\sigma_x = \epsilon E = \frac{E}{r} y$$

This indicates fiber stress is linearly dependent on distance y from neutral axis. This is only applicable when we are dealing with a case of pure bending and well within the elastic limit of the beam material where Hook's law is valid. We can locate the neutral axis position on the cross-section by satisfying the condition that the resultant force produced by the stress distribution over the cross-sectional area must be equal to zero.

Let dA be a small elemental area of the cross section at a distance y from the neutral axis (see Figure 10.3). Then force acting on this differential element is $\sigma_x dA$. Using the earlier derived relationship between stress and strain, we can write,

$$\sigma_x dA = \frac{E}{r} y dA$$

Since there cannot be any resultant force acting on the cross-section as the beam is under static equilibrium, the integral of the above differential relationship over the entire area must be zero. So,

$$\sum F_x = 0$$

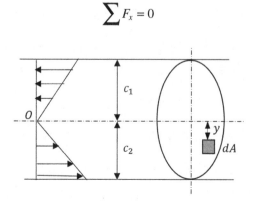

FIGURE 10.3 Stress distribution and location of neutral axis.

or,

$$\int \frac{E}{r} y\, dA = 0$$

From this, we can say that,

$$\int y\, dA = 0$$

$$Ay_c = 0$$

where A is the total area of the beam cross section and y_c is the distance of neutral axis from the CG of the beam. And since $A \neq 0$, hence, we conclude that, $y_c = 0$. This means that neutral axis is passing through the centroid of the beam. Total moment (bending moment) acting in the section due to external loading is given by,

$$M = \int y\sigma_x\, dA = \frac{E}{r} \int y^2 dA$$

The integral in the left-hand side is the second moment of area or moment of inertia of the beam section. Hence, we form an important relationship related to the bending moment developed inside the beam and its moment of inertia and radius of curvature of its bending, which is given by the following relationship:

$$M = \frac{E}{r} I$$

or,

$$\frac{1}{r} = \frac{M}{EI}$$

Using the differential relationship of elastic line derivatives and radius of curvature, we can write the following equation that we have already studied in the previous chapter:

$$\frac{d^2 y}{dx^2} = \frac{M}{EI}$$

The famous equation of elastic line we have again derived. Thus, the equation of elastic line is a corollary of state of pure bending. Once in the state of pure bending, the strain energy stored in the beam, per unit length (or per unit volume for three-dimensional object), can be expressed as:

$$\text{Strain energy } (U) = \frac{1}{2}(\text{stress} \times \text{strain})$$

or, within elastic limit, applying Hook's law, we can write,

$$U = \frac{1}{2} \frac{\text{stress}^2}{E}$$

From earlier discussion, we have related stress developed in an element with that of the bending moment developed inside it. Suppose the length of the neutral axis, *CD* is *l*, then we have (refer Figure 10.2),

$$r\theta = l$$

or,

$$\theta = \frac{l}{r} = \frac{Ml}{EI}$$

So, for the case of pure bending, the net work done in bending the beam by an angle θ is given by,

$$U = \frac{1}{2}M\theta$$

Using the above two equations we get,

$$U = \frac{1}{2}M\theta = \frac{1}{2}M\left(\frac{Ml}{EI}\right)$$

or,

$$U = \frac{1}{2}M\theta = \frac{1}{2}\frac{M^2l}{EI}$$

From the above equation, we can determine the strain energy due to pure bending. If instead of total length, we investigate the strain energy element wise, i.e., by taking small length of beam element *dx*, we have the following integral relating strain energy and bending moment developed inside the beam,

$$U = \int_0^l \frac{M^2 dx}{2EI}$$

or, replacing the above by equations of elastic line, we get,

$$U = \int_0^l \frac{M^2 dx}{2EI}$$

For linear elongation of prismatic element of length *l*, subjected to an axial load *P*, having cross sectional area *A*, and volume *Al*, the strain energy stored in the beam due to the elongation is given by:

$$U = \frac{1}{2}\left(\text{stress} \times \text{strain}\right) \times \text{volume} = \frac{1}{2} \times \sigma \times \frac{\Delta l}{l} \times (Al) = \frac{1}{2} \times \frac{P}{A} \times \frac{Pl}{AEl} \times (Al) = \frac{P^2l}{2EA}$$

The above relationship has been derived from Hook's law of elasticity which is as follows:

$$\frac{\text{Stress}}{\text{Strain}} = E$$

or,

$$\frac{P/A}{\Delta l/l} = E$$

From the above, we have,

$$\Delta l = \frac{Pl}{AE}$$

By inserting this expression in the strain energy relationship, we get the earlier derived result. For a curious-minded reader, here is a hint on how to calculate deflection or elongation of member using the strain energy expression. If we differentiate partially the strain energy expression for prismatic member of length l, we have the following expression:

$$\frac{\partial U}{\partial l} = \frac{Pl}{AE} = \Delta l$$

which is the elongation length we have found out earlier by the application of Hook's Law. Thus, we get an important relationship between strain energy stored in the beam and deflection of the beam; partial derivative of the strain energy stored in a beam with respect to the length variable of the beam gives the deflection of beam in the same direction of the applied force or moment induced by the applied external loads. We will develop this concept further in subsequent sections of this chapter.

10.3   PRINCIPLE OF VIRTUAL WORK

Let us first discuss the concept of work. In mechanics, a force or a couple does work when it displaces a body dx or rotates a body $d\theta$ amount, respectively, in the same direction as the force or couple. The amount of external work done is a scalar quantity and can be expressed as, $dU_e = Fdx$, in case of applied force; $dU_e = Md\theta$, in case of applied moment. If the total displacement is Δ, the work becomes,

$$U_e = \int_0^\Delta Fdx$$

If the total angular displacement is θ radian, the work becomes,

$$U_e = \int_0^\theta Md\theta$$

Suppose the force gradually increases from zero to a final value $F = P$. In this process, the displacement reaches its limiting value Δ. If the material behaves in a linear elastic manner, the force will be directly proportional to the displacement, i.e., $F = \frac{P}{\Delta} x$. If we replace this in the earlier integral expression, we obtain: $U_e = \frac{1}{2} P\Delta$. Similarly, for the case of angular rotation, we can write, $U_e = \frac{1}{2} M\theta$. If the P or M remains constant during the displacement or rotation, the amount of work is given by $U_e = P\Delta$ or $U_e = M\theta$, respectively.

Now, let us discuss about the *virtual work*. The principle of virtual work was first introduced by John Bernoulli in 1717. This principle does not have any physical significance, but it provides a powerful analytical tool for many problems in structural mechanics. The word virtual means imaginary or not real. We already understood that work is a dot product of force and displacement. The 'work' will become virtual, if one of the product elements is imaginary. Both cannot be imaginary simultaneously as we need to analyze and correlate an actual structure. So, considering the product elements to be virtual, we can develop two virtual work principles, namely, *the principle of virtual displacements for rigid bodies* and *the principle of virtual forces for deformable bodies*.

10.3.1 PRINCIPLE OF VIRTUAL DISPLACEMENTS FOR RIGID BODIES

The principle states that, *if a rigid body is in equilibrium under a system of forces and if it is subjected to any small virtual rigid body displacement, the virtual work done by the external forces is zero.* This principle can be easily proved with the help of a simple illustration as shown in Figure 10.4.

The free body diagram of the simply supported beam is shown in Figure 10.4 (b). Now, suppose this beam is given an arbitrary small virtual rigid body displacement from its initial equilibrium position ACB to $A'C'B'$. The total virtual rigid body displacement can be decomposed into translation Δ_{vx} and Δ_{vy} in the x and y direction, respectively, and a rotation θ_v about A. When the beam undergoes this

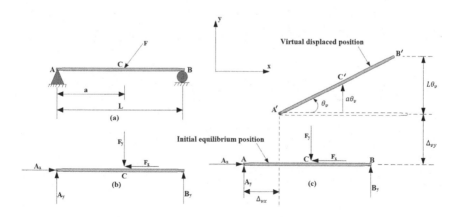

FIGURE 10.4 Virtual displacement of a rigid body.

virtual displacement from ACB to $A'C'B'$, the forces acting on it perform virtual work as:

$$W_{ve} = W_{vx} + W_{vy} + W_{vr}$$

where W_{vx} and W_{vy} are the virtual work done during translation in x and y direction, respectively, and W_{vr} is the virtual work done during rotation. The virtual work done during translation in x and y can be written as follows, respectively:

$$W_{vx} = A_x\Delta_{vx} - F_x\Delta_{vx} = \left(A_x - F_x\right)\Delta_{vx} = \left(\sum F_x\right)\Delta_{vx}$$

$$W_{vy} = A_y\Delta_{vy} - F_y\Delta_{vy} + B_y\Delta_{vy} = \left(A_y - F_y + B_y\right)\Delta_{vy} = \left(\sum F_y\right)\Delta_{vy}$$

And the virtual work done by all the forces during the small virtual rotation θ_v can be expressed as:

$$W_{vr} = -F_y\left(a\theta_v\right) + B_y\left(L\theta_v\right) = \left(-aF_y + LB_y\right)\theta_v = \left(\sum M_A\right)\theta_v$$

Now, we can write the total virtual work as:

$$W_{ve} = \left(\sum F_x\right)\Delta_{vx} + \left(\sum F_y\right)\Delta_{vy} + \left(\sum M_A\right)\theta_v$$

As the beam is in equilibrium under the action of all the forces, we can write: $\sum F_x = 0$, $\sum F_y = 0$ and $\sum M_A = 0$. Therefore, $W_{ve} = 0$. Hence, it proves the principle. To this end, we will study Example 10.1 to understand applying the principle of virtual displacement for rigid bodies in solving engineering problems.

Example 10.1: Each rod in the following frame is uniform of mass m and rigid. There is an acting force P at joint C in the total frame system as shown in Figure 10.5. Calculate the angle θ between the two members under equilibrium condition.

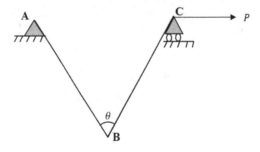

FIGURE 10.5 Example problem on principle of virtual displacement for rigid bodies.

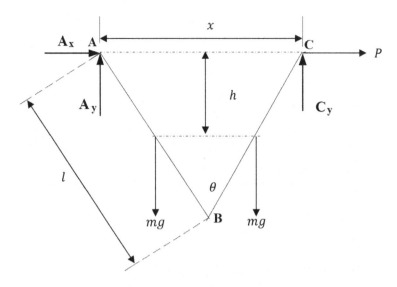

FIGURE 10.6 Free body diagram of the frame under equilibrium.

SOLUTION: We will solve this problem using principle of virtual displacement for rigid bodies, which states that under any virtual displacement, total work done by external forces is zero. Let us draw the free body diagram under equilibrium condition with the following configuration as shown in Figure 10.6.

From Figure 10.6, under equilibrium condition, the following geometric relationships can be made,

$$x = 2l \sin\frac{\theta}{2}$$

so,

$$\delta x = l \cos\frac{\theta}{2} \delta\theta$$

similarly,

$$h = \frac{l}{2} \cos\frac{\theta}{2}$$

or,

$$\delta h = -\frac{l}{4} \sin\frac{\theta}{2} \delta\theta$$

Now, for the entire system, applying principle of virtual work for acting forces we get:

$$P \delta x + 2mg \delta h = 0$$

Substituting the virtual displacements with known parameters already derived, we get,

$$Pl \cos\frac{\theta}{2} \, \delta\theta - 2mg \, \frac{l}{4} \sin\frac{\theta}{2} \, \delta\theta = 0$$

From which, cancelling the common terms we get,

$$\tan\frac{\theta}{2} = \frac{2P}{mg}$$

or,

$$\theta = 2\tan^{-1}\frac{2P}{mg}$$

10.3.2 PRINCIPLE OF VIRTUAL FORCES FOR DEFORMABLE BODIES

The principle states that, *if a deformable structure is in equilibrium under a virtual system of forces and couples and if it is subjected to any small real deformation consistent with the support and continuity conditions of the structure, then the virtual external work done by the virtual external forces and couples acting through the real external displacements and rotations is equal to the virtual internal work done by the virtual internal forces and couples acting through the real internal displacements and rotations.* This principle can be easily proved with the help of a simple illustration as shown in Figure 10.7.

Suppose a truss is in equilibrium under the action of a virtual external force P_v. Consider the free body diagram of joint C as shown in Figure 10.5 (b). As the whole structure is in equilibrium, the joint C will be also in equilibrium. Since the problem

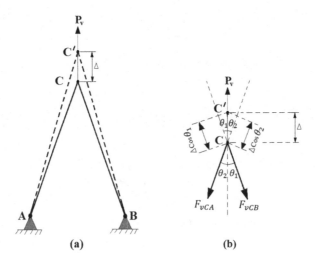

FIGURE 10.7   (a) Virtual forces acting on a deformable body, (b) equilibrium of joint C.

is two-dimensional, we can write the equations of equilibrium in x and y directions as below:

$$\sum F_x = 0; \quad -F_{vCA}Sin\theta_2 + F_{vCB}Sin\theta_1 = 0$$

$$\sum F_y = 0; \quad P_v - F_{vCA}Cos\theta_2 - F_{vCB}Cos\theta_1 = 0$$

F_{vCB}, and F_{vCA} represent the virtual internal forces in members BC and AC, respectively, and angle θ_1, θ_2 are the angles of inclination of these members with respect to the vertical axis. Now, let us assume that the joint C of the truss is given a small real displacement Δ along the positive direction of the y axis. This small deformation Δ, is consistent with the support conditions, i.e., the supports provided at joints A and B are not displaced. In this truss, only joint C can move due to the application of external virtual force P_v. So, the total virtual work for this truss can be written as:

$$W_v = P_v\Delta - F_{vCB}\left(\Delta Cos\theta_1\right) - F_{vCA}\left(\Delta Cos\theta_2\right)$$

or,

$$W_v = \left(P_v - F_{vCB}Cos\theta_1 - F_{vCA}Cos\theta_2\right)\Delta = 0$$

or,

$$P_v\Delta = \left(F_{vCB}Cos\theta_1 + F_{vCA}Cos\theta_2\right)\Delta$$

or we can write, $W_{ve} = W_{vi}$; where W_{ve} is the virtual external work done by the virtual external force, P_v acting through the real displacement Δ, and W_{vi} is the virtual internal work done by the virtual internal forces acting through real internal displacements.

10.4   DEFLECTION OF TRUSSES BY VIRTUAL WORK METHOD

Virtual work method can be applied to trusses to determine deflection of any member under the application of the external loading. To understand the process, let us consider the determinate truss as shown in Figure 10.8, with the external loads P_1 and P_2 as applied at joint C and calculate vertical deflection at node B of the truss by this method.

Since above truss is statically determinate, hence, all member forces due to the applied loading can be calculated by any of the methods we have learnt in the truss analysis chapter. Let us denote the force in any member, say AC, by F_{AC}. Also let us take this force as tensile in nature. Hence, the elongation of the member under this axial load will be:

$$\delta_{AC} = \frac{F_{AC}l}{AE}$$

where A and E is the cross-sectional area and elastic modulus of the member under study, respectively. Now, to determine the vertical deflection at point B, we apply a unit load in the direction of our interest as shown in Figure 10.9.

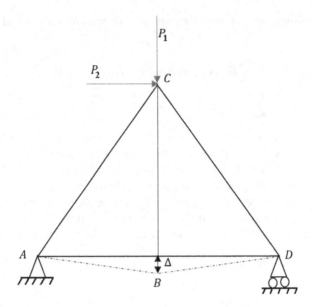

FIGURE 10.8 Finding deflection of truss by virtual work method.

Under the application of this unit load in the direction of vertical deflection at point B (denoted by Δ), the net external virtual work done W_{ve} by the unit load is given by,

$$W_{ve} = \Delta \times 1$$

On the other hand, the total virtual internal work done W_{vi} for all members of the truss is given by the following summation,

$$W_{vi} = \sum F_{vj} \delta_j$$

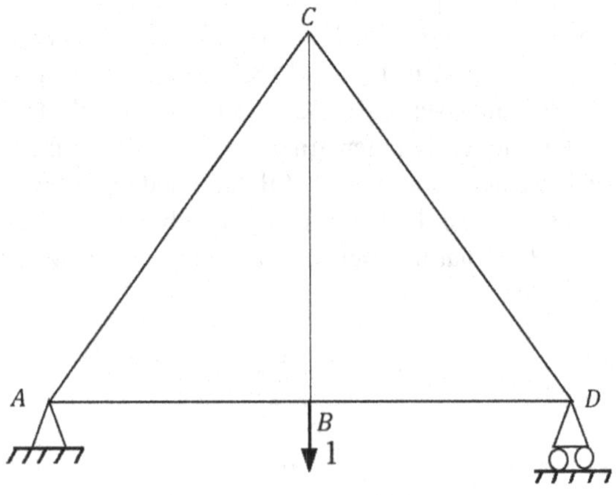

FIGURE 10.9 Finding deflection of truss by virtual work method: unit load application.

where it is to be understood that the summation is over all members of the truss and F_{vj} is the member force in j^{th} member due to the unit load at point B. Since total energy is conserved, hence, net external virtual work done due to unit load should be same as that of the total internal work done by the truss elements. Hence, we may write the following equation:

$$\Delta \times 1 = \sum F_{vj} \delta_j$$

or,

$$\Delta = \sum F_{vj} \frac{F_j L_j}{AE}$$

where F_j is the actual member force in j^{th} member of the truss under the application of original loading.

Since all the terms in the right side of above equation is known, we can determine the vertical deflection at point B due to the applied force on the truss from the above equation. We can solve this equation in an excellent tabular form, as shown in Table 10.1, to calculate the virtual internal work done by all the truss members and, after that, can obtain the required deflection values.

Now we can equate the virtual external work to the virtual internal work and can solve for the desired deflection, Δ easily.

i.e.,

$$1 \times \Delta = \sum F_{vj} \frac{F_j L_j}{AE}$$

Hence, the deflection at node B will be,

$$\Delta = F_{vAB} \frac{F_{AB} L_{AB}}{AE} + F_{vBC} \frac{F_{BC} L_{BC}}{AE} + F_{vAC} \frac{F_{AC} L_{AC}}{AE} + F_{vCD} \frac{F_{CD} L_{CD}}{AE} + F_{vBD} \frac{F_{BD} L_{BD}}{AE}$$

TABLE 10.1
Internal Work Done by Various Members of the Truss

Member	Length (L)	Real Member Force (F_j)	Virtual Member Force Due to Unit Load (F_{vj})	$\sum F_{vj} \frac{F_j L_j}{AE}$
AB	L_{AB}	F_{AB}	F_{vAB}	$F_{vAB} \frac{F_{AB} L_{AB}}{AE}$
BC	L_{BC}	F_{BC}	F_{vBC}	$F_{vBC} \frac{F_{BC} L_{BC}}{AE}$
AC	L_{AC}	F_{AC}	F_{vAC}	$F_{vAC} \frac{F_{AC} L_{AC}}{AE}$
CD	L_{CD}	F_{CD}	F_{vCD}	$F_{vCD} \frac{F_{CD} L_{CD}}{AE}$
BD	L_{BD}	F_{BD}	F_{vBD}	$F_{vBD} \frac{F_{BD} L_{BD}}{AE}$

10.5 DEFLECTION OF BEAMS BY VIRTUAL WORK METHOD

For beams, deflection due to application of external loading can be calculated using principle of virtual work with unit load method as we have done for trusses in the previous section. To understand the method, let us draw an example of calculating deflection at the midpoint C of a simply supported beam with load as shown in Figure 10.10.

Now, from elastic line equation, we know that the slope of elastic line is related to the internal bending moment of the beam by the following relationship:

$$\frac{d\theta}{dx} = \frac{M}{EI}$$

where M is the bending moment at a section x, from the left support point A. So, we have,

$$d\theta = \frac{Mdx}{EI}$$

At this point, let the bending moment at the same point from left support A, due to unit point load, be given by M_v. So, the virtual internal work done due to the small angular virtual displacement $d\theta$, is given by,

$$M_v d\theta = M_v \frac{Mdx}{EI}$$

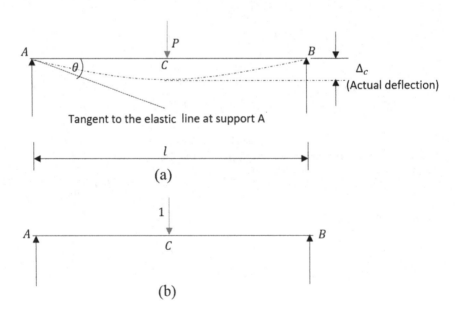

FIGURE 10.10 (a) Beam with original load, elastic line and (b) applied unit load at the desired point.

However, the external work done by the unit load due to actual displacement Δ_C is given by,

$$1 \times \Delta_C$$

Following principle of virtual forces for deformable bodies, we have the following relationship:

$$1 \times \Delta_C = \int_0^l M_v \frac{M dx}{EI}$$

Since all the terms are known in the right-hand side of the above expression, hence, evaluating the integral we can determine the actual deflection at the required point on the beam.

We can summarize the deflection calculation by this method as follows:

1. Determine the actual bending moment at the required point on the beam due to the actual applied load.
2. Apply unit load at the required point on the beam and calculate bending moment due to unit load at the same point of the beam. Unit load direction will be in the same as that for desired deflection direction of the beam.
3. Evaluate the integral as deduced above for the entire length of beam.
4. Value of the integral will be the actual desired deflection at the required point of the beam.

10.6   DEFLECTION OF FRAMES BY VIRTUAL WORK METHOD

To determine the deflection, Δ, and rotation, θ, at a certain point of a frame, we need to apply virtual unit load, or unit moment at that point. After that we can easily find out the virtual external work done just by multiplying the unit load or unit moment with the deflection, Δ and rotation, θ of that point, respectively. As the portions of the frame may undergo axial deformation as well as bending deformation, the total virtual internal work done is the summation of virtual internal work due to bending and due to the axial deformation. Therefore, the total internal virtual work (W_{vi}) for a frame can be written by the following expression:

$$W_{vi} = \sum F_v \left(\frac{FL}{AE} \right) + \sum \int \frac{M_v M}{EI} dx$$

where F_v, F are the axial forces generated in the frame members due to virtual and real forces, respectively, and M_v, M are the bending moments generated due to virtual and real loads, respectively. Now, if we equate this total internal virtual work to the virtual external work, we can get the desired deflection or rotation values of a particular point, respectively, as shown in the following:

$$1(\Delta) = \sum F_v \left(\frac{FL}{AE} \right) + \sum \int \frac{M_v M}{EI} dx$$

or,

$$1(\theta) = \sum F_v\left(\frac{FL}{AE}\right) + \sum \int \frac{M_v M}{EI} dx$$

The axial deformations of the frame members are much smaller than the bending deformations and are generally neglected for common engineering materials. That is why the first part of the right-hand side of the previous expressions is not considered in the calculations, and we can obtain the deflection or rotation values of a particular point, respectively as:

$$1(\Delta) = \sum \int \frac{M_v M}{EI} dx$$

or,

$$1(\theta) = \sum \int \frac{M_v M}{EI} dx$$

Let us consider a frame as shown in Figure 10.11, with a system of external loading acting on it. Here we will neglect the axial deformations of the frame members. Suppose here that our objective is to find out the horizontal deflection at point B due to application of these loads. To be able to do that, first we need to calculate the actual bending moments and support reactions. Since we are dealing with statically determinate frames, hence, this should not be a major issue.

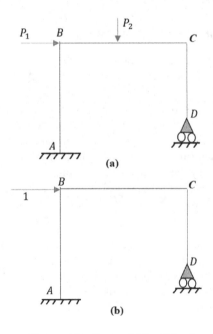

FIGURE 10.11   (a) Frame with actual loading and (b) unit load applied at the desired location and direction of deflection.

Once we complete the analysis of the frame with external actual loading, we apply a unit load at the point B of the frame in the desired direction of deflection and again calculate the support reactions and bending moments in all members of the frame. Suppose we denote the actual bending moment in each member of the frame as follows:

- Actual bending moment due to external loading in member $AB = M_{AB}$
- Actual bending moment due to external loading in member $BC = M_{BC}$
- Actual bending moment due to external loading in member $CD = M_{CD}$

Also, let us denote the bending moment in each member of the frame due to unit load at point B as follows:

- Virtual bending moment due to point load at node B in member $AB = M'_{AB}$
- Virtual bending moment due to point load at node B in member $BC = M'_{BC}$
- Virtual bending moment due to point load at node B in member $CD = M'_{CD}$

Hence, if we denote the actual horizontal deflection at node B by Δ_B, from principle of virtual forces for deformable bodies we can write,

$$1 \times \Delta_B = \int_A^B M'_{AB} \frac{M_{AB}dx}{EI} + \int_B^C M'_{BC} \frac{M_{BC}dx}{EI} + \int_C^D M'_{CD} \frac{M_{CD}dx}{EI}$$

or,

$$\Delta_B = \int_A^B M'_{AB} \frac{M_{AB}dx}{EI} + \int_B^C M'_{BC} \frac{M_{BC}dx}{EI} + \int_C^D M'_{CD} \frac{M_{CD}dx}{EI}$$

Since all the terms in the right-hand side of the above relationship are known, hence, evaluating the integral within the limit of the member lengths, we can evaluate the actual deflection at node B due to the system of actual loads acting on the frame. Thus, for frames, we can apply the same principles as that for the beams to evaluate deflection at the desired direction and desired location.

10.7 CASTIGLIANO'S THEOREM

We already have seen an interesting connection between strain energy and deflection of structural elements. As pointed out toward the end of Section 10.2, partial derivative of internal strain energy with respect to load is equal to the deflection of structural element toward the same force at its point of application. Although we have seen a simple case for point loads, this statement is true for any kind of loading. Only thing we need to be careful about is the correct expression for internal strain energy. At this point, we would introduce Castigliano's 2nd theorem, which is something like following:

Castigliano's 2nd theorem: For linearly elastic structures, the first partial derivative of the strain energy of the structure with respect to any particular force or couple gives the displacement or rotation of the point of application of that force or moment in the direction of its line of action.

Mathematically we can state the theorem as:

$$\frac{\partial U}{\partial P_i} = \Delta_i \text{ or } \frac{\partial U}{\partial M_i} = \theta_i;$$ in which U is the strain energy; Δ_i is the deflection of the point of application of the force P_i in the direction of P_i; and θ_i is the rotation of the point of application of the couple M_i in the direction of M_i.

Another version of above theorem stipulates that partial derivative of strain energy with respect to deflection provides the force acting in the direction of the deflection at the same point where deflection has been measured. It is somewhat reciprocal to the above theorem. People often consider this theorem as the 1st theorem and the above theorem as the 2nd one. However, from application point of view, the measurement of deflection is the major objective and that is the one we will dive deeper into.

To understand 2nd theorem and its importance, let us consider the example of a cantilever beam with point load at its end as shown in Figure 10.12, and let us calculate the deflection of the beam under the action of point load at its free end.

First, the bending moment at any section x from left support point which is:

$$M_x = -Px$$

The internal strain energy due to bending of the beam will be,

$$U = \int_0^l \frac{M_x^2 dx}{2EI}$$

or, putting the value of M_x into the above equation we get,

$$U = \int_0^l \frac{(-Px)^2 dx}{2EI}$$

Now, to determine the deflection under point load at the free end of cantilever, we take partial derivative of the above expression with respect to P and arrive at:

$$\frac{\partial U}{\partial P} = \frac{\partial}{\partial P} \int_0^l \frac{(-Px)^2 dx}{2EI}$$

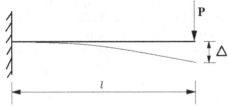

FIGURE 10.12 Cantilever beam with point load at its end and deflection line.

or,

$$\Delta = \frac{\partial U}{\partial P} = \int_0^l \frac{Px^2 dx}{EI}$$

Evaluating the integral, we get,

$$\Delta = \frac{Pl^3}{3EI}$$

Suppose we want to calculate the deflection of the same cantilever beam with uniformly distributed load instead of point load at its free end. In such a situation, we have to apply a fictitious point load R at the free end and then we will carry out the strain energy with the original UDL load, including the fictitious load R.

So, the bending moment at any section x from the left support will be given by,

$$M_x = -Rx - \frac{wx^2}{2}$$

So, strain energy due to bending will be:

$$U = \int_0^l \frac{M_x^2 dx}{2EI} = \int_0^l \frac{\left(-Rx - \left(wx^2/2\right)\right)^2 dx}{2EI}$$

Now the deflection will be the following partial derivative,

$$\Delta = \frac{\partial U}{\partial R} = \frac{\partial}{\partial P} \int_0^l \frac{\left(-Rx - \left(wx^2/2\right)\right)^2 dx}{2EI}$$

or,

$$\Delta = \int_0^l \frac{2\left(Rx + \left(wx^2/2\right)\right)x dx}{2EI}$$

Now, as R was a fictitious force, hence, equating $R = 0$ in the above equation and completing the integral we get,

$$\Delta = \frac{wl^4}{8EI}$$

In addition to the above, if we want to calculate the slope of deflection line at the free end, we can do so by taking partial derivative of the strain energy with respect to moment. Considering the same example of cantilever beam in Figure 10.13, we want to calculate the slope of the deflection line at the free end of the cantilever. To do so, since there is no actual external moment acting at the free end, we will apply a

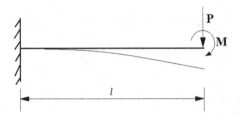

FIGURE 10.13 Cantilever beam with point load and fictitious moment at free end.

fictitious moment M at the free end and calculate the strain energy under the application of both point load and fictitious moment M. So, the cantilever beam with fictitious moment applied at its free end will look something as shown in Figure 10.13.

So, bending moment at any section x from support is given by:

$$M_x = -Px + M$$

Strain energy due to bending will be,

$$U = \int_0^l \frac{M_x^2 dx}{2EI} = \int_0^l \frac{(-Px + M)^2 dx}{2EI}$$

Evaluating the integral completely we get,

$$U = \frac{P^2 l^3}{6EI} + \frac{MPl^2}{2EI} + \frac{M^2 l}{2EI}$$

So, the slope of deflection line at free end is given by,

$$\theta = \frac{\partial U}{\partial M} = \frac{Pl^2}{2EI} + \frac{Ml}{EI}$$

Since, M is the fictitious moment, hence, putting $M = 0$ in the above expression we get,

$$\theta = \frac{Pl^2}{2EI}$$

Methods presented above for cantilever beam can be applied and extended to any type of statically determinate beams and frames to calculate deflection at any point of structure due to external loading conditions. Students should learn and understand the method of applying this theorem to any statically determinate problems. This method is completely independent of the geometry and shape of the bending moment and shear force diagrams, and thus, it can be also named a geometry-independent method.

Castigliano's theorem is somewhat generalized in nature. In a single equation, it combines all kinds of forces (point loads, moment, torsion etc.) to its related conjugate displacements. If we take force as moment, the corresponding deflection or displacement will be angular displacement. If we take force as point loads, deflection or displacement will be linear displacement. Since the equation is related to energy, which is scalar in nature, its application is more generalized than vector equations like Newton's Laws of motion. That is why we introduced the concept of generalized coordinates and generalized forces in Chapter 2 of this book. Familiarity with the generalized coordinates and force helps us to unify the concepts of mechanics in an elegant way.

Also, compared to geometric methods, energy principles provide a much easier way to determine the deflection of complex structures without much trouble. In geometric methods, students need to draw bending moment and shear force diagram accurately and then apply the methods to determine the deflections. In contrary, energy does not require any such drawings to be made a priori. One needs to calculate the strain energy of individual members as per the acting loads and then simply apply principle of conservation of energy to determine the unknown deflections. In any case, it is up to the analyst which method he or she adopts to analyze the structure.

10.8   MAXWELL-BETTI LAW OF RECIPROCAL DEFLECTIONS

Maxwell's law is a special case of more general Betti's law. Initially, in 1864, Maxwell developed it. After that, in 1872, Betti presented a generalized form of it. Both laws apply to any type of structure, whether beam, truss, or frame. These ideas will be developed by considering a simple beam, as shown in Figure 10.14.

Suppose that this beam is subjected to two separate and independent sets of forces, the system of forces P and the system of forces Q. The P system develops the internal bending moment M_P in the beam, while the Q system develops the internal bending moment M_Q. Let us imagine two situations. First, suppose that the beam is at rest under the action of the P system of forces, and then we further

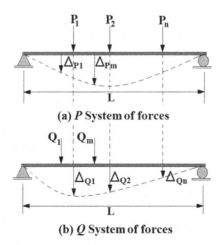

(a) P System of forces

(b) Q System of forces

FIGURE 10.14   Derivation of Betti's reciprocal law: (a) P system of forces and (b) Q system of forces.

deform the beam by applying the Q system of forces. As a second situation, suppose that the reverse is true. i.e., the Q system is acting on the beam, and then we further deform the beam by the P system of forces. In both situations, we may apply the law of virtual work and thereby come to a very useful conclusion known as Betti's law, which states that: *In any linear elastic structure in which the supports are unyielding and the temperature is constant, the external virtual work done by a system of forces P during the deformation caused by a system of forces Q is equal to the external virtual work done by the Q system during the deformation caused by the P system of forces.* From Figure 10.14, the virtual external work done (W_{ve}) can be expressed as:

$$W_{ve} = P_1\Delta_{Q1} + P_2\Delta_{Q2} + \cdots + P_n\Delta_{Qn}$$

or,

$$W_{ve} = \sum_{i=1}^{n} P_i\Delta_{Qi}$$

The virtual internal work done in the beam can be expressed as:

$$W_{vi} = \int_0^L \frac{M_P M_Q}{EI} dx$$

Now applying the principle of virtual forces of deformable bodies, we can write,

$$\sum_{i=1}^{n} P_i\Delta_{Qi} = \int_0^L \frac{M_P M_Q}{EI} dx$$

Next, we assume that the beam is subjected to a Q set of forces and the deflections caused by the P set of forces. By equating the external virtual work to the internal virtual work, we can write:

$$\sum_{j=1}^{m} Q_j\Delta_{Pj} = \int_0^L \frac{M_Q M_P}{EI} dx$$

The right-hand side of both the above expressions is the same. Thus, we can write:

$\sum_{i=1}^{n} P_i\Delta_{Qi} = \sum_{j=1}^{m} Q_j\Delta_{Pj}$; this is the mathematical statement of Betti's law. Betti's law is a very useful principle and sometimes called generalized Maxwell's law.

Maxwell's law of reciprocal deflection states that: *In any linear elastic structure in which the supports are unyielding and the temperature is constant, the deflection at a point i due to a unit load applied at a point j is equal to the deflection at j due to a unit load at i.* In this, the terms deflection and load are given in general sense,

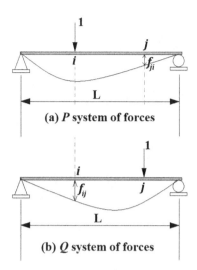

FIGURE 10.15 Development of Maxwell's reciprocal law: (a) P system of forces and (b) Q system of forces.

to include rotation, and couple, respectively. Maxwell's theorem can be realized by observing Figure 10.15, as shown below.

As mentioned earlier, Maxwell's law can be considered as a special case of Betti's law. Applying the Betti's law, we can easily show that: $f_{ij} = f_{ji}$, for the case shown in Figure 10.15, and this is the mathematical statement of Maxwell's theorem. In this figure, f_{ij} represents the deflection at i due to the unit load at j, and f_{ji} represents the deflection at j due to the unit load at i. These deflections per unit load are referred to as *flexibility coefficients*.

We will find the application of this theory in several chapters that follow. Even in indeterminate structural analysis, we will frequently apply this principle to form additional equations to eliminate the unknown forces and moments.

11 Rolling Loads and Influence Lines and Their Applications

11.1 INTRODUCTION

In previous chapters, we have considered loads acting on the structure, which is static in nature. It was assumed that the point of application of the load would remain the same throughout the entire life span of the structural element. But in real-life problems, loads shift their position in a structure. Thus, the location of maximum internal forces and deflections due to the applied loads will also vary depending on the position of the load at that instant. In this chapter, we will develop some methods to analyze the structure under moving loads. For moving loads, placing unit load at different beam locations and drawing subsequent bending and shear force diagrams will be very useful. These diagrams will be used, at a later stage, on other structures like frames to analyze internal forces and moments induced due to moving load conditions. After completing this chapter, students are expected to acquire detailed knowledge of the analysis process when a structure is subjected to moving loads.

11.2 INFLUENCE LINES FOR BEAMS AND FRAMES BY EQUILIBRIUM METHOD

Consider a simply supported beam with unit point load at a distance x from the left support as shown in Figure 11.1. At this location of the load, the support reactions at the two ends of the beam can be calculated using global moment equilibrium conditions (recall from Chapter 2, to determine unknown support reactions, we have applied global equilibrium conditions on the overall structure, and to determine internal forces such as bending moment and shear forces, we applied local equilibrium conditions).

Taking moment of all forces about point B, we get,

$$\sum M_B = 0$$

$$R_A = 1 \times \frac{(l-x)}{l}$$

or,

$$R_A = 1 - \frac{x}{l}$$

DOI: 10.1201/9781003081227-13

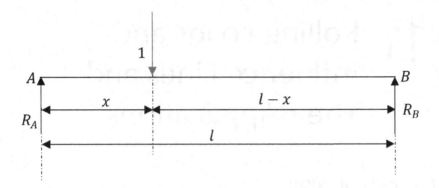

FIGURE 11.1 Simply supported beam with unit point load at a distance x from left support A.

So, R_B will be,

$$R_B = 1 - \left(1 - \frac{x}{l}\right) = \frac{x}{l}$$

We developed two equations for support reactions, for the arbitrary location of external load, and both the equations are linear functions of x. So, by varying x, we can determine the support reactions for any location of the point load acting on the beam. It is interesting to put $x = 0$ in the abovementioned equation. In that case, $R_A = 1$ and $R_B = 0$. This result is quite reasonable as, in this case, point load is sitting just at the left support point, and hence, total downward force will be balanced by the support A, and thus support reaction at B will be zero. Similarly, by setting $x = l$ in the abovementioned equation, we find that, $R_A = 0$, $R_B = 1$ and the reason is quite self-explanatory. So, the influence line for support reactions for a simply supported beam with point load can be drawn as Figure 11.2.

Main advantage of drawing influence line is that we can determine the support reactions for any load other than unit one from the above triangles by applying principle of similar triangles and multiplying the result by the actual load applied on the structure. To understand this concept, let us calculate the support reaction at A when

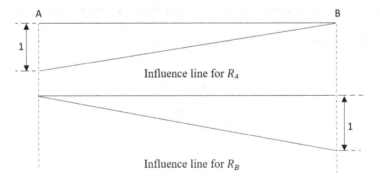

FIGURE 11.2 Influence line diagram for support reactions of a simply supported beams.

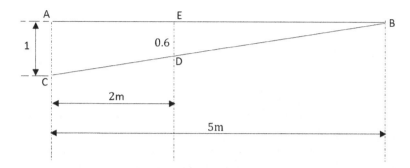

FIGURE 11.3   Influence line diagram for support reaction of example problem.

a load of 5 kN is acting at a distance 2 m from left support. Total length of beam is 5 m. We take help of the first influence line to determine support reaction R_A.

From Figure 11.3, considering similar triangle ABC and BED, we have,

$$\frac{BE}{ED} = \frac{BA}{AC}$$

or,

$$ED = \frac{BE \times AC}{BA} = \frac{3 \times 1}{5} = 0.6$$

Since the actual load acting on the structure is 5 kN, the actual support reaction at support A will be $0.6 \times 5 = 3$ kN.

Thus, the beauty of the influence line diagram can be well appreciated from the abovementioned example. In similar way, one can find out the support reaction at B due to any position of load on the beam using the relevant influence line diagram for the same.

Having equipped with the concept of influence line diagrams for support reactions, we can form the influence line diagrams for bending moment and shear force also. Using the same beam as shown in Figure 11.1, let us derive the expression for change in bending moment diagram at a distance l_1 from the left support if a unit load moves across the beam from left support to right support. Let the point of concern be C as shown in Figure 11.4.

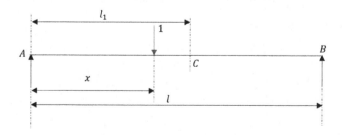

FIGURE 11.4   Influence line for bending moment at section C.

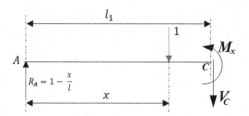

FIGURE 11.5   Free-body diagram when unit load is at $0 \le x \le l_1$.

Now, for the range $0 \le x \le l_1$ (as shown in Figure 11.5):

$$M_x = \left(1 - \frac{x}{l}\right)l_1 - 1 \times (l_1 - x)$$

For, $x = l_1$, $M_x = \left(1 - \frac{l_1}{l}\right)l_1$, and for $x = 0$, $M_x = 0$. So, when unit load is placed at $x = l_1$, bending moment at section C will be $\left(1 - (l_1/l)\right)l_1$ and when unit load is placed at left support point A, the bending moment at section C will be zero.

Now, for the range $l_1 \le x \le l$ (as shown in Figure 11.6).

$$M_x = R_B\ (l - l_1) - 1 \times (x - l_1)$$

or, $M_x = \frac{x}{l}(l - l_1) - 1 \times (x - l_1) = x\left(1 - \frac{l_1}{l}\right)$

So, for $x = l_1$, we get,

$$M_x = \left(1 - \frac{l_1}{l}\right)l_1$$

And for $x = l$, $M_x = 0$. So, when unit load is placed at right support point B, the bending moment at section C will be zero and when unit load is placed at section C, the bending moment at the same location will be $\left(1 - \frac{l_1}{l}\right)l_1$, which is in perfect agreement with the earlier calculation.

So, the influence line diagram for the moment at point C will be linear in nature, and its values are zero at both supports. It is customary to note the difference between the bending moment diagram and the influence line diagram for bending moment. In the bending moment diagram, we draw the bending moment of a beam at various points along its span due to a fixed external load acting on it. On the other hand, in

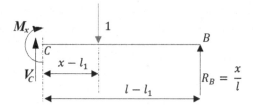

FIGURE 11.6   Free-body diagram when unit load is at $l_1 \le x \le l$.

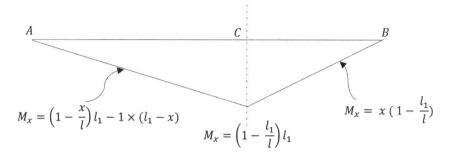

$$M_x = \left(1 - \frac{x}{l}\right) l_1 - 1 \times (l_1 - x)$$

$$M_x = \left(1 - \frac{l_1}{l}\right) l_1$$

$$M_x = x\left(1 - \frac{l_1}{l}\right)$$

FIGURE 11.7 Influence line diagram of bending moment at point C.

the influence line for bending moment, we draw change in bending moment at any particular section on a beam due to the movement of a unit concentrated load over its span. The influence line for bending moment at point C due to unit point load moving across the beam is shown in Figure 11.7.

For shear force at point C, let us write the first equation of shear force corresponding to Figure 11.5.

For the range $0 \le x \le l_1$, we can find out the shear force by using the free-body diagram and applying local force equilibrium equation.

$$R_A - 1 - V_C = 0; \; V_C = 1 - \frac{x}{l} - 1 = -\frac{x}{l}$$

So, at $x = 0$, means when unit load is placed at left support point A, then shear force at C, $V_C = 0$. Also, for $x = l_1$, the shear force at C will be $V_C = -(l_1/l)$.

For the other portion of the beam, when the unit load is beyond point C, i.e., $l_1 \le x \le l$, then we can use the right section about C (refer to Figure 11.6 for proper unit load location).

Constructing the shear force equation by using the free-body diagram shown in Figure 11.6, and applying local force equilibrium equation,

$$V_C - 1 + R_B = 0; \; V_C = 1 - \frac{x}{l}$$

So, for $x = l_1$, we have $V_C = 1 - (l_1/l)$ and for $x = l$, we have $V_C = 0$. And also, it is to be noted that all the equations for shear forces are linear in x. So, influence line for shear force at C due to varying position of unit load can be drawn as shown in Figure 11.8.

As the unit load moves from A to C, the shear at C decreases linearly until it becomes $-(l_1/l)$, when the unit load reaches just to the left of C. As the unit load crosses point C, the shear at C increases abruptly to $1 - (l_1/l)$. It then decreases linearly as the unit load moves toward B until it becomes zero when the unit load reaches the right support B.

For frames, similar mathematical equations can be formed depending on the geometry of the frame and the chosen point of reference for which the influence

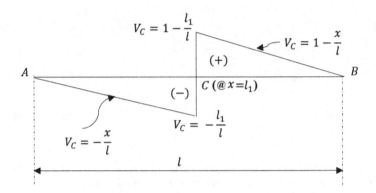

FIGURE 11.8   Influence line diagram of shear force at point C.

lines need to be drawn. To understand this concept, let us take an example problem for a frame as shown in Figure 11.9.

Now, we want to draw the influence line for vertical reaction and support moment at point A of the above frame. To be able to do this, let us first draw the free-body diagram as shown in Figure 11.10 of the above frame with the applied unit load.

Now for support reactions, applying global force equilibrium condition, we get,

$$R_A - 1 = 0$$

or,

$$R_A = 1$$

And, applying global moment equilibrium conditions about point A, we get,

$$M_A - 1 \times (x - 6) = 0$$

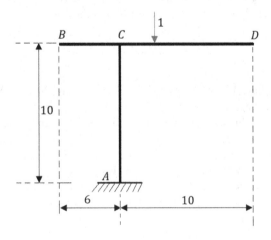

FIGURE 11.9   Example problem for frame structure.

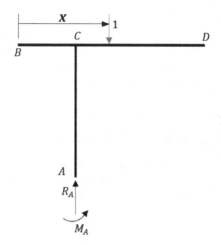

FIGURE 11.10   Free-body diagram of frame.

or,

$$M_A = 1 \times (x - 6)$$

So, for $x = 0$, $M_A = -6$, and when $x = 6$, $M_A = 0$. Also, for $x = 18$, $M_A = 12$

So, with the above-calculated values and since all the equations are linear functions of x, the influence line diagrams for support reaction and support moment at point A can be shown in Figure 11.11 (a) and (b), respectively.

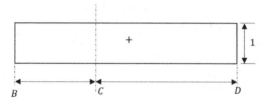

(a) Support reaction influence line diagram corresponding to support point A

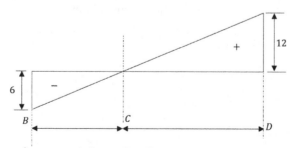

(b) Support moment influence line diagram corresponding to support point A

FIGURE 11.11   Influence line diagram for the example frame.

So, by following this method, we can draw influence line diagrams of any response function (reaction, shear, bending moment etc.) for any frame for any reference point selected on the frame. Having gained enough knowledge about the construction of influence lines, now, we will learn another important topic known as Müller-Breslau principle for drawing qualitative construction of influence line diagrams.

11.3 QUALITATIVE INFLUENCE LINES AND MÜLLER-BRESLAU'S PRINCIPLE

For quantitative construction of influence lines, the method based on equilibrium equations is found to be very useful. In contrast to quantitative analysis, and in real life problems, we need to understand the nature and behavior of influence lines without calculating in detail the numerical values. It is the shape and nature of influence line that are of utmost important for analyzing any structure qualitatively and instantly. Müller-Breslau principle provides the best tool for qualitative analysis of structures to construct influence lines for shear, support reactions, and bending moment. However, this principle cannot be used to construct deflection influence line diagrams as will be apparent when we will learn in detail about this principle.

11.3.1 MÜLLER-BRESLAU PRINCIPLE

The influence line equations for force or moment can be derived using the deflected shape of the released structure. The released structure is obtained by removing the restraint related to the point for which the influence lines are to be drawn. Then, we need to apply unit displacement or rotation at that location and in the direction of the force or moment at the reference point of the structure so that only the force or moment at the reference point and the unit load do the external work.

To understand the actual working procedure, as per the above principle, we need to go through an example. Let us consider the same simply supported beam as was shown in Figure 11.4 and try to form the influence line equations for bending moment at point C of that beam by applying this principle. As our objective is to find the equation for bending moment influence line at point C, we remove the restraint at C by introducing a hinge connection at that point (remember in hinge there will be no moment, and thus, we can eliminate the bending moment at C by this method). The beam with a new hinge connection at point C will look something like as shown in Figure 11.12.

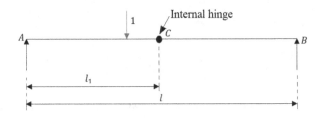

FIGURE 11.12 Released structure of simply supported beam with hinge at point C.

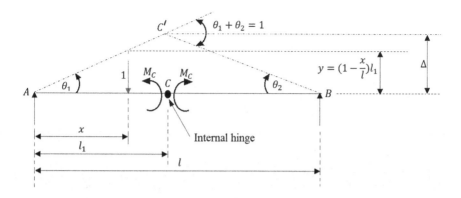

FIGURE 11.13 Released structure moment and virtual displacement diagrams.

As we have provided a hinge at point C, the beams AC and CB are now free to rotate about this point. To keep the released beam in equilibrium, we apply a moment at C, which is shown in Figure 11.13.

Then, we apply a virtual rotation at point C by unit value, and due to this rotation, AC is rotated by say θ_1 and BC is rotated by say θ_2, so that, $\theta_1 + \theta_2 = 1$. Refer to Figure 11.13 for the displaced portion of the beams.

Now, applying the principle of virtual work, we get,

$$W = M_C\theta_1 + M_C\theta_2 - 1 \times y = 0$$

which implies,

$$M_C(\theta_1 + \theta_2) = 1 \times y$$

or,

$$M_C = y$$

which indicates that the deflected shape of the beam is equal to the influence line diagram for the beam, which was then the claim of Müller-Breslau principle. Now from the radius and angular displacement relationship, so, the ordinate, Δ is given by,

$$\Delta = l_1\theta_1 = (l - l_1)\theta_2$$

or,

$$l_1 \times \theta_1 = (l - l_1) \times \theta_2$$

So,

$$\theta_1 = \frac{(l - l_1)}{l_1} \times \theta_2$$

Now, using the relationship, $\theta_1 + \theta_2 = 1$, we get,

$$\theta_2 = \frac{l_1}{l}$$

So, we get,

$$\Delta = \frac{(l-l_1)l_1}{l} = l_1\left(1-\frac{l_1}{l}\right)$$

This is in exact agreement with earlier derived expression for bending moment influence line equation for the portion CB of the beam by applying equilibrium methods. By following the same steps, we can form the other influence line equations for reaction and shear forces as well.

Although Müller-Breslau principle can determine the influence lines quantitatively, but in real-life problems, this principle is mostly used to determine the influence lines qualitatively. Qualitative nature of the influence lines helps us to analyze nature and qualitative failure mode analysis for large structures and individual structural elements as well. For detailed analysis purpose, one may adopt Müller-Breslau principle to obtain the qualitative influence line diagrams and then apply equilibrium methods to obtain the complete influence line diagrams with numerical values.

11.4 INFLUENCE LINES FOR FLOOR GIRDERS

In the previous section, influence lines for beams and frames for reactions, shear, and bending moments have been provided. We have developed the equations for influence lines for the case when unit load is directly moving on the same system. In most bridges and building, there are main structural elements on which direct application of loading is not possible. For example, let us consider the case of bridge girders. In most of the structural bridge girders, deck is supported on smaller steel members known as the stringers. Stringers are then supported on structural beams that are also supported on the main beams known as girders. Thus, for girders, no matter what type of loading has been applied on the deck level, it will be always transferred as point loads. For structural analysis and design, it is customary to determine the actual shear and bending moment values acting on the girder due to load applied on the deck.

To illustrate the necessary steps to draw influence lines for bridge girders, we assume that the girders are supported at the ends as a simply supported beam. Refer Figure 11.4 (a) for better understanding of the stringer/floor beam/girder system. Influence line for support reactions will be same as that for the simply supported beam we have already discussed in the first section of this chapter and can be derived as next.

$$\circlearrowleft + \sum M_F = 0$$

i.e.,

$$-R_A \times l + 1 \times (l-x) = 0$$

or,

$$R_A = 1 - \frac{x}{l}$$

$$\circlearrowleft + \sum M_A = 0$$

i.e.,

$$R_F \times l - 1 \times x = 0$$

or,

$$R_F = \frac{x}{l}$$

Hence, influence lines are drawn without much ado for the support reactions as shown in Figure 11.14 (b) and (c).

Next, let us consider shear force at a point G in the panel BC as shown in Figure 11.14 (d). When the unit load is located to the left of point B, then shear force at any point within the panel BC will be:

$$V_{BC} + R_F = 0, \; V_{BC} = -\frac{x}{l} \text{ for } 0 \le x \le \frac{l}{5}$$

When unit load is placed within the panel BC, then force at support B (as shown in Figure 11.15) on the girder needs to be incorporated in the equation of the shear force for panel BC. So, in this case the shear force equation will be:

$$V_{BC} = R_A - R_B = \left(1 - \frac{x}{l}\right) - \left(2 - \frac{5x}{l}\right) = -1 + \frac{4x}{l}; \frac{l}{5} \le x \le \frac{2l}{5}$$

Similarly, when the unit load is located to the right of the panel point C, the shear at any point within the panel BC is:

$$V_{BC} = R_A = 1 - \frac{x}{l}; \frac{2l}{5} \le x \le l$$

Thus, from these equations, it is clear that influence line for shear force does not actually depend on the location of unit load on the panel. Thus, shear force remains constant for the entire panel of girder. Hence for girders, shear force is drawn for panels and not point to point of the entire length and for that reason, it is called as panel shear.

Now, let us find out the influence line for the bending moment at point G, located at panel BC. When the unit load is to the left of point B, bending moment at point G can be written as:

$$M_G = R_F(l - a) = \frac{x}{l}(l - a) \text{ for } 0 \le x \le \frac{l}{5}$$

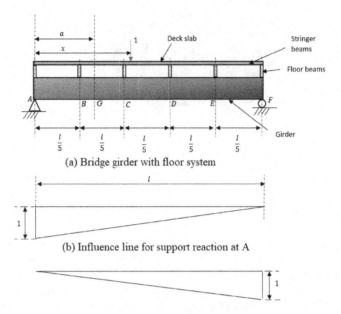

(a) Bridge girder with floor system

(b) Influence line for support reaction at A

(c) Influence line for support reaction at F

(d) Influence line for shear at panel BC, V_{BC}

FIGURE 11.14 Influence line diagram for panel BC of a bridge girder.

When unit load is located to the right of panel point C, then bending moment at G is given by:

$$M_G = R_A \times a = \left(1 - \frac{x}{l}\right)a \text{ for } \frac{2l}{5} \leq x \leq l$$

When unit load is placed in between panel points B and C, then bending moment at F is given by:

$$M_G = R_A \times a - R_B\left(a - \frac{l}{5}\right) \text{ for } \frac{l}{5} \leq x \leq \frac{2l}{5}$$

Upon simplification, we get $M_G = (2l/5) - a - x\left(1 - \frac{4a}{l}\right)$ for $\frac{l}{5} \leq x \leq \frac{2l}{5}$.

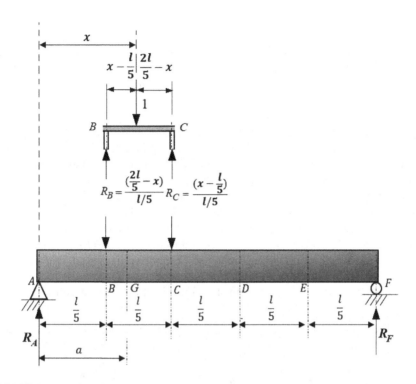

FIGURE 11.15 Finding shear force for panel BC when the unit load is within this panel.

Thus, unlike shear, bending moment for panels varies along point to point on the panel and is dependent on the location of the chosen point (point G in our case) on the panel. Since all the equations are linear in x, the bending moment influence line will also consist of straight-line segments as that found for shear force also. Using the abovementioned equations, we can draw the influence line for bending moment for panel BC as shown in Figure 11.16.

Following the same procedure, one can complete the influence lines for bending moment diagrams for every panel of the girder and same has been left as an exercise.

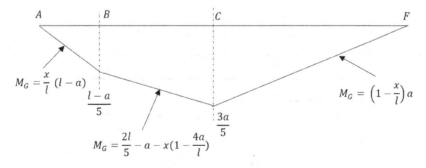

FIGURE 11.16 Influence line diagram for bending moment at point G in panel BC of a bridge girder.

11.5   INFLUENCE LINES FOR TRUSSES

The flooring system used for bridge girders is also used for trusses. The floor or deck slab is supported by stringer beams and stringer beams are supported on floor beams. Floor beams are connected to the nodes of longitudinal trusses. Hence, no matter the type of load moving on the deck, force transmitted to the truss at the nodes are always point loads. A typical arrangement of longitudinal truss with decking arrangements is shown in Figure 11.17 for ease of understanding.

Like bridge girders, trusses are assumed to be supported freely at the ends. Hence, influence lines for trusses will also consist of straight-line segment as it was derived and drawn for bridge girders in the previous section.

To illustrate the mechanism of constructing influence lines for trusses, let us consider the truss shown in Figure 11.17. On the floor slab, let us assume a unit load is moving from left to right. Load from the floor will be transferred to the stringer beams, from stringer beam to floor beams, and from floor beam to the nodal points of the truss. Our target is to draw influence lines for vertical support reactions at points A and D and to draw influence lines for member force in the member EB. Influence lines for reactions are obtained by following the same procedure as that for

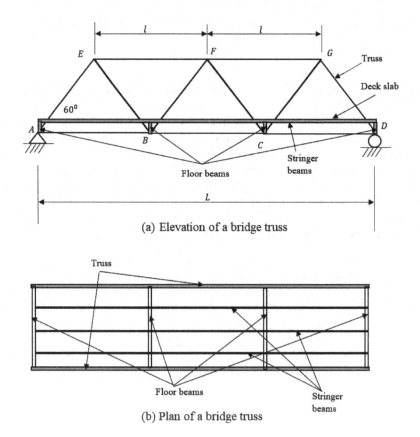

(a) Elevation of a bridge truss

(b) Plan of a bridge truss

FIGURE 11.17   Typical deck/floor slab supporting arrangements for truss.

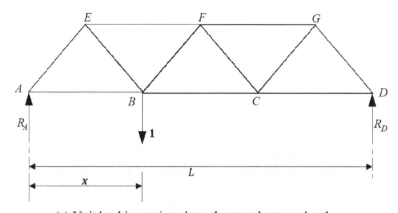

(a) Unit load is moving along the truss bottom chord

(b) Influence line for support reaction at A

(c) Influence line for support reaction at D

FIGURE 11.18   Influence line diagrams for support reactions of the truss.

simply supported beams. Thus, applying the equations of equilibrium, the equations of influence lines for the vertical reactions R_A and R_D can be obtained as shown in Figure 11.18.

$$\circlearrowright + \sum M_D = 0$$

i.e.,

$$-R_A \times l + 1 \times (L - x) = 0$$

or,

$$R_A = 1 - \frac{x}{L}$$

$$\circlearrowright + \sum M_A = 0$$

i.e.,

$$R_D \times L - 1 \times x = 0$$

or,

$$R_D = \frac{x}{L}$$

After completing the support reaction influence lines, now, we can move ahead to determine the equations for influence line of member force for the member *EB*. When the unit load is at the left of point *B*, then applying the method of sections as shown in Figure 11.19, we can form the local equilibrium force equations as follows:

Applying local force equilibrium equation in *y* direction, we get,

$$R_A - F_{EB} \sin 60 - 1 = 0 \text{ for } 0 \leq x \leq \frac{L}{3}$$

or,

$$F_{EB} = -\frac{2x}{L\sqrt{3}} \text{ for } 0 \leq x \leq \frac{L}{3}$$

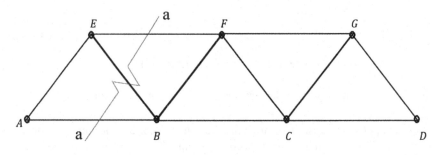

(a) Section line a-a

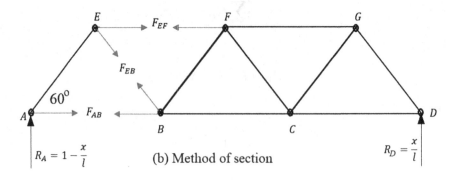

(b) Method of section

FIGURE 11.19 Free-body diagram of truss members applying method of sections.

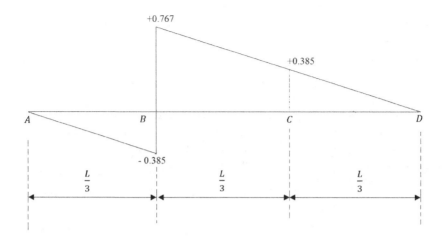

FIGURE 11.20   Influence line diagram for member force F_{EB}.

When unit load is placed beyond point B i.e., right of point B, then it is convenient to work with the other half of the section. From that, we can form the equation of equilibrium easily as shown next:

$$F_{EB} \sin 60 + R_D - 1 = 0 \text{ for } \frac{L}{3} \le x \le L$$

$$\text{or, } F_{EB} = \frac{2}{\sqrt{3}} \left(1 - \frac{x}{L}\right) \text{ for } \frac{L}{3} \le x \le L$$

From the abovementioned equations, we can draw the influence lines for member force F_{EB} as shown in Figure 11.20.

The member force F_{EB} was assumed to be tensile (Figure 11.19 (b)) while derivation of the influence line equations. So, a positive ordinate of the influence line indicates that the unit load applied at that point causes a tensile force in the member EB and vice versa. Thus, the influence line for F_{EB} (Figure 11.20) indicates that member EB will be in tension when unit load is located between B and D, whereas it will be in compression when unit load is located between A and B. It should be noted that when the unit load just crosses point B and moves toward the right, then the load in member EB will become tensile from compressive.

By following the above footsteps, we can form the influence lines of all other members of the truss.

11.6   MAXIMUM INFLUENCE AT A POINT DUE TO A SERIES OF CONCENTRATED LOADS

First, we need to find out the influence lines of any response functions (reaction force, axial force, shear force, bending moment etc.) for a particular point in a structure. Now, the maximum effect caused by a live concentrated force can be obtained by multiplying the peak ordinate of the influence line by the magnitude of the force. So, for any arbitrary point load P, the following two important points need to be understood clearly:

1. To obtain the influence lines for force P, we multiply the ordinates of the unit load influence lines with the numerical value P.
2. To obtain maximum positive value of ordinate, we need to multiply the load P with the maximum positive unit load influence line ordinate. Similarly, to obtain the maximum negative influence line value, we need to multiply the maximum negative value of the unit load influence line ordinate with the numerical load value of P.

To understand this concept, let us consider the following beam with overhang portion. We wish to calculate the maximum positive and negative bending moment values due to the moving load P. To be able to do that, we first draw the influence line for bending moment at a chosen point (say point D) for the given beam applying moving unit load. Upon constructing the unit load influence line for bending moment, we can now conveniently decide the location of the actual load P to obtain maximum values as illustrated graphically in Figure 11.21.

For most of the real-life bridge and flyover problems, during analysis, all traffic loads are considered as a series of many concentrated loads placed at a fixed distance from each other. Most of the time, these points are wheel locations of the moving vehicles like trucks or big carriage vehicles. For these types of movements, it is essential to determine the critical locations of these series point loads for which maximum adverse shear or bending moment is generated on the bridge structures. Influence lines provide the easiest and accurate way to determine the locations of these series loads for which maximum adverse conditions are met. Design against these adverse data provides a safe and reliable design of bridges and flyovers for the entire design life span of these structures. In the following sections, we will learn these techniques in detail and prepare ourselves to tackle such problems in real life.

Let us consider a series of moving point loads as shown in the following diagram. This is the pictorial representation of a large truck or vehicle class AA as per IRC norms. However, for our analysis purpose, we only need to know the center-to-center distance of the wheels and load acting at those points. Suppose, we want to calculate the maximum shear force at any specific point C on the span of the beam due to this moving load. The process of determining the maximum shear due to the moving load is calculated by trial-and-error method.

As shown in Figure 11.22, first we place the first wheel at the reference point C on the beam. Since we already know the shear force influence line diagram for the unit load, from similar triangles, we can determine the ordinates of the influence line diagram under the load positions, and then we multiply the same with corresponding load magnitudes to get the actual shear force that will be induced by the truck at the position of interest, i.e., C. Hence, the shear force at C due to the first load position will be:

$$V_C = w_1 s_1 + w_2 s_2 + w_3 s_3 + w_4 s_4$$

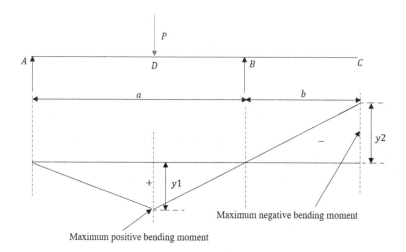

(a) Influence line diagram for M_D

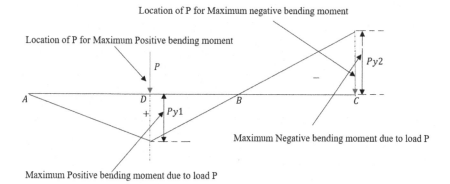

(b) Location for maximum positive and negative bending moment

FIGURE 11.21   Maximum positive and negative bending moment locations for load P.

Now, let us move the vehicle toward left support A so that now the second wheel rests at the reference point C. At this position, we can calculate the shear force at C following the same logic explained previously. Similarly, we will place the other two rear wheels, respectively, at the point C and calculate the shear forces.

Proceeding as per the earlier method, we calculate the shear force value for each different position of the wheel loads on the reference point C of the beam. Once we calculate all these values, we can determine the maximum value of the shear force and wheel position for the same.

When many concentrated loads act on the span, the trial-and-error method of finding maximum shear force at a point of interest becomes very tedious. Then by

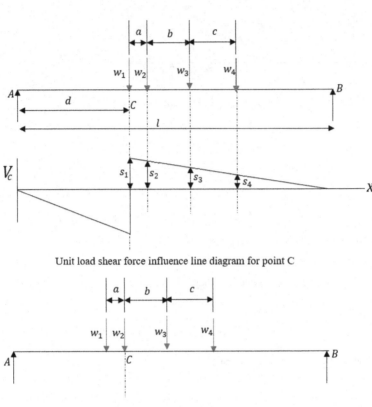

Unit load shear force influence line diagram for point C

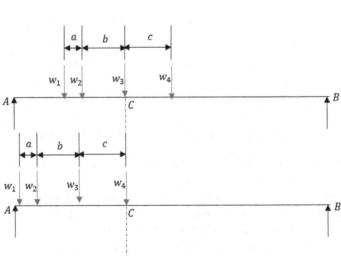

FIGURE 11.22 Different position of wheel load on the reference point C of the beam.

finding the change in shear force, ΔV from case to case, it becomes easy. If each computed ΔV is positive, the new position will yield a larger shear in the point of interest than the previous one. Each movement is investigated until a negative change in shear is found. When this occurs, the previous position of loads will give the critical value.

If the slope of the influence line is S, then we can write,

$$S = \frac{y_2 - y_1}{x_2 - x_1}$$

where $y_2 - y_1$ is the change in the ordinate of influence line, let say ΔV, and $x_2 - x_1$ is the change in the x-coordinate. So, we can write, $\Delta V = S(x_2 - x_1)$. Therefore, the change in shear, ΔV for a load P that moves form position x_1 to x_2 over a beam can be determined by:

$$\Delta V = PS(x_2 - x_1)$$

Now, if the load moves past a point where there is discontinuity of jump in the influence line, then change in shear can be written as:

$$\Delta V = P(y_2 - y_1)$$

Similarly, for a horizontal movement $(x_2 - x_1)$ of a concentrated load P, the change in moment, ΔM, is equivalent to the magnitude of the load times the change in influence line ordinate under the load written as:

$$\Delta M = PS(x_2 - x_1)$$

Like shear here also, as long as each computed ΔM is positive, the new position will yield a larger bending moment in the point of interest than the previous one. Each movement is investigated until a negative change in bending moment is found. When this occurs, the previous position of loads will give the critical value.

At this end, we should appreciate the importance of drawing influence line for unit load for internal force and moments, which enables us to analyze the same for any kind of loading as per actual requirements. Also, it is to be noted that following the same scheme, we can find out the position of the wheel load that will cause maximum bending moment in the beam using the unit load bending moment influence line. We will show how to calculate maximum bending moment for a moving uniformly distributed load (*UDL*) using influence line diagram in the next section.

Example 11.1: Find the maximum influence of shear at a point C due to series of concentrated loads as shown in Figure 11.23.

SOLUTION: Method 1: Let us first solve the problem by trial-and-error method. Three cases have been investigated as shown in Figure 11.24.

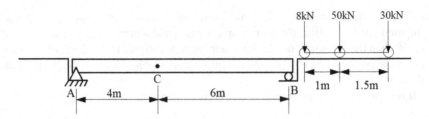

FIGURE 11.23 Example problem on finding the location for maximum influence.

Case 1:

In Case 1, let us assume that the first wheel is located at point C. The shear force at point C due to this case,

$$(V_C)_{Case\ 1} = 8(0.6) + 50(0.5) + 30(0.35) = 40.3\ \text{kN}$$

Case 2:

In Case 2, let us assume that the second wheel is just to the right of point C. The shear force at point C due to this case,

$$(V_C)_{Case\ 2} = 8(-0.3) + 50(0.6) + 30(0.45) = 41.1\ \text{kN}$$

Case 3:

In Case 3, let us assume that the third wheel is just to the right of point C. The shear force at point C due to this case,

$$(V_C)_{Case\ 3} = 8(-0.15) + 50(-0.25) + 30(0.6) = 4.3\ \text{kN}$$

Hence, Case 2 yields the largest value for V_C and, therefore, represents critical loading position.

Method 2: In this method, finding the change in shear force, ΔV from case to case we will obtain the load position for maximum shear at point C.

First consider the loads of Case 1, moving 1 m to Case 2. When this occurs, 8-kN load jumps down (−1) and all loads move up the slope of the influence line. This causes a change in shear force.

$$\Delta V_{1-2} = 8(-1) + (8 + 50 + 30)(0.1)(1) = +0.8\ \text{kN}$$

Since ΔV_{1-2} is positive, Case 2 will yield a larger value of V_C than Case 1. While calculating ΔV_{1-2}, we should note that after jumping down (−1) in the influence line diagram, the 8-kN load has moved forward and positioned 1 m ahead from point C. In this region also (AC), the 8-KN load has climbed a positive slope of 0.1.

Now, investigating ΔV_{2-3} which occurs when Case 2 moves to Case 3. Here the loads of Case 2, moving 1.5 m to Case 3. We must account for the downward (−1) jump of the 50-kN load, and all loads move up the slope of the influence line.

$$\Delta V_{2-3} = 50(-1) + (8 + 50 + 30)(0.1)(1.5) = -36.8\ \text{kN}$$

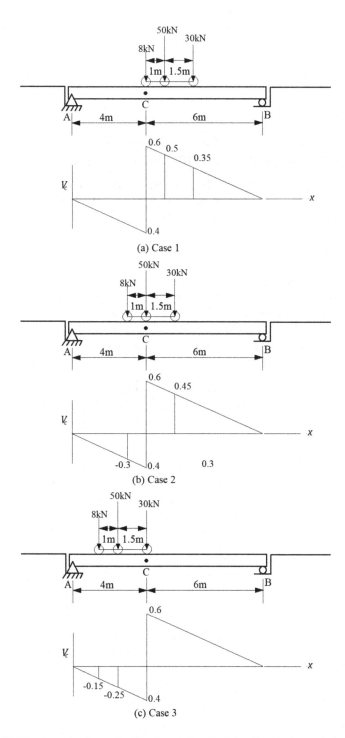

FIGURE 11.24 Investigations of various cases for obtaining the maximum influence at *C*.

Since ΔV_{2-3} is negative, Case 2 will yield a larger value of V_C than Case 3. Hence the critical load position is that of Case 2 as investigated previously by trial-and-error method earlier.

Similarly, one can investigate the influence of bending moment at a particular point after constructing the bending moment influence line for a point of interest, as depicted in Figure 11.7.

11.7 MAXIMUM INFLUENCE AT A POINT DUE TO A UNIFORMLY DISTRIBUTED LIVE LOAD

Let us consider a simply supported beam with a moving *UDL* of magnitude *w* acting on it. At any instant, let the starting and the end point of the *UDL* be at a distance of *a* and *b*, respectively, from the left support. We wish to calculate the bending moment at point *C* of the beam, as shown in Figure 11.25. Influence line diagram for bending moment at point *C* is also shown at the bottom of the main beam for the analysis purpose.

For a small length *dx* at a distance *x* from support *A*, the equivalent point load acting on the beam is *wdx*. Hence, total bending moment generated at point *C*, due to the above differential load is $(wdx)y$, where *y* is the ordinate of the influence line under the differential load. Now, we can find the total bending moment at point *C* due to the whole *UDL* as given by the following integral:

$$M_C = \int_a^b wy\,dx = w\int_a^b y\,dx$$

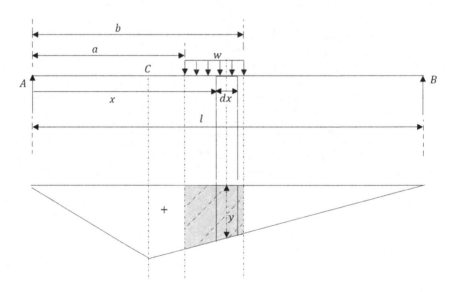

FIGURE 11.25 Maximum influence at a point due to moving uniformly distributed load (*UDL*).

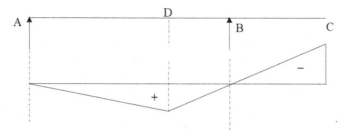

(a) Influence line diagram for M_D

(b) Arrangement of *UDL* for maximum negative bending moment at point D

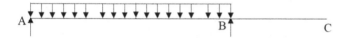

(c) Arrangement of *UDL* for maximum positive bending moment at point D

FIGURE 11.26   Arrangement of *UDL* for maximum positive and negative bending moment at a point *D*.

The integral in the right-hand side in the abovementioned expression represents the area under the influence line diagram between point *a* and *b* (shaded portion). This integral also tells us that for maximum positive bending moment, the *UDL* needs to be placed in that portion of the beam where the ordinates of the influence line diagram are all positive. To determine the maximum negative bending moment, if there was any overhang portion of the beam, then the *UDL* needs to be placed in that portion where the ordinates of influence line diagram are all negative. The arrangement for maximum negative and positive bending moment at a point of interest *D* on the beam is shown in Figure 11.26.

11.8   ABSOLUTE MAXIMUM SHEAR AND MOMENT

So far, we have discussed maximum and minimum bending moment and shear force values with respect to a reference point on the beam. Now, we want to develop a concept to determine the absolute maximum value of the same quantities for the entire beam for a given loading and its position. The location and magnitude of the absolute maximum shear force or bending moment for a beam are challenging to formulate. But we can easily find them constructing influence lines for shear force or bending moment at chosen points along the entire length of the beam and then computing the maximum shear or moment for each point in the beam. These maximum values of shear and moment, when plotted together with respect to their positions along the beam, yield an 'envelope of maximums'. From this envelope of maximums, we can easily find out the absolute maximum value of shear or moment and its location.

As usual, we can start this analysis by taking a moving unit load on the beam. We already knew that once the ordinates for the influence line diagrams are obtained, we can determine the values for any magnitude of loads just by multiplying the ordinates at that point of influence line diagrams. From our earlier discussion, for single unit point load on the beam, if we superimpose both influence line diagrams for support reactions, then we can easily determine from this the envelope of the maximum value of shear and its location.

We know for any arbitrary section at a distance a from left support on the simply supported beam subjected to a point load P, the maximum positive and negative shear, respectively, are:

$$V_a = +P \left(1 - \frac{a}{l} \right)$$

$$V_a = -\frac{Pa}{l}$$

These equations show that maximum positive and maximum negative shear at any section a varies linearly with the distance from left support A. Suppose, we plot the values of maximum positive and negative shear according to the abovementioned two equations simultaneously on the same beam axis along the entire length of the beam and join all of them on both sides of the beam axis. In that case, an envelope will be obtained as shown in Figure 11.27. From this envelope, we can determine the location and value of absolute maximum shear force. It is at the support points where maximum positive or negative shear force values are P, which is clearly seen from the envelope itself.

The envelope for maximum bending moment can be easily evaluated by considering the maximum bending moment at any section a from left support point A expressed as,

$$M_a = Pa \left(1 - \frac{a}{l} \right)$$

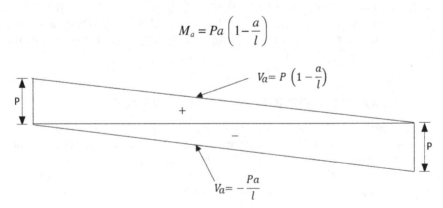

FIGURE 11.27   Envelope of maximum shear for single point load P.

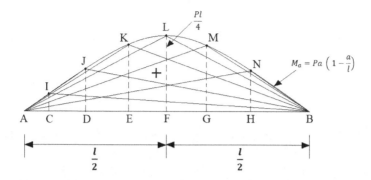

FIGURE 11.28 Envelope of maximum bending moment for single point load *P*.

Clearly, the bending moment also varies quadratically with respect to distance from the left support point. Hence, unlike shear force, the envelope for bending moment for the simply supported beam with single concentrated load will be parabolic in nature as shown in Figure 11.28. And the absolute maximum bending moment from the envelope itself can be found to be at midspan with value *Pl/4*.

Same can be obtained for *UDL*s also. For *UDL*, the maximum positive and negative shear at any section *a* from left support is given, respectively, by:

$$V_a = +\frac{w}{2l}(l-a)^2$$

$$V_a = -\frac{w}{2l}(a)^2$$

The abovementioned values can be obtained by placing the *UDL* load at the beam segment with maximum positive or negative ordinate values of unit load influence line for shear force and calculating the area under between the reference points. As from the abovementioned expression, it is clear that the shear force varies quadratically with respect to the distance of the load from the left support. Hence, the envelope for maximum positive and negative shear force will be parabolic in nature, which is shown in Figure 11.29.

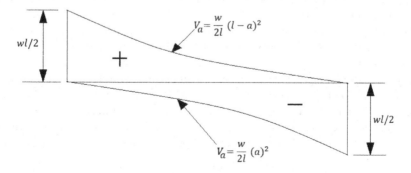

FIGURE 11.29 Envelope of maximum shear for uniformly distributed load.

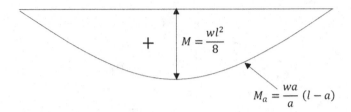

FIGURE 11.30 Envelope of maximum bending moment for uniformly distributed load.

Similarly, the maximum positive bending moment values for *UDL* from a distance a from the left support will be:

$$M_a = \frac{wa}{a}(l-a)$$

So, the bending moment equations also vary quadratically with respect to distance from the left support. Hence, bending moment envelope will be parabolic in nature and absolute maximum bending moment from the envelope itself can be found to be at midspan with value $wl^2/8$. The maximum bending moment envelope can be as seen in Figure 11.30.

11.8.1 ABSOLUTE MAXIMUM BENDING MOMENT FOR SERIES OF CONCENTRATED LOADS

For a series of concentrated loads, influence line envelopes can be drawn along the length of the member to determine the maximum response values. But to be able to that, we have to place the loads at various locations along the length of the beam, which will ultimately attract a large number of computational efforts. By inspection also the critical position of loads and the associated absolute maximum moment cannot be determined even for a simple beam. For this reason, these analyses are done in computer with the help of various software. However, for our analysis purpose, we will provide the method of obtaining the maximum response for simple beams. For a simply supported beam, a series of concentrated loads P_1, P_2, and P_3 are applied as shown in Figure 11.31. Since the moment diagram for a series of concentrated loads consists of straight-line segments having peaks at each load, the absolute maximum moment will also occur under one of these loads. Let us assume the maximum moment occurs under the load P_2, and P_2 is x distance apart from the beam's centerline. By positioning the load P_2 from a fixed distance x from the centerline of the beam, we can fix the location of other loads in the series as well. Now, to determine the specific value of x, we need to find the resultant force of this load series, P_R, and its distance $\bar{x}$ measured from P_2. After obtaining this, if we now sum up moments with respect to support B, we can easily find out the reaction force A_y as follows:

$$\circlearrowleft + \sum M_B = 0$$

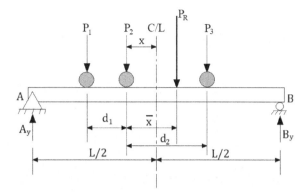

FIGURE 11.31 Condition for the absolute maximum moment in a simply supported beam for series of concentrated loads.

or,

$$-\left(A_y \times L\right) + P_R\left[\frac{L}{2} - (\bar{x} - x)\right] = 0$$

Therefore,

$$A_y = \frac{1}{L}(P_R)\left[\frac{L}{2} - (\bar{x} - x)\right]$$

We have assumed the absolute maximum bending moment will occur under the load P_2. So first, we need to find the internal moment, M_2 generated under this load and after that need to find the condition for, M_2 to be maximum. If the beam is sectioned just to the left of P_2 as shown in Figure 11.32, we can write the moment equilibrium equation as under:

$$M_2 = A_y\left(\frac{L}{2} - x\right) - P_1 d_1$$

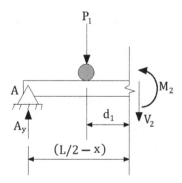

FIGURE 11.32 Section of the beam just to the left of load P_2.

Now, replacing the value of A_y in the abovementioned expression, we obtain,

$$M_2 = A_y \left(\frac{L}{2} - x \right) - P_1 d_1$$

or,

$$M_2 = \frac{1}{L}(P_R)\left[\frac{L}{2} - (\bar{x} - x)\right] \times \left(\frac{L}{2} - x\right) - P_1 d_1$$

therefore,

$$M_2 = \frac{P_R L}{4} - \frac{P_R \bar{x}}{2} - \frac{P_R x^2}{L} + \frac{P_R x \bar{x}}{L} - P_1 d_1$$

For M_2 to be maximum from this expression, we require,

$$\frac{dM_2}{dx} = 0$$

i.e.,

$$\frac{-2P_R x}{L} + \frac{P_R \bar{x}}{L} = 0; \therefore x = \frac{\bar{x}}{2}$$

This is a very important relationship and from this relationship, we can say that, for a simply supported beam with series of concentrated loads, maximum bending moment occurs under a load, when midspan of the beam equally divides the distance between the said load and the resultant of all loads acting on the beam. Thus, after noting the resultant of the applied series loads, we can place each load about mid-span of the beam and calculate bending moment, satisfying the abovementioned condition, to calculate the maximum absolute bending moment acting on the beam. This method is widely adopted to calculate the maximum bending moment for bridge girders during analysis and design works.

Example 11.2: Find the absolute maximum bending moment in the simply supported beam for the series of as shown in Figure 11.23.

SOLUTION: First let us determine the magnitude and position of the resultant force of the system as next,

$$P_R = (8 + 50 + 30) = 84 \text{ kN}$$

Suppose the resultant force (84 kN) is $\bar{x}$ distance apart from the 8-kN load. Now we need to find $\bar{x}$ from the following equilibrium equation:

$$84\bar{x} = 50(1) + 30(2.5)$$

or,

$$\bar{x} = 1.49 \text{ m}$$

Now we can locate the resultant force as shown in Figure 11.33.

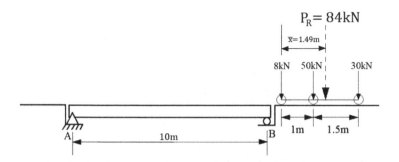

FIGURE 11.33 Location and magnitude of resultant force for the load series.

Let us assume the absolute maximum bending moment occurs under the 50-kN load. The load and the resultant force are positioned equidistant from the beam's centerline as shown in Figure 11.34.

Now in this load position, let us sum up all the moments about point B,

$$\circlearrowleft + \sum M_B = 0$$

$$-A_y(10) + 84(4.755) = 0$$

or,

$$A_y = 39.94 \text{ kN}$$

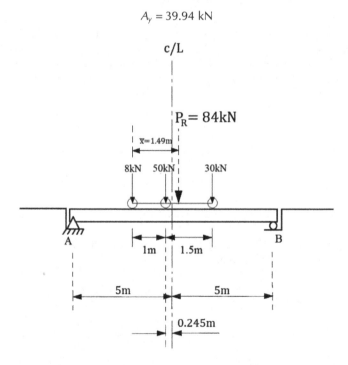

FIGURE 11.34 Position of load series loads assuming the absolute maximum moment occurs under 50-kN load.

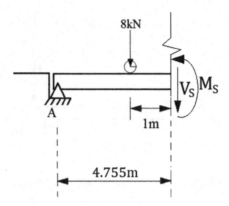

FIGURE 11.35   Section of the beam just to the left of load 50 kN.

If the beam is sectioned just to the left of 50 kN as shown in Figure 11.35, we can write the moment equilibrium equation as under:

$$-39.94\left(4.755\right)+8\left(1\right)+M_s = 0$$

or,

$$M_s = 181.91\,\text{kN}-\text{m}$$

Now let us check another possibility before concluding 181.91 kN m as the absolute maximum bending moment. For that let us assume the absolute maximum bending moment occurs under the 30-kN load. The load and the resultant force are positioned equidistant from the beam's centerline as shown in Figure 11.36.
   Now in this load position, let us sum up all the moments about point B,

$$\circlearrowleft+\sum M_B = 0$$

$$-A_y\left(10\right)+84\left(5.505\right) = 0$$

or,

$$A_y = 46.24\,\text{kN}$$

If the beam is sectioned just to the left of 30 kN as shown in Figure 11.37, we can write the moment equilibrium equation as under:

$$-46.24\left(5.505\right)+8\left(2.5\right)+50\left(1.5\right)+M_s = 0$$

or,

$$M_s = 159.55\,\text{kN}-\text{m}$$

Now, we can conclude that the absolute maximum bending moment will occur under the 50-kN load and its value is 181.91 kN m.

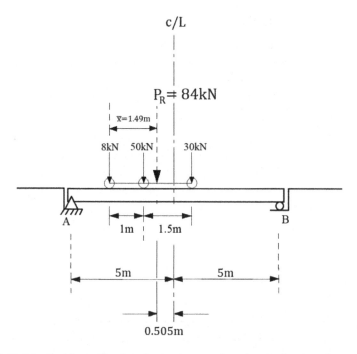

FIGURE 11.36 Position of load series loads assuming the absolute maximum moment occurs under 30-kN load.

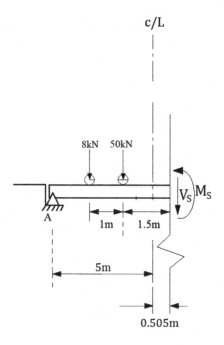

FIGURE 11.37 Section of the beam just to the left of load 30 kN.

11.9 INFLUENCE LINES FOR DEFLECTIONS

Influence line for deflections provides the information of deflection due to a moving concentrated unit load on a beam structure. To obtain influence line for deflection, Maxwell's reciprocal law of deflection is found to be most useful. Consider a simply supported beam on which we want to determine the deflection influence line for the reference point C on the beam as shown in Figure 11.38.

Suppose when unit load is placed at a distance x from left support (point X), the deflection at point C due to this case is δ_{CX} as shown in Figure 11.38 (a). Also, when

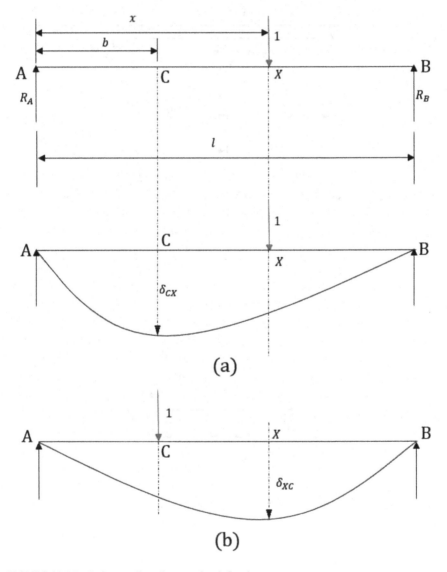

(a)

(b)

FIGURE 11.38 Influence line diagram for deflection.

we place the unit load at point C itself, the vertical deflection at point X is given by δ_{XC} as shown in Figure 11.38 (b). Now, following Maxwell's theorem, we can say that:

$$\delta_{XC} = \delta_{CX}$$

Since the point X is arbitrary in nature, this relationship is applicable for any point on the beam. From the abovementioned relationship, we can conclude that influence line for deflection for any point on a beam can be determined by drawing the elastic curve/deflection line for the same beam when unit load is applied at that point itself. We have already learned how to draw deflection lines for any determinate beams using different principles in earlier chapters. Hence, by placing unit load at the required point on the beam and doing the same calculation for drawing elastic curve will help us to draw the influence line diagram of deflection for the beam.

12 Cables, Arches, and Suspension Bridges

12.1 INTRODUCTION

In major structures, cables and arches are found to be the principal load-carrying elements, and in this chapter, we will explore some salient features of the same and their structural analysis. The chapter starts with a general discussion of cables, leading to the analysis of cables subjected to concentrated and uniformly distributed loads (UDL). Since the most common type of arches is statically indeterminate, only the special case of three-hinged arch will be discussed in this chapter. The knowledge of this type of structural analysis will provide some insight into the core behavior of all arched structures.

12.2 CABLES

Cables are used in structures to support and transfer loads from one member to another. When cables are used to support suspension roofs, bridges, etc., they form the main load-bearing element in the structure. During force analysis of such structures, the weight of the cable is mostly neglected; however, when cables are used as guy wires for radio antennas, electrical transmission lines, etc., the cable weight may become an important factor, and it has to be considered in the structural analysis. Separate cases will be discussed in the sections as follows: a cable subjected to concentrated loads and another one to a UDL. It is worthwhile to mention that these loadings are coplanar with the cable, and corresponding requirements for equilibrium are formulated accordingly.

While deriving the equations and relations between the force in the cable and its slope, it will be assumed that the cable is perfectly flexible and inextensible. Due to flexibility, the cable does not induce shear force and bending moment, and thus, the force acting in the cable will always be tangential to the cable at the same points along its length. As it is inextensible, the cable has invariant length both before and after the external load is applied. As a result, once the external load is applied, the geometry of the cable remains fixed, and the cable or a segment of it can be treated as a rigid body.

12.3 CABLES SUBJECTED TO CONCENTRATED LOADS

When a cable (neglecting self-weight) supports several concentrated loads, the cable takes the form of different straight-line segments, each of which is acted upon by a constant tensile force. Let us consider, for example, the cable shown in Figure 12.1. Here, θ indicates the angle of the cable's chord AB, and L is its span. If the distances

DOI: 10.1201/9781003081227-14

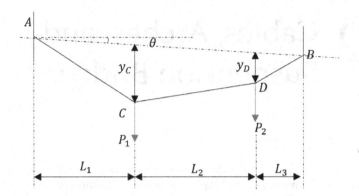

FIGURE 12.1 Cable with concentrated loads.

L_1, L_2, L_3 and the loads P_1, P_2 are known, our goal is to calculate the nine unknowns consisting of the tension in three segments of the cable, the four components of reaction at support points A and B, and the vertical displacements y_C, y_D at the two points C and D. For the calculation, we can write two equations of force equilibrium at each of points A, B, C, and D. This results in a total of eight equations. To complete the analysis and calculations, it will be necessary to know about the geometry of the cable to obtain the necessary ninth equation.

For the sake of simplicity, if the cable's total length L is specified, the Pythagorean theorem can be used to relate each of the three segmental lengths, in terms of L_1, L_2, L_3, y_C, y_D, and θ to the total length, L. Practically, this type of problem cannot be solved by hand with ease. On the other hand, we can analyze the same problem by specifying either of the vertical deflections y_C, y_D, instead of the total length of the cable. Whatever way we may proceed, we can form the equilibrium equations and complete the calculations to determine the unknown tension forces acting at different segments of the loaded cable and the support reactions.

While carrying out equilibrium analysis for a problem like this one, the unknown forces in the cable can also be determined by developing the equilibrium equations for the entire cable or any portion. Example 12.1 provides these necessary concepts.

Example 12.1: Determine the tension in each segment of the following cable subjected to the point loads as shown in Figure 12.2.

SOLUTION: By the nature of this loading pattern, we can declare that there are four unknown support reactions (H_A, V_A, H_B, V_B) and three unknown tension forces (T_{DB}, T_{CD}, T_{CA}) acting in the cable and a sag h. So, there is total eight unknowns. These eight unknowns can be determined from eight available equilibrium equations $\left(\sum F_x = 0, \sum F_y = 0\right)$ applied through A to D. Since the geometry is known in detail from the given diagram, we can first determine the unknown tensile force in the cable DB (Figure 12.3).

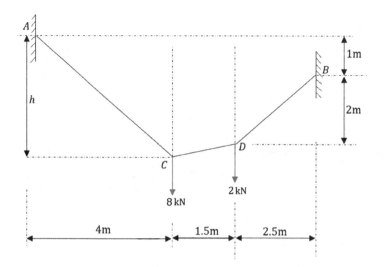

FIGURE 12.2 Example problem on cable under concentrated loads.

Taking moment about A,

$$T_{DB}\left(\frac{2.5}{3.2}\right)1 + T_{DB}\left(\frac{2.0}{3.2}\right)8 - 8 \times 4 - 2 \times 5.5 = 0$$

$$T_{DB} = 7.44 \text{ kN}$$

Now, let us consider the equilibrium of point D as shown in Figure 12.4.

$$\rightarrow + \sum F_x = 0$$

$$7.44 \times \left(\frac{2.5}{3.2}\right) - T_{CD}\cos\theta_{CD} = 0$$

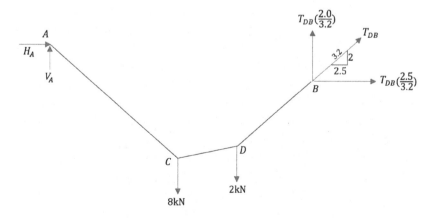

FIGURE 12.3 Free-body diagram to determine the tension force T_{DB}.

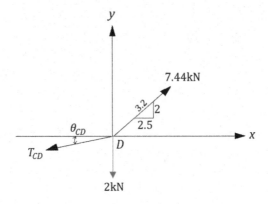

FIGURE 12.4 Equilibrium of point D.

$$\uparrow + \sum F_y = 0$$

$$7.44 \times \left(\frac{2.0}{3.2}\right) - T_{CD} \sin \theta_{CD} - 2 = 0$$

Solving the abovementioned two equilibrium equations, we get $\theta_{CD} = 24.51°$, $T_{CD} = 6.37$ kN.

Now, let us consider the equilibrium of point C as shown in Figure 12.5.

$$\rightarrow + \sum F_x = 0$$

$$6.37 \times \cos 24.51° - T_{AC} \cos \theta_{AC} = 0$$

$$\uparrow + \sum F_y = 0$$

$$6.37 \times \sin 24.51° + T_{AC} \sin \theta_{CA} - 8 = 0$$

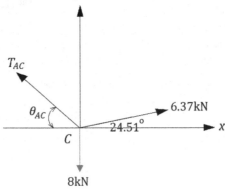

FIGURE 12.5 Equilibrium of point C.

Solving the abovementioned two equilibrium equations, we get, $\theta_{CA} = 42.75°$, $T_{AC} = 7.89$ kN.

From the geometry we get, $h = 4(\tan 42.75°) = 3.7$ m

12.4   CABLE SUBJECTED TO A UNIFORMLY DISTRIBUTED LOAD

Cables provide a very efficient means of withstanding and transferring the dead weight of girders or bridge decks having very long spans. A suspension bridge is an important example in which the bridge deck is supported by the cable using a series of equally spaced hangers. In Figure 12.6 (a), we have shown a cable under

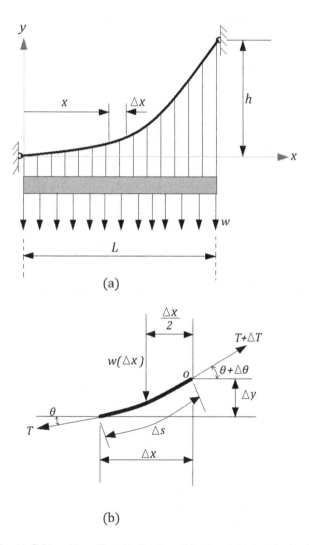

(a)

(b)

FIGURE 12.6   (a) Cable with uniformly distributed load and (b) free-body diagram of the small element Δx.

the application of a UDL of intensity w. The cable will take a curvilinear shape as shown in the figure. Exact nature of the curve will depend upon the load intensity and support conditions, and based on the same, exact equation for the deflected shape of the cable can be drawn. To analyze the problem, we will take an arbitrary small element of loaded cable of length Δs at a distance x from the left support as shown in Figure 12.6. The projection of this small element Δs, on the x-axis is suppose Δx. Free-body diagram of this small element is also shown in Figure 12.6 (b) for the ease of understanding. Here the origin of the x, y coordinates has been chosen at the lowest point of the cable as shown. The distributed force is represented as an equivalent concentrated force of magnitude $w(\Delta x)$, and acting at a distance of $\Delta x/2$ from point O.

Applying force equilibrium equations in horizontal direction, we get,

$$\rightarrow + \sum F_x = 0 - T\cos\theta + (T + \Delta T)\cos(\theta + \Delta\theta) = 0$$

Also, applying force equilibrium equation for vertical direction, we get,

$$\uparrow + \sum F_y = 0 - T\sin\theta - w\Delta x + (T + \Delta T)\sin(\theta + \Delta\theta) = 0$$

Also, taking moment about point O of the free-body diagram we get,

$$+ \circlearrowleft \sum M_o = 0$$

$$(w\Delta x)\frac{\Delta x}{2} - T\cos\theta\,\Delta y + T\sin\theta\,\Delta x = 0$$

Dividing each of the abovementioned equations by Δx and taking the limit $\Delta x \rightarrow 0$, and hence, by $\Delta\theta \rightarrow 0$ and $\Delta T \rightarrow 0$, we obtain,

$$\frac{d(T\cos\theta)}{dx} = 0$$

$$\frac{d(T\sin\theta)}{dx} = w$$

$$\frac{dy}{dx} = \tan\theta$$

Integrating the first equation, where $T = F_H$ at $x = 0$, we have,

$$T\cos\theta = F_H$$

which indicates horizontal component of tension force at any point along the cable subjected to UDL remains constant.

Integrating the second equation, with the initial condition $T \sin \theta = 0$ at $x = 0$, we get,

$$T \sin \theta = wx$$

Dividing these last two equations, we get,

$$\tan \theta = \frac{dy}{dx} = \frac{wx}{F_H}$$

Performing the integration with $y = 0$ at $x = 0$ yields,

$$y = \frac{w}{2F_H} x^2$$

The preceding equation is the equation of parabola. The constant force F_H can be obtained by putting the boundary condition,

$$y = h, \text{ at } x = L$$

Putting this value in the parabolic equations, we get,

$$F_H = \frac{wL^2}{2h}$$

Replacing the value of F_H in the master equation, we finally get the equation of parabolic shape of the cable as follows:

$$y = \frac{h}{L^2} x^2$$

Also, from the equation, $T \cos \theta = F_H$, and $T \sin \theta = wx$; maximum tension in the cable occurs when θ is maximum, i.e., at $x = L$. So, from these equations, we get,

$$T_{max} = \sqrt{F_H^2 + (wL)^2}$$

or,

$$T_{max} = \sqrt{\left(\frac{wL^2}{2h}\right)^2 + (wL)^2}$$

i.e.,

$$T_{max} = wL\sqrt{\left(\frac{L}{2h}\right)^2 + 1}$$

As stated earlier, we have neglected the self-weight of the cable during derivation of the abovementioned relationship. In fact, when a cable is suspended and only acting load on it is the self-weight, the cable assumes the shape of catenary. From the outcome of the abovementioned analysis, it indicates that a cable will assume a parabolic shape, provided the dead load of the deck for a bridge will be uniformly distributed on the horizontal projection length of the cable. Hence, if the bridge girder is supported by a series of hangers, which are uniformly spaced and not far from each other, the load in each hanger must be the same to ensure that the cable has a parabolic shape.

Taking help of this assumption, we can complete the structural analysis of the girder or any other framework that is freely suspended by the cable. We will assume that the girder is simply supported, and as a result, this will be a statically indeterminate problem of degree one. Example 12.2 will elaborate on this concept for cable-stayed bridge girder analysis.

Example 12.2: Determine the tension at points A, B, and C of the following cable-stayed bridge girder as per Figure 12.7. Self-weight of the girder is 70 kN/m.

SOLUTION: The origin has been selected at point B, which is the lowest point of the cable. Now, from our theoretical analysis, we can insert these geometrical values to calculate the exact equation.

$$y = \frac{w}{2F_H}x^2$$

or,

$$y = \frac{70}{2F_H}x^2 = \frac{35}{F_H}x^2$$

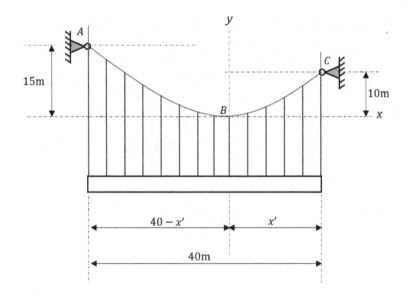

FIGURE 12.7   Example problem of cable-supported bridge girder.

Assuming point C at a distance of x' from point B,

$$10 = \frac{35}{F_H} x'^2$$

or,

$$F_H = 3.5 x'^2$$

Also, for point A,

$$15 = \frac{35}{F_H}\left[-\left(40 - x'\right)\right]^2$$

or,

$$15 = \frac{35}{3.5x'^2}\left[-\left(40 - x'\right)\right]^2$$

Solving the quadratic equation, we get,

$$x' = 17.98 \text{ m}$$

So, once we get the value for x', we can substitute it to get the value of F_H. So,

$$F_H = 3.5 \times 17.98^2 = 1131.48 \text{ kN}$$

Now, at point A, $x = -\left(40 - 17.98\right) = -22.02$ m

$$\tan\theta_A = \frac{dy}{dx}_{x=-22.02} = \frac{35}{1131.48} \times 2 \times \left(-22.02\right) = -1.36$$

Hence, $\theta_A = -53.67°$
So,

$$T_A = \frac{F_H}{\cos\theta_A} = \frac{1131.48}{\cos\left(-53.67°\right)} = 1909.88 \text{ kN}$$

Now, at point B, $x = 0$ m

$$\tan\theta_B = \frac{dy}{dx}_{x=0} = \frac{35}{1131.48} \times 2 \times \left(0\right) = 0$$

Hence, $\theta_B = 0°$
So,

$$T_B = \frac{F_H}{\cos\theta_B} = \frac{1131.48}{\cos\left(0°\right)} = 1131.48 \text{ kN}$$

Now, at point C, $x = 17.98$ m

$$\tan \theta_C = \frac{dy}{dx_{x=17.98}} = \frac{35}{1131.48} \times 2 \times (17.98) = 1.11$$

Hence, $\theta_C = 47.98$
   So,

$$T_C = \frac{F_H}{\cos \theta_C} = \frac{1131.48}{\cos(47.98°)} = 1690.32 \text{ kN}$$

12.5   ARCHES

Arches can be constructed to control and reduce the bending moments in long-span bridges, airplane hangars, etc. because one of the main distinguishing features of an arch is the development of vertical reactions as well as horizontal thrusts at the supports, even in the absence of a horizontal load. Operationally, an arch act just like an inverted cable, so it transmits its load mainly in compression. However, due to its rigidity, it also has to resist bending and shear depending upon external loading patterns and its geometric shape. Specifically, if the arch is parabolic in nature and loaded by a uniform horizontally distributed vertical load, from our earlier knowledge on the analysis of cables it follows that only compressive forces will be resisted by the arch. Under such circumstances, the arch shape is called a funicular arch because no bending or shear forces will be induced within the arch due to external loading.

A typical arch is shown in Figure 12.8, which specifies some of the names of various parts of an arch used to define geometry.

Depending upon the requirements, different types of arches, as shown in Figure 12.9, can be modeled to support different types of external loading. A fixed arch (Figure 12.9 (a)) is often made from reinforced concrete. Although it may require less material to construct than other types of arches, it must have a solid foundation

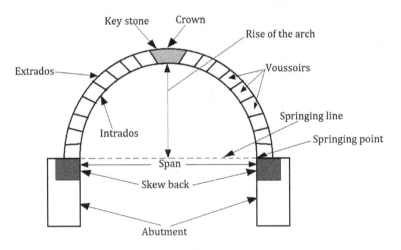

FIGURE 12.8   Different components of an arch.

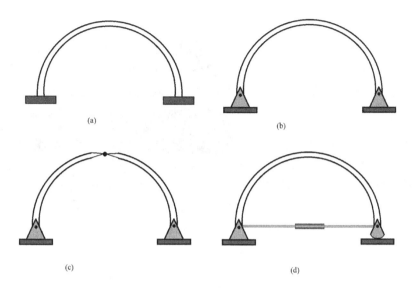

FIGURE 12.9   Different types of arches: (a) fixed arch, (b) two-hinged arch, (c) three-hinged arch, and (d) tied arch.

since it is indeterminate to the third degree, and additional stresses can be introduced into the arch due to relative settlement of its supports. A two-hinged arch (Figure 12.9 (b)) is generally made from metal or timber. It is indeterminate to the first degree. Although it is not as rigid as a fixed arch, it is to some extent insensitive to settlement. We could make this structure statically determinate by replacing one of the hinges with a roller. But this will affect the capacity of the structure to resist bending along its span, and, as a result, it would serve as a curved beam and not as an arch. A three-hinged arch (Figure 12.9 (c)), which is also made from metal or timber, is statically determinate. It is not affected by settlement or temperature changes. Finally, we can attach a tie rod (Figure 12.9 (d)) at the supports, so that the arch will behave like a rigid body and thus we can avoid the need for larger foundation abutments. This will also remain unaffected even under the relative settlement of supports.

12.6   THREE-HINGED ARCHES

We will now consider the analysis of a three-hinged arch such as the one shown in Figure 12.10 (a) and (c), to provide some insight into how arches transmit loads. In this case, the third hinge is located at the crown, C and the supports are located at different heights as shown in the diagram. To determine the reactions at the supports, the arch is disassembled to construct the free-body diagram of each member as shown in Figure 12.10 (b). It is clear from the free-body diagram that there are six unknowns for which six equations of equilibrium are available. One method of solving this problem is to apply the moment equilibrium equations about points A and B. The simultaneous solution will yield the reactions at the internal hinge at C. From the force equations of equilibrium and then the support reactions at A and B are determined. Once obtained, the internal normal force, shear, and moment at any

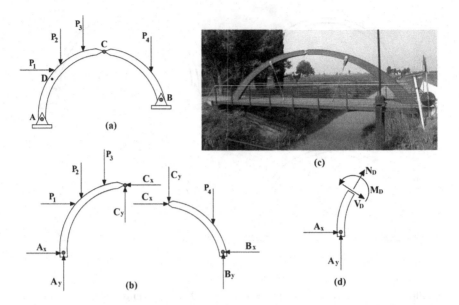

FIGURE 12.10 (a), (c) Three-hinged arch (source of 'c': https://structurae.net/en/media/315533-via-guglielmo-marconi-cycle-bridge), and (b) and (d) free-body diagrams for internal loads and external support reactions.

point along the arch can be found using the method of sections. Here, of course, the section should be taken perpendicular to the axis of the arch at the point considered. Here, for example, the free-body diagram of segment AD is shown in Figure 12.10 (d).

By applying moment equilibrium equations and remembering the fact that at hinge locations, net moment will be zero, we can calculate the unknown forces acting in the arch. Once the support reactions are obtained, we can determine the bending moment and shear force at any typical section taken perpendicular to the arch at that point. These concepts will be easier to understand once we go through Examples 12.3 and 12.4 (Figures 12.11 and 12.15).

Example 12.3: Show that any point (like point D) of the uniformly loaded parabolic arch as shown in Figure 12.11 will always be under compression. The arch is hinged at locations A, B, and C.

SOLUTION: Since the UDL is acting on the horizontal bridge deck above the arch, net point load acting on the arch at the crown is,

$$P = 3.5 \times (20 + 10 + 10) = 140 \text{ kN}$$

The free-body diagram will be like the one in Figure 12.12.
Taking moment about point A, we will get,

$$140 \times 20 - C_y \times 40 = 0$$

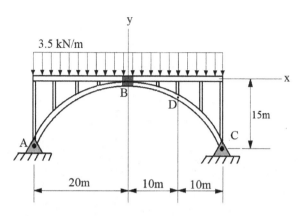

FIGURE 12.11   Three-hinged arch example problem-1.

or,

$$C_y = 70 \text{ kN} \uparrow$$

Since at point B, there is a hinge support, the algebraic sum of moment at B will be zero. The free-body diagram of the right half portion is shown in Figure 12.13.
   Arch segment BC,
   Taking moment about point B, we get,

$$\circlearrowleft + \sum M_B = 0$$

or,

$$70 \times 20 - C_x \times 15 - 70 \times 10 = 0$$

$$C_x = 46.67 \text{ kN} \leftarrow$$

$$\rightarrow + \sum F_x = 0$$

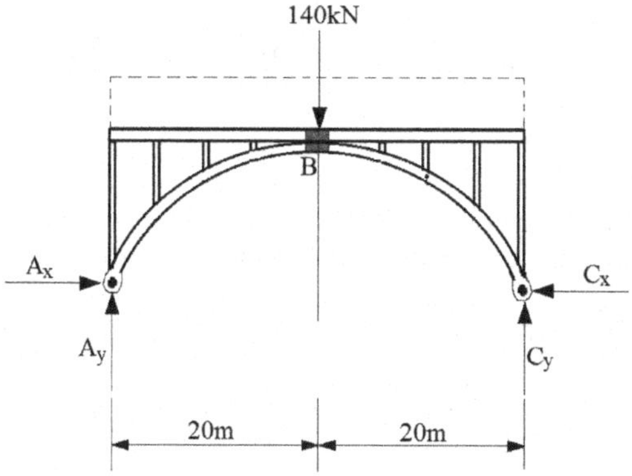

FIGURE 12.12   Free-body diagram of the entire arch for support reactions.

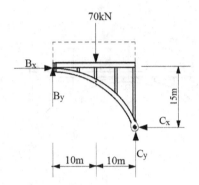

FIGURE 12.13   Free-body diagram of the right half portion of arch.

or,

$$B_x = 46.67 \text{ kN} \rightarrow$$

$$\uparrow + \sum F_y = 0$$

$$B_y + C_y = 70$$

or,

$$B_y = 0$$

Taking a section along the point D, from cable equation of parabolic shape, we get,

$$y = \frac{h}{L^2} x^2$$

where $h = 15$ m, $L = 20$ m. Hence, the elevation of point D will be,

$$y = -\frac{15}{20^2}(10^2) = -3.75 \text{ m}$$

And the slope of the tangent at point D is given by,

$$\tan \theta_D = \frac{dy}{dx}_{x=10 \text{ m}} = \frac{-15 \times 2}{20^2} \times 10 = -0.75$$

$$\theta_D = -36.87°$$

The free-body diagram at point D for calculating the forces is shown in Figure 12.14. Thus, force and moment equilibrium equation will produce,

$$\rightarrow + \sum F_x = 0 \Rightarrow 46.67 - N_D \cos 36.87° - V_D \sin 36.87° = 0$$

$$\uparrow + \sum F_y = 0 \Rightarrow -35 + N_D \sin 36.87° - V_D \cos 36.87° = 0$$

$$\circlearrowleft + \sum M_D = 0 \Rightarrow M_D + 35 (5) - 46.67(3.75) = 0$$

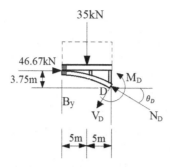

FIGURE 12.14   Free-body diagram of section D.

Solving the abovementioned equations, we get,

$$N_D = 58.41 \text{ kN}$$

$$V_D = 0$$

$$M_D = 0$$

Please note that when arch is having a different shape and with unsymmetrical loading, the bending moment and shear force values will be nonzero.

Example 12.4: A three-hinged parabolic arch is loaded as shown in Figure 12.15. Calculate the support reactions and internal stresses for the sections shown. The arch is hinged at locations A, B, and C.

SOLUTION: The ordinate (y) at any point along a parabolic arch is given by:

$$y = \frac{4y_c\left(Lx - x^2\right)}{L^2}$$

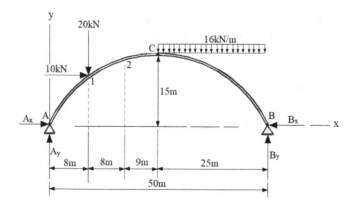

FIGURE 12.15   Three-hinged Arch example problem 2.

where y_c is the height of the crown of the arch from the base = 15 m, L is the length of the arch = 50 m, x is the horizontal coordinate of interest.
The coordinate is chosen at point A as shown in figure.
Hence,

$$y = \frac{4 \times 15 \left(50x - x^2\right)}{50^2} = \left(\frac{6}{5}\right)x - \left(\frac{3}{125}\right)x^2$$

Differentiating the abovementioned equation with respect to x.

$$\frac{dy}{dx} = \tan \theta = \left(\frac{6}{5}\right) - \left(\frac{6}{125}\right)x$$

The values of y-ordinates, and $\tan \theta$ for corresponding different x-coordinates of interest are calculated as from the abovementioned equations and shown in Table 12.1.

Support reactions

$$\circlearrowleft + \sum M_B = 0$$

$$\left(-A_y \times 50\right) + \left(16 \times 25 \times 12.5\right) + \left(20 \times 42\right) - \left(10 \times 8.064\right) = 0$$

or,

$$A_y = 115.18 \text{ kN}$$

$$\circlearrowleft + \sum M_A = 0$$

$$\left(B_y \times 50\right) - \left(16 \times 25 \times 37.5\right) - \left(20 \times 8\right) - \left(10 \times 8.064\right) = 0$$

or,

$$B_y = 304.81 \text{ kN}$$

Now, from span BC, we will find out the horizontal reaction B_x, considering the moment equilibrium at crown C. In this portion about point C, the anticlockwise

TABLE 12.1
Values of Different y-Coordinates and Slopes for Corresponding x-Coordinates of the Arch

Point	$x(m)$	$y(m)$	$\tan \theta$
A	0	0	1.2
1	8	8.064	0.816
2	16	13.056	0.432
C	25	15	0

moment will sag the arch and clockwise moment will hog the arch. So, the anti-clockwise moment is considered positive here.

$$\circlearrowleft + \sum M_C^R = 0$$

$$(304.81 \times 25) - (B_x \times 15) - (16 \times 25 \times 12.5) = 0$$

or, $\qquad B_x = 174.68 \text{ kN} \leftarrow$

Now, from span AC, we will find out the horizontal reaction A_x, considering the moment equilibrium at crown C. In this portion about point C, the anticlockwise moment will hog the arch and clockwise moment will sag the arch. So, the clockwise moment is considered positive here.

$$\circlearrowleft + \sum M_C^L = 0$$

$$(115.18 \times 25) - (A_x \times 15) - (20 \times 17) - 10(15 - 8.064) = 0$$

or, $\qquad A_x = 164.68 \text{ kN} \rightarrow$

Internal stresses

Now we will determine the internal stresses (normal force, shear, and bending moment) of the arch at the section of interests. For this, let us consider one free-body diagram up to some arbitrary section (x_i, y_i) as shown in Figure 12.16.

Let us consider the moment equilibrium at that section,

$$\circlearrowleft + \sum M = 0$$

or, $\qquad M = A_y x_i - A_x y_i - P_1 x_1$

From the free-body diagram, we can obtain,

$$A_x + N \cos \theta + Q \sin \theta = 0$$

$$A_y - P_1 + N \sin \theta - Q \cos \theta = 0$$

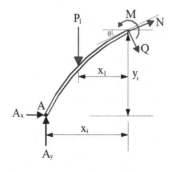

FIGURE 12.16 Free-body diagram of the arch up to some arbitrary section.

Solving these two equations for different sections, we can get the values of N, and Q. Let us find these internal forces for different sections of arch as follows:

Calculation for the left side of the hinge

For point A,

$$\tan \theta = 1.2, \theta = 50.19°$$

$$M_A = 0 \text{ (hinged support)}$$

$$164.68 + N \cos 50.19° + Q \sin 50.19° = 0$$

$$115.18 + N \sin 50.19° - Q \cos 50.19° = 0$$

Solving this,

$$N = -193.91 \text{ kN, and } Q = -52.76 \text{ kN}$$

For point 1,

$$\tan \theta = 0.816, \theta = 39.21°$$

$$M_1 = 115.18 \times 8 - 164.68 \times 8.064 = -406.54 \text{ kN} - m$$

$$164.68 + 10 + N \cos 31.21° + Q \sin 31.21° = 0$$

$$115.18 - 20 + N \sin 31.21° - Q \cos 31.21° = 0$$

Solving this,

$$N = -198.71 \text{ kN, and } Q = -9.11 \text{ kN}$$

For point 2,

$$\tan \theta = 0.432, \theta = 23.36°$$

$$M_2 = 115.18 \times 16 - 164.68 \times 13.056 - 10 \times (13.056 - 8.064) - 20 \times 8$$

$$= -517.1 \text{ kN} - m$$

$$164.68 + 10 + N \cos 23.36° + Q \sin 23.36° = 0$$

$$115.18 - 20 + N \sin 23.36° - Q \cos 23.36° = 0$$

Solving this,

$$N = -198.11 \text{ kN, and } Q = 18.12 \text{ kN}$$

For point C,

$$\tan \theta = 0, \ \theta = 0°$$

$$M_C = 115.18 \times 25 - 164.68 \times 15 - 10 \times (15 - 8.064) - 20 \times 17 = 0 \text{ kN} - m$$

$$164.68 + 10 + N = 0$$

$$115.18 - 20 - Q = 0$$

Solving this,

$$N = -174.68 \text{ kN, and } Q = 95.18 \text{ kN}$$

12.7   THREE-HINGED STIFFENING GIRDERS

The curvature of the cable of an unstiffened bridge always changes as the live load moves on the deck, because the cable of the suspension bridge is the main load-bearing element. To avoid this, the bridge deck is stiffened by either providing two-hinged or three-hinged stiffening girders. The stiffening girders are assumed to transfer a uniform or equal load to each suspender, irrespective of the load positions on the bridge deck. When bridges are stiffened with three-hinged stiffening girders, it is assumed that the cable retains its parabolic shape when subjected to loads from moving traffic or from other sources and hence the load on the cable will remain uniform throughout the entire length. When the bridge is subjected to UDL, the total load is directly taken care of by the cable itself and does not affect the stiffening girders. On the other hand, in the case of moving loads, the cable will be assumed to carry uniform load, and, as a result, the stiffening girders will be subjected to bending moments and shear forces as that happens for normal transversely loaded beams.

Now, let us consider a suspension bridge with three-hinged stiffening girders as shown in Figure 12.17. Let two external loads W_1, and W_2 act on the deck. Let w_e per unit run UDL transferred to the cables through the suspenders.

Now, let us separately consider the equilibrium of cable and the stiffening girder as shown in Figure 12.18.

For the cable part, we can easily prove that the vertical reactions, $V = w_e l/2$, and the horizontal thrusts, $H = w_e l^2/8d$, where d is the dip of the cable at center. The maximum tension in the cable can be written as, $T_{\max} = \sqrt{V^2 + H^2}$.

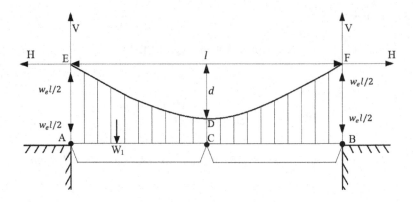

FIGURE 12.17   Three-hinged stiffening girder.

Now consider the girder which is subjected to two load systems (Figure 12.19), such as part (a) the applied external load (Figure 12.19 (b)) system, W_1, and part (b) upward UDL of w_e per unit run from the suspenders (Figure 12.19 (c)).

Let us consider a section at a distance x form support A of the girder. The bending moment, M_x at this section due to this two loading systems on the girder can be expressed as,

$$M_x = \left[V_A \cdot x - W_1(x-a) \right] - \left[\frac{w_e l}{2} \cdot x - w_e \cdot x \cdot \frac{x}{2} \right]$$

or,

$$M_x = M_{\text{beam}} - \frac{w_e \cdot x}{2}[l-x]$$

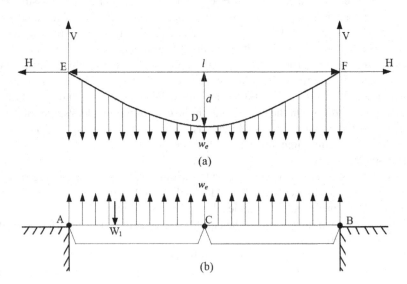

FIGURE 12.18   Equilibrium of (a) cable and (b) stiffening girder shown separately.

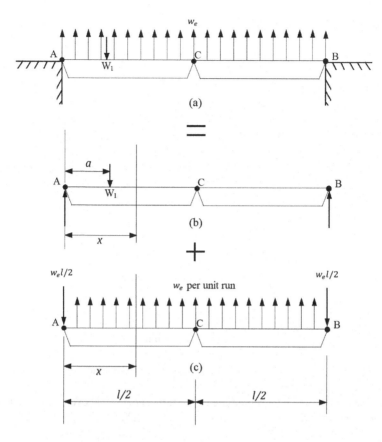

FIGURE 12.19 (a) Girder subjected to two load system, (b) the applied external load system, W_1, and (c) upward UDL of w_e per unit run from the suspenders.

Rewriting the abovementioned expression as,

$$M_x = M_{\text{beam}} - \frac{w_e l^2}{8d} \cdot \frac{4dx}{l^2}[l-x]$$

or,

$$M_x = M_{\text{beam}} - H.y$$

where M_{beam} is the bending moment at the section due to girder load considering the span as that of a simply supported girder.

Now, let us find the moment at hinge C using the abovementioned expression,

$0 = M_C - Hd$ (as dip of the cable at center is d)

Therefore, horizontal thrust, $H = M_C/d$

Similarly, the shear force at this section can be expressed as,

$$S_x = [V_A - W_1] - \left[\frac{w_e l}{2} - w_e x\right] = [V_A - W_1] - \frac{w_e}{2}(l-x)$$

Rewriting the abovementioned expression as,

$$S_x = [V_A - W_1] - \left[\frac{w_e l^2}{8d} \times \frac{4d(l-2x)}{l^2} \right]$$

or,

$$S_x = [V_A - W_1] - \left[\frac{w_e l^2}{8d} \times \frac{d}{dx} \left(\frac{4dx}{l^2} [l-x] \right) \right]$$

or,

$$S_x = [V_A - W_1] - \left[\frac{w_e l^2}{8d} \times \frac{dy}{dx} \right]$$

or,

$$S_x = [V_A - W_1] - [H \times \tan\theta]$$

or,

$$S_x = S_{beam} - [H \times \tan\theta]$$

where S_{beam} is the shear force at the section due to girder load considering the span as that of a simply supported girder.

The shear force and bending moment diagram as per the abovementioned expressions of three-hinged girders are shown in Figure 12.20.

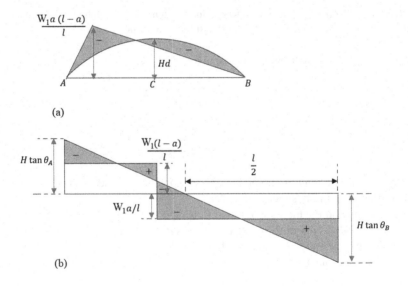

FIGURE 12.20 Three-hinged stiffening girder (a) bending moment and (b) shear force diagram.

The bending moment diagram for the girder can be drawn by superposing the same for a simply supported beam over the bending moment diagram due to Hy. The value of H for a particular loading type is constant, and hence, the product Hy will be a parabola. The diagram is obtained by taking the parabolic shape of the cables and multiplying its ordinates by H as shown in the bending moment diagram in Figure 12.20 (a). Similarly, the shear force diagram is obtained by superposing the same of a simply supported beam over the $H \tan \theta$ diagram as shown in Figure 12.20 (b). The $\tan \theta$ varies linearly with x, and hence, the diagram will be straight line. As H is constant, $H \tan \theta$ will also vary linearly from $H \tan \theta_A$ at A to $H \tan \theta_B$ at B.

Example 12.5: A suspension bridge of 110-m span has a three-hinged stiffening girder supported by two cables having a central dip of 10 m as shown in Figure 12.21. The roadway has a width of 5 m. The dead load on the bridge is 10 kN/m² while the live load is 20 kN/m² that acts right half of the span. Determine the shear force and bending moment in the girder at 30 m from the right end. Find the maximum tension in the cable for the current position of the live load.

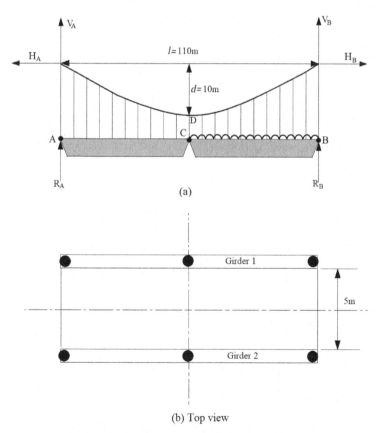

(a)

(b) Top view

FIGURE 12.21 Example problem on three-hinged stiffening girder.

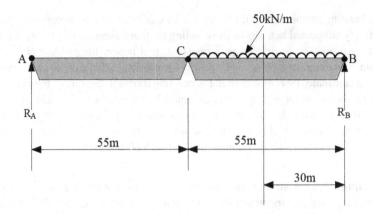

FIGURE 12.22 Finding reaction forces due to live load on the girder.

SOLUTION: Dead load = 10 kN/m² × 5 m = 50 kN/m, live load = 20 kN/m² × 5 m = 100 kN/m. Dead load per girder = 50 kN/m/2 = 25 kN/m, live load per girder = 100 kN/m/2 = 50 kN/m,

Only considering the live load on each girder as shown in Figure 12.22, let us find out the reaction forces.

Taking moment about *B*,

$$R_A \times 110 - 50 \times 55 \times \frac{55}{2}$$

or,

$$R_A = 687.5 \text{ kN}; R_B = (50 \times 55) - 687.5 = 2062.5 \text{ kN}$$

Horizontal thrust,

$$H = \frac{M_C}{d} = \frac{687.5 \times 55}{10} = 3781.25 \text{ kN}$$

Actual bending moment at 30 m from the right side due to the live load $= M_{beam} - H.y = \left(2062.5 \times 30 - \frac{50 \times 30^2}{2} \right) - 3781.25y$
where,

$$y = \frac{4dx}{l^2}(l-x) = \frac{4 \times 10 \times 30}{110^2} \times (110 - 30) = 7.93 \text{ m}$$

Now, putting the value of *y* in the abovementioned expression, we get the actual bending moment at 30 m from support *B*, as 9389.69 kN m.

Likewise, we can calculate the actual shear force at 30 m from support *B* due to the live load as,

$$S_{30} = S_{beam} - \left[H \times \tan \theta \right] = (2062.5 - 50 \times 30) - 3781.25 \times \tan \theta$$

where,

$$\tan\theta = \frac{dy}{dx} = \frac{4d}{l^2}(l-2x) = \frac{4\times10\times(110-2\times30)}{110^2} = 0.1653$$

Now, putting the value of $\tan\theta$ in the abovementioned expression, we get the actual shear at 30 m from support B, as -62.5 kN m.

Now to get the maximum tensile force in the cable, we need to consider both the effects of dead loads and live loads. For this, we need to find the equivalent UDL, w_e acting in the suspenders due to dead load and live load.

We know horizontal thrust in cables under UDL, $H = \frac{wl^2}{8d} = \frac{w\times110^2}{8\times10} = 151.25w$.

Again, we have obtained the horizontal thrust due to the live load as 3781.25 kN. If we equate these two, we will get the value of w, which is basically the UDL due to the live loads transferred from girder to cable through suspender.

Therefore,

$$w = 25 \text{ kN/m}$$

so, the total UDL, w_e, that will be transferred from girder to cable through suspender is = UDL from dead load + UDL from live load = (25 + 25) = 50 kN/m.

Now from Figure 12.21 (a),

$$V_A = V_B = V = \frac{w_e l}{2} = \frac{50\times110}{2} = 2750 \text{ kN}$$

$$H_A = H_B = H = \frac{w_e l^2}{8d} = \frac{50\times110^2}{8\times10} = 7562.5 \text{ kN}$$

Therefore, the maximum tension in the cable will be $T_{max} = \sqrt{(2750^2 + 7562.5^2)} = 8046.98$ kN.

13 Analysis of Symmetric Structures

13.1 INTRODUCTION

In this chapter, we will introduce the concept of symmetric structures. Symmetry may be related to loading conditions or geometry or both. Having learned this method of analysis, one may acquire quick detection of structural methods to be adopted for part of the structures and applying the results for the entire one to complete the analysis. However, it is advisable to check the results of symmetric analysis with complete analysis of the same structure ignoring the symmetry in case of any confusion. In this way, one can check if there is any mismatch between the outcomes of both the processes and detect any calculation or conceptual error thereof. In short, structural geometry can be determined based on (a) symmetric supports, (b) symmetric loading, (c) symmetric geometry. If any one of these parameters is nonsymmetric, the structure as a whole becomes antisymmetric.

13.2 SYMMETRIC AND ANTISYMMETRIC COMPONENTS OF LOADINGS

In the case of structural analysis of any highly indeterminate structure, even for a statically determinate structure, it can be simplified, provided the designer or analyst can recognize those structures that are symmetric and support either symmetric or antisymmetric loadings. In a general sense, a structure can be termed as being symmetric, provided that half of it develops the same internal loadings and deflections as its mirror image reflected about its central axis. Generally, symmetry requires the material properties, geometry, supports, and loading to be the identical on each side of the structure. However, this does not always have to be the case as shown in Figure 13.1. It is to be noted that for horizontal stability, a pin is required to support the truss horizontally. This truss is violating the symmetry condition due to its supports, but as only vertical load is applied on this truss, the horizontal reaction at the pin is zero, and so, this structure will deflect and produce the same internal loading as its reflected counterpart. As a result, they can be classified as being symmetric. It is to be understood that this would not be the case for the frame shown in Figure 13.2. If the fixed support at A is replaced by a pin or roller, the deflected shape and internal forces would not be the same on its left and right sides, even if the vertical loading is only applied.

As can be seen from Figure 13.2, the structure is perfectly symmetrical in this situation. But as explained earlier, when we replace one fixed support at location A, this change in support condition will lead to asymmetry in the structural analysis.

DOI: 10.1201/9781003081227-15

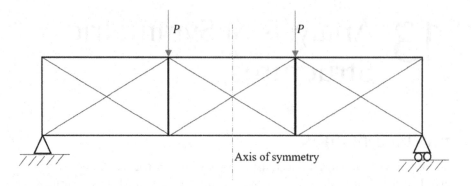

FIGURE 13.1 Symmetrical structure – truss.

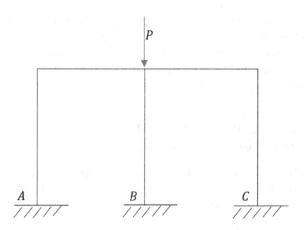

FIGURE 13.2 Frame structure with loading.

13.3 SYMMETRIC AND ANTISYMMETRIC COMPONENTS OF LOADINGS

Sometimes, a symmetric structure supports an antisymmetric loading, (i.e., the loading is considered to be antisymmetric with respect to an axis in its plane if the negative of the reflection of the loading about the axis is identical to the loading itself) such as shown in Figures 13.3 and 13.4. Provided the structure is symmetric in geometry and its loading is either symmetric or antisymmetric, the structural analysis will only have to be carried out on half of the structure since the same (symmetric) or opposite (antisymmetric) results will be generated on the other half of it. If a structure is geometrically symmetric and its applied loading is unsymmetrical, then it is possible to transform this loading into symmetric and antisymmetric components. To do this, the loading is first divided in half, then it is reflected to the other side of the structure

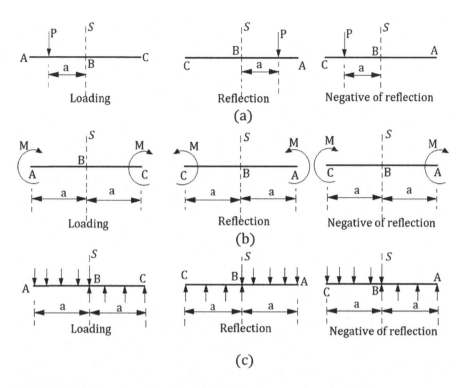

FIGURE 13.3   Examples of antisymmetric loading.

and both symmetric and antisymmetric components are produced. For example, the loading on the beam in Figure 13.5 is divided by two and reflected about the beam's axis of symmetry. From this, the symmetric and antisymmetric components of the load are produced as shown in the same figure. When combined all together, these components produce the original loading. A separate structural analysis can now be performed using the symmetric and antisymmetric loading components and the results superimposed to obtain the actual behavior of the structure.

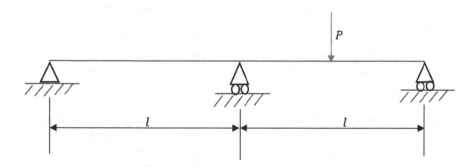

FIGURE 13.4   Continuous beam with antisymmetric loading.

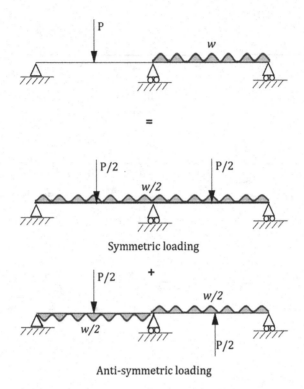

FIGURE 13.5   Superposition of loading on continuous beam.

13.4   BEHAVIOR OF SYMMETRIC STRUCTURES UNDER SYMMETRIC AND ANTISYMMETRIC LOADINGS

When a symmetric structure is subjected to a loading with respect to the structure's axis of symmetry, the response of the structure is also symmetric.

Displacement behavior along the axis of symmetry for symmetric loading results neither in rotation (unless there is a hinge at such a point) nor any deflection perpendicular to the axis of symmetry.

Force behavior along the axis of symmetry for symmetric loading results in zero force along the axis of symmetry.

For a symmetric structure, which is subjected to a loading that is antisymmetric with respect to the structure's axis of symmetry, the response of the structure is also antisymmetric. Displacement behavior along the axis of symmetry for antisymmetric loading conditions results in no displacement along the axis of symmetry. Force behavior along the axis of symmetry for antisymmetric loading results in zero force normal to the axis of symmetry and zero bending moment. In Figure 13.6, the deflection pattern of a symmetric frame is shown for both symmetric (Figure 13.6 (a)) and antisymmetric (Figure 13.6 (c)) loading condition. It can be also noted that, considering the symmetric and antisymmetric boundary conditions on the half frame along the location of axis of axis of symmetry, we can easily determine the nature of response of the whole structure.

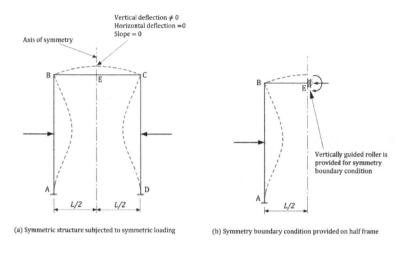

(a) Symmetric structure subjected to symmetric loading   (b) Symmetry boundary condition provided on half frame

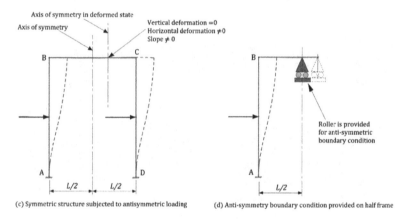

(c) Symmetric structure subjected to antisymmetric loading   (d) Anti-symmetry boundary condition provided on half frame

FIGURE 13.6 Symmetric frame subjected to symmetric and antisymmetric loading and respective boundary conditions for the half frame.

For general loading on a symmetric structure, the loading can be decomposed into symmetric and antisymmetric components. Displacement and force boundary conditions for symmetric and antisymmetric loadings along the axis of structural symmetry apply. To obtain the total response, use superposition of the symmetric and antisymmetric result.

Example 13.1: Find the substructures for the analysis of symmetric and antisymmetric responses for the statically indeterminate beam shown in Figure 13.7.

SOLUTION: The procedure for determining the substructures for analyzing symmetric and antisymmetric responses of the given indeterminate beam for this loading condition is shown in Figure 13.8.

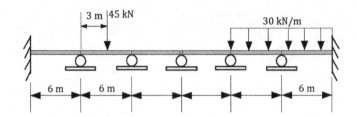

FIGURE 13.7 Indeterminate beam for the example problem 13.1.

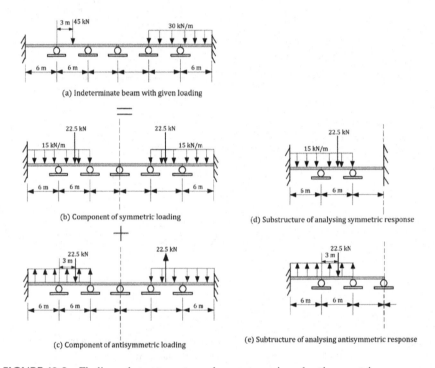

FIGURE 13.8 Finding substructures to analyze symmetric and antisymmetric response.

From this figure it is clear that if we add up the responses of symmetric loading component as shown in Figure 13.8 (b) with the antisymmetric loading component as shown in Figure 13.8 (c), we will get the original response of the given structure. Putting the appropriate boundary conditions along the axis of symmetry on the half structure, we will be able to find the response of the whole beam as well.

Part III

Analysis of Statically Indeterminate Structures

14 Introduction to Statically Indeterminate Structures

14.1 INTRODUCTION

In Part III (Chapters 14–24) of this text, our principal attention will be on the analysis of statically indeterminate structures. As discussed earlier, the support reactions and internal forces (member force, shear force, bending moment) of statically determinate structures can be determined from the equations of equilibrium (including equations of conditions, if required). However, since indeterminate structures have more supports and/or members (called redundant members) than required for static stability and equilibrium, the equilibrium equations alone are insufficient to determine the unknown reactions and internal forces of such structures. There must be some other relationships based on the geometry of deformation of structures or the nature of constraints imposed on the structure. These additional relationships or equations, which are termed as the compatibility conditions or equations, ensure that the continuity of the displacements is maintained throughout the structure and that the structure's various parts remain connected together without any damage to the stability of the structure. For example, at a rigid joint, all the members' deflections and rotations related to the joint must be the same.

Thus, analysis of an indeterminate structure requires, in addition to the dimensions and geometric arrangement of members/elements of the structure, its cross-sectional and material properties (such as cross-sectional areas, moments of inertia – both geometric and mass moment of inertia, moduli of elasticity, etc.), which in turn, also depend on the internal forces of the same. The design of an indeterminate structural element is, therefore, carried out in an iterative manner, whereby the (relative) sizes of each structural member is initially assumed and used to analyze the structure, and the internal forces, thus, obtained are used to revise the member sizes and orientations. If the revised member sizes are not same as to those initially assumed sections, then the structure is reanalyzed using the latest member sizes. The iteration keeps continuing till the member sizes based on the results of analysis are close to those assumed for that analysis. Due to this iterative nature of structural analysis, computer-aided programmers are called for to do the iterations correctly and within short time span.

Despite the difficulty in designing indeterminate structures, a great majority of structures being built in today's modern world are statically indeterminate in nature; for example, most modern reinforced concrete buildings are statically indeterminate, and all the beams are supported at more than two supporting columns. These types of beams are called continuous beams, for which we shall study the analysis procedure soon.

In this chapter, we will discuss some important advantages and disadvantages of indeterminate structures as compared to determinate structures and will develop the fundamental concepts for the analysis of indeterminate structures.

DOI: 10.1201/9781003081227-17

14.2 ADVANTAGES OF INDETERMINATE STRUCTURE

The advantages of statically indeterminate structures over determinate structures include the following:

1. *Lesser Stresses and greater stiffness* – The maximum stress intensity and deflection in statically indeterminate structures is lower than those in similar determinate structures. For example, let us consider the statically determinate and indeterminate beams shown in Figure 14.1 (a) and (b), respectively. Bending moment diagrams and deflections for the beams due to a uniformly distributed load intensity, w, are also shown. The methods for analyzing indeterminate structures will be introduced in subsequent chapters. It can be understood from the figures that the maximum bending moment – and, consequently, the maximum bending stress intensity as well as deflection – in the indeterminate beam is much lesser than in the determinate beam. In other way we can say, statically indeterminate structures generally have more stiffness (i.e., smaller deflection), than those of similar determinate structures.
2. *Redundancies* – In statically indeterminate structures, if correctly designed, acting loads are redistributed when certain structural elements become overloaded or collapse in cases of overload due to earthquakes, wind, snow, impact, and other events. In practical situations, indeterminate structures have more members and/or supporting points than required for static stability. If a part of such a structure fails, the entire structure will not collapse immediately, and the loads will be redistributed to the adjacent members of the structure. In transmission line towers, this is frequently seen where more members are connected with the main leg elements to reduce the slenderness ratio of the main members and, thus, ensure the overall structural stability even under worst

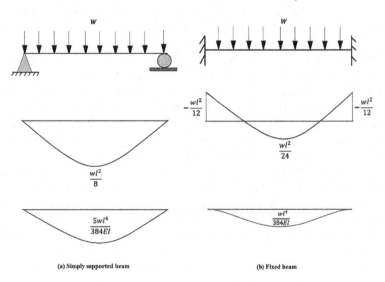

(a) Simply supported beam (b) Fixed beam

FIGURE 14.1 (a) Simply supported beam and (b) fixed beam bending moment and deflection diagrams.

loading situations. These additional members are called redundant members and play a crucial role in the overall structural stability. We will investigate the effect of redundant members in other chapters of this section and establish the analysis procedure for the same. Even in some redundant elements after carrying out analysis, it will be found that there is no force playing at all in those elements. These elements are called zero force elements.

14.3   DISADVANTAGES OF INDETERMINATE STRUCTURE

The disadvantages of statically indeterminate structures over determinate structures include the following:

1. *Stresses due to support settlements* – In many practical situations, supports may get settled as shown in Figure 14.2, due to faulty design, weak soil strata, or any unforeseen consequences due to excessive loading originated from seismic or wind, or any other adverse conditions.

   Under such circumstances, for determinate structures, no additional stresses are developed due to such settlements. But in case of indeterminate structures, support settlements induce additional internal force and moments due to which structural elements undergoes higher stress concentrations than they are designed to withstand for. We will analyze the effects of support settlements and additional force and moments coming from them in subsequent chapters. But for now, it is to be understood that for indeterminate structures, support settlements cause adverse effect, which is a major drawback of designing members as indeterminate in nature.

2. *Stresses due to temperature change and fabrication errors* – Change in temperature and fabrication errors do not cause additional stresses in determinate structures but may induce significant stresses in indeterminate ones as shown in Figure 14.3, and 14.4, respectively.

   Using the simple relationship for linear elongation/contraction of solids under change in temperature $(\alpha \Delta TL)$, due to support fixity, axial loads will come into play in the member or structural element. This will cause additional direct stress generation in addition to the bending stress due to external loading. This effect produces most adverse stress condition for fixed beams. In such beams, the axial load due to temperature change will induce axial compressive or tensile force, and thus, direct stress will be induced in

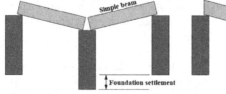 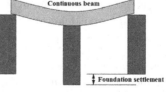

(a) No moment generates due to settlement as no curvature formed for simple beam

(b) Moment generates due to settlement as curvature formed for continuous beam

FIGURE 14.2   Effect of foundation settlement to determinate and indeterminate structures.

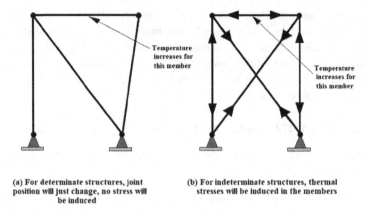

(a) For determinate structures, joint position will just change, no stress will be induced

(b) For indeterminate structures, thermal stresses will be induced in the members

FIGURE 14.3 Effect of temperature change to determinate and indeterminate structures.

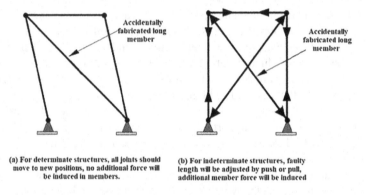

(a) For determinate structures, all joints should move to new positions, no additional force will be induced in members.

(b) For indeterminate structures, faulty length will be adjusted by push or pull, additional member force will be induced

FIGURE 14.4 Effect of fabrication errors to determinate and indeterminate structures.

the fixed beam along with the bending stress due to other external applied loading. Thus, without proper care, the members will fail due to this combined (direct as well as bending) excessive stress conditions. In case of fabrication errors, if any member length accidentally becomes short or long, that will be adjusted by the structural system by elongating or compressing the particular member, respectively, for indeterminate structures. In contrast, for determinate structures, joints will be shifted to new positions to accommodate the situation. Thus, material properties and their effect on change in temperature, good fabrication practices need to be considered while analyzing and designing indeterminate structures.

This chapter has provided a basic qualitative introduction to indeterminate structures and their advantages and disadvantages over determinate structures. We will learn in detail in subsequent chapters about the analysis procedures for the indeterminate structures, which is the first step toward carrying out structural design engineering work. We will also learn how to form additional equations depending upon the geometry and constraint imposed by different supports.

15 Approximate Analysis of Statically Indeterminate Structures

15.1 INTRODUCTION

The analysis of statically indeterminate structures applying the force and displacement methods is considered as exact in the sense that the compatibility and equilibrium conditions of the structure are exactly satisfied in such an analysis. However, the results of such an exact analysis represent the actual response to the extent that the mathematical model of the structure reflects the actual structure under study. Experimental analysis has established the fact, that the response of most common types of structures under various acting loads can be correctly predicted by the force and displacement methods, provided an accurate mathematical model of the structure is considered at the beginning of analysis.

An approximate method of analysis proves to be quite a convenient way to apply in the preliminary planning phase of any project when several alternative design philosophies of the structure are usually evaluated for optimized economic results. The results of approximate analysis can also be used to assume the dimensions of various structural elements needed to begin the exact analysis. The approximate dimensions of various structural elements are then changed iteratively, using the results of successive exact design and analyses, to achieve their final dimensions. Moreover, approximate analysis is seldom used to roughly estimate the results of exact analysis program, which due to its complexity and time-consuming calculations can be prone to erroneous results. Finally, in recent times, there has been an increased interest toward renovating and retrofitting older and heritage structures. So, a knowledge and understanding of approximate methods used by the original designers is usually helpful in a renovation of work progress.

15.2 ASSUMPTIONS FOR APPROXIMATE ANALYSIS

As discussed in the earlier chapter, statically indeterminate structures have more support reactions and/or members/elements than are required for static equilibrium; therefore, all the reactions and internal forces, including moments, of such structures cannot be determined from the equations of equilibrium alone. The additional reactions and internal forces of an indeterminate structure are referred to as redundants, and the number of redundants (i.e., the difference between the total number of unknowns and the number of equilibrium equations available) is termed the degree of indeterminacy of the structure. So, in order to calculate the reactions and internal forces of an indeterminate structure, the equilibrium equations must be aided with

DOI: 10.1201/9781003081227-18

additional equations, the number of which must be the same as that of the degree of indeterminacy of the structure. In approximate analysis, these additional equations are formed by applying engineering judgment to make simplifying assumptions about the response of the structure under the action of external loads. The total number of equations corresponding to assumptions must be equal to the degree of indeterminacy of the structure. Each assumption leads to an independent relationship between the unknown reactions and/or internal forces. The equations based on the simplifying assumptions are then solved in line with the equilibrium equations of the structure to determine the approximate values of its reactions and internal forces.

Two assumptions usually applied to carry out the approximate analysis are detailed in the following sections.

15.2.1 ASSUMPTIONS ABOUT THE LOCATION OF POINTS OF INFLECTION

As a starter, a qualitative deflected shape of the indeterminate structure is sketched and applied to determine the location where the curvature of elastic line changes its direction/sign or becomes zero. These points are known as the point of inflection. At the inflection points, bending moments also become zero. Thus, to model zero bending moment points, internal hinges are inserted in the indeterminate structure at the assumed inflection points to obtain a general determinate structure. By inserting internal hinge points, each hinge provided additional condition, thereby providing additional equations. Moreover, the inflection points should be placed in such a way, so that the resulting determinate structure, thus formed, must be statically and geometrically stable. The simplified determinate structure thus obtained is then analyzed to determine the approximate values of the reactions and internal forces of the original indeterminate structure.

To understand the above concept, let us consider an example of a statically indeterminate portal frame. See Figure 15.1 for the actual portal frame and load acting on it.

We start by drawing possible deflected shape of the frame due to the applied loading. The deflection diagram has been shown in Figure 15.1 for reference. From this diagram, it seems that the inflection point is situated near the center of the span CD. Although to determine the exact location of inflection point, we need to carry out the exact analysis. With this assumption, we insert a hinge at the middle of the span CD and applying the concept of that moment at hinge or inflection point is zero; thus, we get the following equations:

$$\sum M_B = 0$$

$$A_y l - Ph = 0$$

or,

$$A_y = \frac{Ph}{l}$$

$$\sum F_y = 0$$

$$-\frac{Ph}{l} + B_y = 0$$

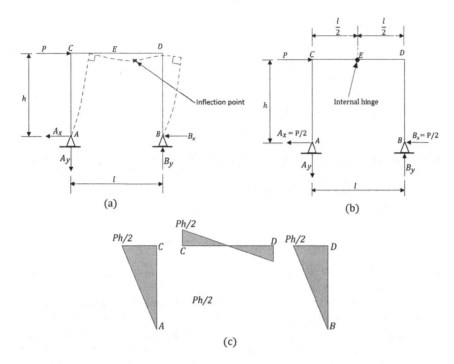

FIGURE 15.1 (a) Indeterminate frame with external loading, (b) corresponding simplified determinate structure, and (c) approximate bending moment diagram.

or,

$$B_y = \frac{Ph}{l}$$

Similarly, calculating bending moment at point B, we get from the left of section with respect to point B:

$$M_x^{BE} = B_y \frac{l}{2} - B_x h$$

Since, at point B, we have the point of inflection, hence,

$$M_x^{BE} = 0$$

Resulting,

$$B_y \frac{l}{2} = B_x h \text{ or, } B_x = \frac{P}{2}$$

Thus, progressing the abovementioned way, we can easily determine all other unknown support reactions at the other support point A. So, approximate analysis is found out to be handy while carrying out the structural analysis work for indeterminate structures.

15.2.2 Assumptions about the Distribution of Forces and Reactions

Approximate analysis of indeterminate structures can also be performed by assuming that the distribution of forces among the members and support reactions of the structures. The number of these assumptions adopted for the analysis of the structure is equal to the degree of indeterminacy of the structure. Each assumption provides an independent equation combining the unknown member forces and support reactions. The equations are then solved simultaneously with the equilibrium equations of the structure to calculate its approximate reactions and internal forces. For example, the same problem in the previous example can be solved by assuming that horizontal support reactions at two supports are equal. By assuming this, one can solve and analyze the entire frame without any difficulty. The same is left as an exercise for the readers.

15.3 VERTICAL LOADS ON BUILDING FRAMES

The most accepted procedure for approximate analysis of building frames subjected to vertical loads (mainly due to gravity) required taking three assumptions about the nature of each girder of the frame. To understand the concept, let us take a building frame subjected to uniformly distributed loads w, as shown in Figure 15.2. From the typical free body diagram of member DE and its deflected shape, the point of inflections is located near the column support points D and E. Since the member DE is fixed and connected with the framing columns, hence, the column provides end

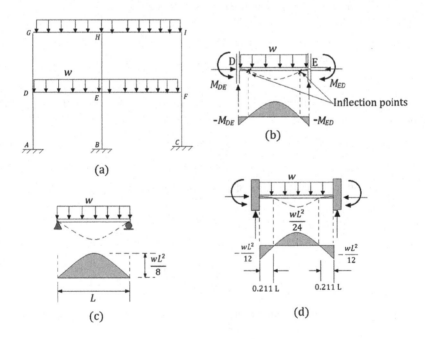

FIGURE 15.2 (a) Building frame subjected to vertical load, (b) typical girder, (c) simply supported girder, and (d) ideally fixed girder.

restraint against rotation of the member due to applied loading. Actual location of the inflection points depends upon the proper analysis of indeterminate structure, which we will discuss later. However, for approximate analysis, we can at least get a feel of the extreme conditions and its effect toward the deflection of the said member under investigation. If the girder was simply supported, bending moment would be zero at the support points and thus inflection points would lie at the support locations too. But since this is not simply supported, hence for practical purposes, we can assume that the point of inflection lies at 0.11L from each support. This is half of the actual distance of inflection point as per exact analysis. However, for approximate analysis, the assumed distance of 0.11L is proved to be very effective one.

The third assumption is a very simple one and it is known from the exact analysis of structures. From exact analysis, for a building frame subjected to vertical loads, member force (force acting along the axis of member) is very negligible and hence can be neglected. We take the same assumption accordingly while analyzing the same frame using approximate analysis procedure.

15.4 LATERAL LOADS ON BUILDING FRAMES: PORTAL METHOD

Response of any rectangular building frames is unique and different under lateral (horizontal) loads than under vertical loads explained in the previous section. Hence, completely different assumptions must be adopted in the approximate analysis for lateral loads. Two methods are popularly applied for approximate analysis of rectangular frames subjected to lateral loads. These are: (1) the portal method and (2) the cantilever method. In this section, we will discuss the portal and cantilever methods.

In portal analysis, it is assumed that point of inflections lies at the midpoint of each member of the portal frame under investigation. To understand this concept, let us take the example of a simple portal frame that is indeterminate to third degree, as shown in Figure 15.3.

As seen in the above diagram, there are three points of inflections in the frame element. These three points of inflection can be modeled as internal hinge points at the same location. By inserting the internal hinges at these members, the frame looks similar to the one shown in Figure 15.4.

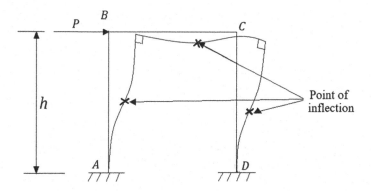

FIGURE 15.3 Building frame analysis by portal method.

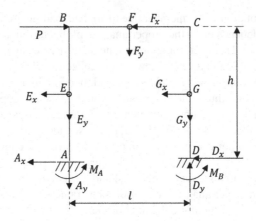

FIGURE 15.4 Building frame with internal hinges.

With the assumption that point of inflection lies at the center of each member, we can solve the indeterminate structure by considering moment at the inflection point, thereby eliminating the unknown support reactions one by one. Since all the internal hinges act like virtual supports, hence, there will be induced support reactions due to the external loading. We will show some steps of evaluating the unknown reaction forces and others will be left as an exercise for the reader.

At each internal hinge points, we have marked unknown horizontal and vertical reaction forces, as shown in Figure 15.4. To solve the above frame, we pass an imaginary section through point E and G of the above frame and draw the free body diagram of the sections with all forces acting at the respective cut locations (for recap of Method of Sections refer Chapter 7).

From Figure 15.5, taking moment about G, we get:

$$P\frac{h}{2} - E_y l = 0$$

or,

$$E_y = P\frac{h}{2l}$$

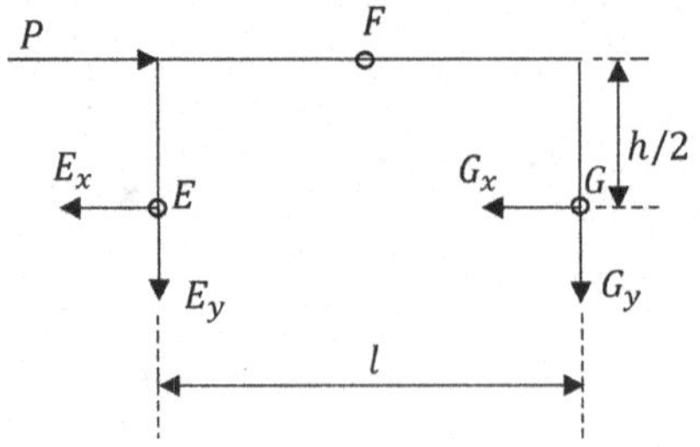

FIGURE 15.5 Sectional force free body diagram of simplified determinate portal frame.

Similarly, by taking moment of all forces about point E, we get:

$$G_y = -P\frac{h}{2l}$$

Now, the bending moment at the point of inflection F, we get:

$$M_F - E_x\frac{h}{2} - E_y\frac{l}{2} = 0$$

Since, bending moment at point F is zero, hence, we get:

$$E_x\frac{h}{2} - P\frac{h}{2l}\frac{l}{2} = 0$$

or,

$$E_x = \frac{P}{2}$$

Now, applying force equilibrium equation in x direction, we get:

$$G_x + \frac{P}{2} - P = 0$$

or,

$$G_x = \frac{P}{2}$$

Hence, by adopting a simple assumption about the location of inflection points, we can determine the unknown support reactions by applying force and moment equilibrium equations.

As we have just learned about the application of the portal method for statically indeterminate single bay frame, the same understanding can be applied to multiple bay portal frames of buildings to analyze the unknown forces and moments acting in the frame due to eternal loading. To understand this, let us consider the following indeterminate frame of a multi-bay building, as shown in Figure 15.6.

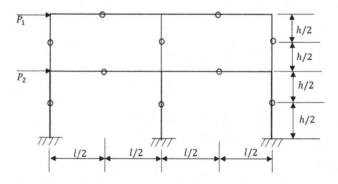

FIGURE 15.6 Multi-bay simplified building frame.

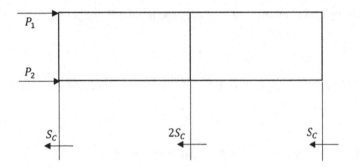

FIGURE 15.7   Multi-bay building frame – shear force distribution.

Following the same procedure, we assume that the point of inflection exists at the midpoint of each member of the above multi-bay frame. Now, before inserting the internal hinges, the total number of unknown support reactions are $= 3 \times 4$ and number of girders $= 12$. Now, with the introduction of internal hinges (ten in total), the structure does not become statically determinate. Hence, with the internal hinges, the net degree of indeterminacy becomes $12 - 10 = 2$. Hence, additional two equations or conditions are required to make the structure statically determinate. Hence, in addition to assumption on location of inflection point, we need another two conditions to make this structure statically determinate, which will enable us to complete the analysis.

As analyzed above, we have found under the application of horizontal loading, the perimeter columns carry shear force equal to the half of the applied loading. Since internal columns represent two legs of the two portal frames placed side by side, hence, it may be assumed that the internal columns can carry twice more load than the peripheral columns. Thus, this assumption provides additional equations for each column in the portal frame. Hence, total available equations become $1 \times 3 = 3$. However, we needed only two additional equations and we got three in turn. However, since all these additional equations are derived based on the same principle, hence, one extra equation than required does not pose any problem for analysis. Hence, following this assumption, we get the following Figure 15.7 related to the analysis program of multiple bays building frames by portal frame method.

Although we have tried to provide the detailed approximate calculation techniques by portal frame methods, we will provide Example 15.1 for better understanding of the concept below.

Example 15.1: Calculate the member forces using portal frame method of the building frame, as shown in Figure 15.8.

SOLUTION: After inserting the hinges at midpoint of each member, we pass a section $a - a$ near the midportion of the frame and another section $b - b$ at the bottom of the total system. After that, we draw the free body diagram for the

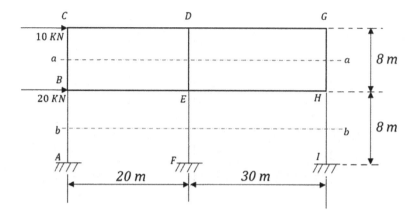

FIGURE 15.8   Multi-bay building frame – example problem.

separated frames along with horizontal shears, as explained earlier. The free body diagrams will be something like as shown in Figures 15.9 and 15.10.

From the free body diagrams as mentioned, applying force equilibrium equation in x direction:

$$\sum F_x = 0$$

Thus,

$$10 - R_C - 2R_C - R_C = 0$$

or,

$$R_C = 2.5 \text{ kN}$$

$$30 - S_C - 2S_C - S_C = 0$$

or,

$$S_C = 7.5 \text{ kN}$$

Once we calculate the shear forces acting in the column, we are now able to analyze the forces in members, as will be described. Let us open a joint located at C and draw free body diagram of the connected member at this node.

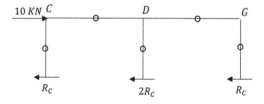

FIGURE 15.9   Free body diagram of portal frame of a–a section.

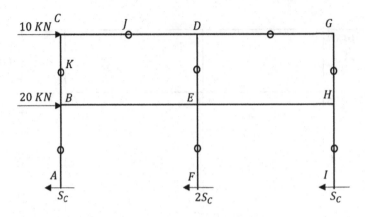

FIGURE 15.10   Free body diagram of portal frame of *b–b* section.

Now, applying force equilibrium equation in y direction, we get:

$$V_K + V_J = 0$$

So,

$$V_K = -V_J$$

Now, considering bending moment at point K of the member CK, we get:

$$M_{CK} - 10 \times \frac{8}{2} = 0$$

or,

$$M_{CK} = 40 \text{ kNm}$$

Now, since at joint C, total summation of moments will be zero to maintain equilibrium condition, hence, we get:

$$M_{CJ} = -M_{CK} = -40 \text{ kNm}$$

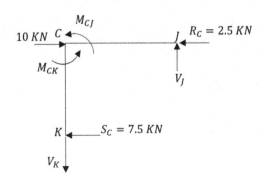

FIGURE 15.11   Free body diagram of joint *C*.

Once the nodal moments are calculated, we can take moment of all forces about any points, J or K, which will help us to determine the other unknown forces acting in the member. As for example, if we take moment of all forces about point J, we get:

$$S_C \times \frac{8}{2} - V_K \times \frac{20}{2} = 0$$

Solving the above equation, we get:

$$V_K = \frac{7.5 \times 4}{10} = 3 \text{ kN}$$

Hence, for the node C and connected members in this node, we have completely analyzed all unknown forces and moments for the same. Proceeding in this way, students are encouraged to complete the analysis of this frame by the same fashion. In the next section, we will introduce another effective approximate method for analyzing multistoried building frames, which is popularly known as cantilever method.

15.5 LATERAL LOADS ON BUILDING FRAMES: CANTILEVER METHOD

This method was initially developed by A. C. Wilson in 1908 and is generally applied mainly for the approximate analysis of tall building frames. Cantilever method assumes that under the application of lateral loads, building frames act like cantilever element, as shown in Figure 15.12. From the strength of materials, we already know that the axial stress on a cross section of a cantilever beam subjected to lateral/transverse loads varies linearly with the distance from the centroidal axis (neutral line), so that the longitudinal fibers of the beam on the concave side of the neutral line are in the state of compression, whereas fibers on the convex side remain in tension.

As we determine the neutral line for the beams under the application of transverse loading, for building, we determine the centroidal line accordingly. This centroidal line behaves in the same way as the neutral line behaves for the building structure. Framing columns are then considered as the different fibers of the beam, and axial stress distribution, at the mid-height of column, is assumed to be linearly proportional to the distance of the column from the centroidal axis of the building. If it is further assumed that the cross-sectional areas of all the columns throughout the building structure, the axial force in each column will be linearly proportional to the distance of the column from the centroidal line of the building.

In addition to the abovementioned assumption, the cantilever method uses the same assumption regarding the location of inflection points, the same as that in the portal method. Thus, the assumptions of cantilever method can be stated as follows:

1. An inflection point is located at the midspan of each member of the frame.
2. In each story, the axial stress in columns is directly proportional to the distance of the column from the centroidal line.

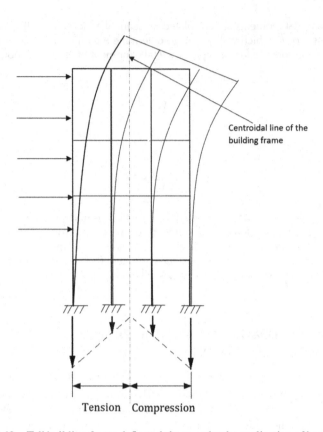

Centroidal line of the
building frame

Tension   Compression

FIGURE 15.12   Tall building frame deflected shape under the application of horizontal load.

3. As this method also assumes the same location of inflection points as that
for the portal method, hence, at midspan of each member there will be
internal hinge indicating the location of the inflection point.

With this two assumptions and subsequent analysis procedure, we will show through
Example 15.2 that this method is very effective toward determining the approximate
force and moments in each and every framing elements of any tall building structure.

Example 15.2: Calculate the member forces using portal frame method of the building frame, as shown in the Figure 15.13.

SOLUTION: To solve this problem using cantilever method, we first insert the
internal hinge points at each member midspan location. Then we pass two imaginary section lines along the hinge points to form the reduced or simplified frame.
In reduced frame thus obtained, we then draw the forces and moments as we
show in any free body diagram. The free body diagram of the frame is shown in
Figure 15.14.

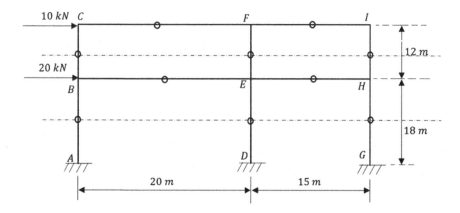

FIGURE 15.13 Building frame for cantilever method.

Assuming all columns are of the same cross-sectional area = A, the centroidal line location can be determined by taking area moments with respect to line CB:

$$\bar{x} \times (3A) = 20 \times A + 35 \times A$$

or,

$$\bar{x} = \frac{(20 \times A + 35 \times A)}{3A} = 18.33 \text{ m}$$

Once the location of CG line is determined, we immediately know that the columns to the left of CG line will be under tension, whereas the columns to the right

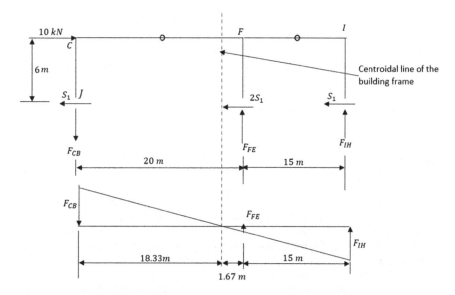

FIGURE 15.14 Free body diagram of reduced building frame.

of CG line will be under compression. Hence, the force triangle will be as shown Figure 15.14. Since the force in the columns is directly proportional to the distance of the same from the centroidal axes, hence, from a similar triangle, we can write:

$$\frac{F_{CB}}{18.33} = \frac{F_{FE}}{1.67}$$

or,

$$F_{FE} = \frac{F_{CB} \times 1.67}{18.33} = 0.091\,F_{CB}$$

Similarly,

$$\frac{F_{IH}}{16.67} = \frac{F_{CB}}{18.33}$$

or,

$$F_{IH} = \frac{F_{CB} \times 16.67}{18.33} = 0.91\,F_{CB}$$

Now calculating the moment at joint J, we get:

$$\sum M_J = 0$$

So,

$$6 \times 10 - 20 \times F_{FE} - 35 \times F_{IH} = 0$$

or,

$$60 - 20 \times 0.091\,F_{CB} - 35 \times 0.91\,F_{CB} = 0$$

or,

$$F_{CB} = 1.782 \text{ kN}$$

So, from earlier derived relationships, we get:

$$F_{FE} = 0.162 \text{ kN}$$

$$F_{IH} = 1.621 \text{ kN}$$

Following the same procedure, we can determine the column forces of the lower segment columns by drawing an imaginary line through the inflection points as shown in Figure 15.12 and taking the drawing of the triangle of forces as per tension and compression states of the member. Column member forces for all other columns are left as exercise for the interested readers.

Once all the column member forces are determined then we can proceed to calculate the girder forces and moments by drawing suitable free body diagrams and applying force or moment equilibrium equation or both suitably as per requirement.

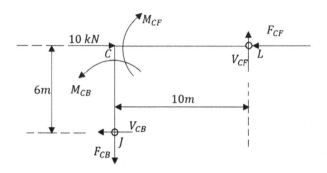

FIGURE 15.15   Free body diagram of girder *CF*.

To understand the concept, let us calculate the member force and moments in the girder CF of the above frame. The free body diagram of the girder with all forces and moment is shown in Figure 15.15.

Taking moment about hinge *J*, we get:

$$M_{CB} - 10 \times 6 = 0$$

or,

$$M_{CB} = 60 \text{ kNm}$$

Since at joint *C*, sum of all moments should be zero to maintain equilibrium, hence,

$$M_{CB} + M_{CF} = 0$$

or,

$$M_{CF} = -60 \text{ kNm}$$

Similarly, by taking moment of all forces about point *C* for the span CJ, we get:

$$V_{CB} \times 6 - M_{CB} = 0$$

or,

$$V_{CB} = 10 \text{ kN}$$

Following these procedures, we can carry out all the forces and moments in all members of the frame using the cantilever method. All other girders are left as an exercise for the readers. It is customary to mention here that by merely studying theories and worked examples will not help one to master the underlying concepts and problem-solving techniques. To be able to have a good hold on any mathematical topic, it is of utmost importance to go through the exercises in detail and carry out all necessary steps to obtain the result.

16 Method of Consistent Deformations

16.1 INTRODUCTION FORCE METHOD OF ANALYSIS: GENERAL PROCEDURE

This chapter will discuss a general analysis procedure of the force (flexibility) method, also called the method of consistent deformations, to analyze statically indeterminate structures. James C. Maxwell introduced this method in 1864, and this involves removing enough restraints and/or constraints from the indeterminate structure to convert it into a statically determinate structure. This determinate structure, thus obtained, must be statically (and geometrically) stable and is known as the primary structure. The excess restraints (and/or constraints) taken away from the indeterminate structure to make it determinate primary structure are called redundant restraints, and the reactions or internal forces related with these restraints are called redundants. The redundants are then applied as unknown forces on the primary structure, and their values are determined by solving the compatibility equations. Compatibility equations are being solved based on the condition that the actual displacement is a linear superposition of displacements of the primary structure with the applied external load and load due to force of constraint. In this method of analysis, the main unknown forces are the redundant reactions, and they must be evaluated at first before evaluating other unknown parameters like deflections and rotations. Since the forces are first determined, this method is known as the force method.

In this chapter, we will also learn how to apply Maxwell-Betti's law to analyze indeterminate structure to determine the unknown forces and moments.

16.2 STRUCTURES WITH A SINGLE DEGREE OF INDETERMINACY

To understand the concept of force method of analysis for single degree of indeterminacy, let us consider the following example of propped cantilever as shown in Figure 16.1.

So, for the shown propped cantilever, we have three unknown support reactions at A (one vertical, one horizontal, one moment) and another unknown support reaction is at support B (since this is a roller support, we have only one unknown vertical support reaction). So, the total unknown support reactions are four, and we have three equilibrium equations, namely, $\Sigma F_x = 0$, $\Sigma F_y = 0$, $\Sigma M_x = 0$. So, degree of indeterminacy = 4 − 3 = 1; hence, the system has one more reaction than required for static stability. Now to analyze the structure, we need to remove one redundant support reaction and make it to determinate primary structure. In primary structure, there will be all external loading acting on the system.

DOI: 10.1201/9781003081227-19

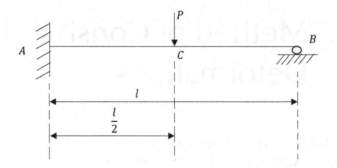

FIGURE 16.1   Propped cantilever.

So, for the applied loading, the bending moment at the support is $M_A = P(l/2)$.
Now from the bending moment diagram, we can apply moment area theorem to calculate the deflection at free end B, by taking moment of bending moment area about
support B and dividing the same by EI (flexural rigidity of the beam).

$$\delta_B = \frac{(1/2) \times P(l/2) \times (l/2)}{EI} \times \frac{5l}{6} = -\frac{5Pl^3}{48EI}$$

Once this deflection at free end has been calculated from the primary structure, we
are in a position to apply the unknown redundant reaction at the propped end to calculate the deflection in the reverse direction. Here the downward deflection at B due
to the external loading is considered as negative.

Let the unknown support reaction at propped end is given by R_B. Now due to this
support reaction, the deflection at end B in vertical upward direction will be:

$$\delta_B = \frac{(1/2) \times R_B l \times l}{EI} \times \frac{2l}{3} = \frac{R_B l^3}{3EI}$$

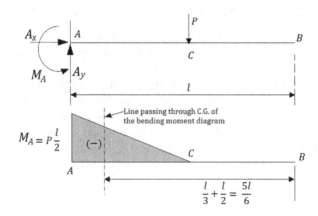

FIGURE 16.2   Primary structure and bending moment diagram due to external loading.

Since there is no actual deflection at the free end, the total algebraic sum of deflections at the propped end should be zero (this is also known as principle of superposition). Thus, we obtain the following equation:

$$\frac{R_B l^3}{3EI} - \frac{5Pl^3}{48EI} = 0$$

or, solving the above equation we get,

$$\frac{R_B l^3}{3EI} = \frac{5Pl^3}{48EI}$$

or,

$$R_B = \frac{5P}{16} \uparrow$$

Hence, we have solved the indeterminate structure and calculated the redundant support reaction by the application of force method and in conjunction with principle of superposition. As stated earlier, we have tackled the problem starting with redundant force; hence, the name force method came into picture. This simple trick for calculating the redundant forces or moments by converting the indeterminate structure into determinate primary structure can be applied in general type of indeterminate structures as well for analysis.

16.3 METHOD OF LEAST WORK

In this portion, we consider an alternative procedure of the force method known as the method of least work. For this analysis process, the compatibility conditions are formed with the help of Castigliano's second theorem instead by deflection superposition method, as in the method we have studied previously. With this difference, the two methods are very similar and demand essentially the same amount of computational work. The method of least work generally proves to be easy for analyzing composite structures that contain both axial force and flexural members (e.g., beams supported by columns or cables). However, the method cannot be treated as general as the method of consistent deformations in a sense that in its original process (as presented here), this method cannot be applied for analyzing the impact of support settlements, temperature changes, and fabrication errors.

To derive the method of least work, let us take a statically indeterminate beam with unyielding supports subjected to an external loading P, as shown in Figure 16.3.

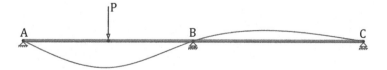

FIGURE 16.3 Indeterminate beam and its deflected shape.

Suppose that we select the vertical reaction B_y at the support point B to be the redundant reaction. By taking the redundant as an unknown load applied to the beam along with the applied loading P, relation for the strain energy can be written as a function of the known load P and the unknown redundant B_y as:

$$U = f\left(P, B_y\right)$$

According to Castigliano's second theorem, the partial derivative of the strain energy with respect force is equal to the deflection of the structure at the point where load has been applied. Also, the direction of deflection will be consistent with the direction of the applied load. Hence, for deflection in the direction of redundant reaction, the following partial differential equation forms:

$$\frac{\partial U}{\partial B_y} = 0$$

Above equation indicates that the value of the redundant that satisfies equilibrium and compatibility conditions, the strain energy of any elastic system attains maximum or minimum value. Since for linearly elastic bodies, there is no upper limit for elastic strain energy, we conclude that, for true value of redundant, strain energy attains minimum value. So, in short, the magnitude of redundant reaction in any indeterminate structure must be such that the strain energy attains its minimum value at that situation.

The method described above is suitable for structures with higher degrees of indeterminacy as well. In such case, there will be n number of partial differential equation combining the elastic stored energy and redundant forces. The number of partial differential equation is same as that of the degrees of indeterminacy as shown below:

$$\frac{\partial U}{\partial B_1} = 0$$

$$\frac{\partial U}{\partial B_2} = 0$$

$$\dots$$
$$\dots$$
$$\dots$$
$$\dots$$
$$\dots$$

$$\dots$$

$$\frac{\partial U}{\partial B_n} = 0$$

To understand the concept of energy principle, let us consider the previous section example with the redundant force as R_y.

Under the application of load as was shown in Figure 16.2, the bending moment at any section x from right support B is given by,

$$M_x = R_y x - P\left(x - \frac{l}{2}\right)$$

So, total strain energy stored in the beam is given by,

$$U = \int_0^l \frac{M_x^2}{2EI} dx$$

or, in full terms, the above relationship becomes:

$$U = \int_0^l \frac{\left[R_y x - P\left(x - \frac{l}{2}\right)\right]^2}{2EI} dx$$

Now, from principle of least work, we know that the partial derivative of internal strain energy with respect to redundant force will be zero. Hence, we get,

$$\frac{\partial U}{\partial R_y} = 0$$

which implies,

$$\frac{\partial U}{\partial R_y} = \int_0^l \frac{M_x \left(\partial M_x / \partial R_y\right)}{EI} dx = 0$$

So,

$$\frac{\partial M_x}{\partial R_y} = \frac{\left(\partial\left(R_y x - P\left(x - (l/2)\right)\right)/\partial R_y\right)}{EI}$$

or,

$$\frac{\partial M_x}{\partial R_y} = \frac{x}{EI}$$

which implies,

$$\frac{\partial U}{\partial R_y} = \int_0^l \frac{M_x \left(\partial M_x / \partial R_y\right)}{EI} dx = \int_0^l \frac{\left(R_y x - P\left(x - (l/2)\right)\right)(x) dx}{(EI)(EI)} = 0$$

or, performing the integration, we get,

$$R_y \frac{l^3}{3} - P\left(\frac{l^3}{3} - \frac{l^3}{4}\right) = 0$$

or, solving this equation, we get,

$$R_y = \frac{5P}{16}$$

It is same as we have derived in the previous section. So, method of virtual work provides another alternative way to effectively calculate the unknown support reactions for indeterminate structures.

16.4 STRUCTURES WITH MULTIPLE DEGREE OF INDETERMINACY

The method of consistent deformations introduced in the previous sections for analyzing structures with single degree of indeterminacy can easily be extended to the analysis of structures with multiple degrees of indeterminacy. To understand the concept, let us consider an example of four-span continuous beam subjected to a uniformly distributed load w, as shown in Figure 16.4 (a). As can be seen, there are six unknown support reactions. Since this is a planar structure, the total available equilibrium equations are three. Hence, the degree of indeterminacy will be $= 6 - 3 = 3$. Following the logic of previous sections, we need to take any three unknown support reaction as redundant. To begin with, let us consider the vertical reactions at support B, C, and D to be the redundants. The roller supports at B, C, and D are then removed from the given indeterminate beam to form statically determinate and stable primary beam, as shown in Figure 16.4 (b). Once the primary beam is formed, we can now

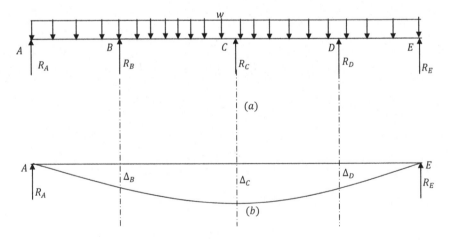

FIGURE 16.4 (a) Indeterminate structure with multiple degrees of indeterminacy and (b) corresponding primary structure.

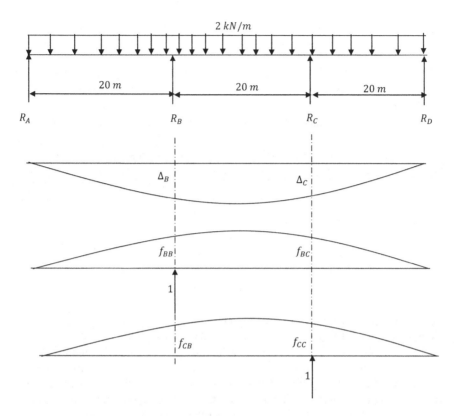

FIGURE 16.5 Three-span continuous beam with redundant reactions.

determine the vertical deflection at the support s using any of the known methods. Since there will be no deflection, we apply redundant forces as loads at these points to calcite the deflection I opposite direction. The algebraic sum of deflection due to these two cases should add up to zero.

To better understand the analysis procedure for the same, let us consider that the uniformly distributed load applied on the three-span continuous beam as shown in the following diagram is 2 kN/m. The span of each segment of the beam is 20 m.

Now we have here four supports with the support A as hinge, and all other supports are roller supports. Hence, total unknown is $2 + 1 + 1 + 1 = 5$. Total available equilibrium equations are 3 $(\Sigma F_x = \Sigma F_y = \Sigma M_z = 0)$. Hence, degree of indeterminacy of the structure is $5 - 3 = 2$.

Now, to solve this problem, we must take two support reactions as redundant. Let us take R_B and R_C. In this situation, we remove roller supports at B and C to make the determinate primary beam as shown in the above diagram. Next, we apply unit load at the redundant location B and C separately and calculate the deflection in the reverse order as shown in the above diagram. As can be seen from the unit load diagrams, there are few new terms introduced, namely, f_{BB}, f_{BC}, f_{CC}, and f_{CB}. These terms are known as flexibility coefficients. These coefficients are calculated based on the deflection due to unit load applied at a certain point on the structure.

f_{BB} indicates the deflection at point B when unit load is applied there. First suffix represents point of application of unit load and second suffix indicates the location at which deflection has been considered. Thus, f_{CB} is the deflection at point B when unit load has been applied at point C. Thus, from the unit force diagrams, we can easily determine the deflection in the direction of force by applying previously acquired knowledge on any one of the deflection calculation methods.

Understanding the concept of flexibility coefficients is quite useful when we are dealing with structures having multiple degrees of indeterminacy. Now, considering the actual redundant force R_B and R_C, the total deflection at support B due to redundant forces is $f_{BB}R_B + f_{BC}R_C$. So, the compatibility equation for deflection at support B is given by,

$$\Delta_B - f_{BB}R_B - f_{BC}R_C = 0$$

Similarly, for support at C, we can have another compatibility equation as given below:

$$\Delta_C - f_{CB}R_B - f_{CC}R_C = 0$$

Since all the parameters except the redundant forces are unknown, we can simultaneously solve these two equations to get two unknown support reactions.

Once redundant support reactions are obtained, we can apply principle of equilibrium conditions alone to determine the bending moment and shear force diagram for the entire beam. In Figure 16.6, the bending moment and shear force diagram is shown. Though the detailed calculation to determine the support reactions and bending moment diagrams is left as an exercise for the readers.

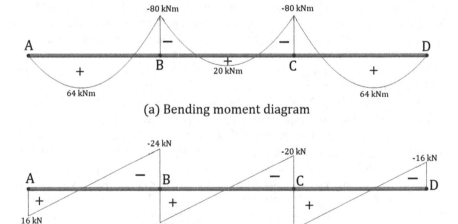

(a) Bending moment diagram

(b) Shear force diagram

FIGURE 16.6 (a) Bending moment and (b) shear force diagrams for three span continuous beam.

16.4.1   SHEAR AND BENDING MOMENT DIAGRAMS
OF THREE-SPAN CONTINUOUS BEAMS

The shear and bending moment diagrams of the previous example beam are shown in Figure 16.6. The shapes of the shear and bending moment diagrams for continuous beams can be drawn only after complete determination of redundant support reactions as explained above. Once all unknown support reactions are known, we can proceed from any one extreme support end, and proceed progressively toward the other end as we have done for statically determinate beams. Once bending moment equations are obtained, we vary the length parameter x along the beam axis to determine magnitude of bending moment at different points on the axis of beam. In general, the following observations will be formed once we complete the analysis of the beam:

1. In case uniformly distributed loading, the bending moment diagrams in general will be parabolic in nature with negative value at the interior supports and positive values near the mid span of the beam. At the extreme supports, if they are hinged, then bending moment will be zero. Otherwise, it will be generally negative.
2. In case of point loading, the bending moment diagrams will be linear segments with negative moment at the supports and positive moments at the mid span of the beam.

However, actual values of the bending moments, of course, depend on the magnitude of the loading as well as on the lengths and flexural rigidities of the spans of the continuous beam. It is strongly suggested that students should carry out the detailed analysis and match the magnitude of shear and bending moment as given in Figure 16.6, along with proper algebraic signs.

16.5   SUPPORT SETTLEMENTS, TEMPERATURE CHANGES, AND FABRICATION ERRORS

So far, we have discussed the analysis of structures with unsettled supports. As discussed in previous chapters, support settlements due to inadequate foundation system and the like may result significant stresses in externally indeterminate structures and must be considered in the designs. Support settlements, however, do not cause any considerable effect on the stress conditions of structures that are internally indeterminate but externally determinate. This is majorly due to the fact that settlements cause structures to move or rotate as a rigid body. The method of consistent deformations, as developed in the previous sections, can be conveniently modified to include the effect of support settlements in the analysis.

Let us consider, as example, the same continuous beam subjected to a uniformly distributed load w, as shown in Figure 16.7 (a). In this situation, let us also consider that the supports B and C of the beam undergo small settlements Δ_B and Δ_C, respectively, as shown in the same figure. To analyze the beam, like previous example, we consider the vertical reactions R_B and R_C as the redundants. The supports B and C

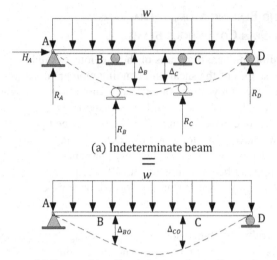

(a) Indeterminate beam

$$=$$

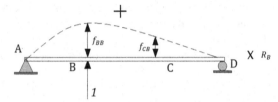

(b) External loading is imposed on primary beam

$$+$$

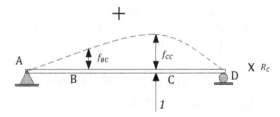

(c) Redundant R_B is imposed on the primary beam

$$+$$

(d) Redundant R_C is imposed on the primary beam

FIGURE 16.7   Three-span continuous beam with settlement of supports.

are then removed from the indeterminate beam to form the primary beam, which is then applied separately to the external load w and the unit values of the redundants R_B and R_C as shown in Figure 16.7 (b), (c), and (d), respectively. By realizing that the deflections of the actual beam at supports B and C are equal to the settlements Δ_B and Δ_C, respectively, we obtain the compatibility equations as given below:

$$\Delta_{BO} + f_{BB}R_B + f_{BC}R_C = \Delta_B$$

$$\Delta_{CO} + f_{CB}R_B + f_{CC}R_C = \Delta_C$$

Since all the parameters except the redundant support reactions are unknown in the above equations, we can easily solve them simultaneously to obtain the unknown redundant forces and thereby complete the analysis of the indeterminate structure.

Although support settlements are usually defined with respect to the undeformed position of the indeterminate structure, the magnitudes of such settlements to be used in the compatibility equations must be calculated from the chord connecting the deflected positions of the supports of the primary structure to the settled positions of the redundant supports. Any such support settlement is positive if it has the same direction as that assumed for the redundant. For our case, the beam of Figure 16.7, since there is no settlement at supports A and D, the chord AD of the primary beam remains same as that for the undeformed position of the indeterminate beam. Thus, the settlements of supports B and C relative to the chord of the primary beam are equal to the prescribed settlements Δ_B and Δ_C, respectively.

Now, let us consider a more severe case when all the supports of a beam undergo settlement as shown in Figure 16.8. If we consider the reactions R_B and R_C as redundants, then the displacements Δ_{BF} and Δ_{CF} to be the of supports B and C, respectively, in respect to the chord of the primary beam should be applied in the compatibility equations instead of the specified displacements Δ_B and Δ_C. This is because only the displacements relative to the chord induce stresses in the beam. In other words, if all the supports of the beam would have settled either by equal amounts or by amounts so that the deformed positions of all of the supports would lie on a straight line, then the beam would remain straight without bending, and no stresses would develop in the beam. In this case, as no external load is present on the beam, the compatibility equations can be written as:

$$f_{BB} R_B + f_{BC} R_C = \Delta_{BF}$$

$$f_{CB} R_B + f_{CC} R_C = \Delta_{CF}$$

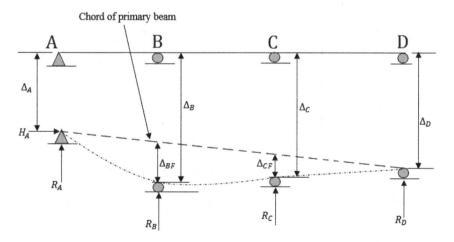

FIGURE 16.8   Three-span continuous beam with settlement of all the supports.

16.5.1 Temperature Changes and Fabrication Errors

Apart from support settlements, which affect only externally indeterminate structures, temperature changes and fabrication errors may affect the stress conditions of both externally and internally indeterminate structures. The method to analyze these structures subjected to temperature changes and/or fabrication errors is the same as used previously for the case of external loads. The only difference is that the primary structure is now applied with the prescribed temperature changes and/or fabrication errors (instead of external loads) to evaluate its deflection at the locations of redundants due to these cases. The redundants are then calculated by applying the usual compatibility conditions that the deflections of the primary structure at the locations of the redundants due to the combined effects of temperature changes and/or fabrication errors and the redundants must equal the known deflections at the corresponding locations on the actual indeterminate structure. The procedure is explained by the Example 16.1.

Example 16.1: Determine the reactions and the force in each member of the truss shown in Figure 16.9 due to a temperature increase of 45°C in member *AB* and a temperature drop of 20°C in member *CD*. Use the method of consistent deformations. Cross-sectional area of each member is 5000 mm² and diagonal members are 3000 mm², *E* = 200 GPa, and coefficient of linear thermal expansion, $\alpha = 1.2\ (10^{-5})/°C$.

SOLUTION: The degree of indeterminacy is $m + r - 2j = 6 + 3 - 2 \times 4 = 1$. Thus, the truss is internally indeterminate to first degree. Hence, let us consider one diagonal element, say *AD*, as the redundant member. Hence, the primary truss is first obtained by removing the redundant member from the original structure. Once it is removed, it will look similar to the one shown in Figure 16.10.

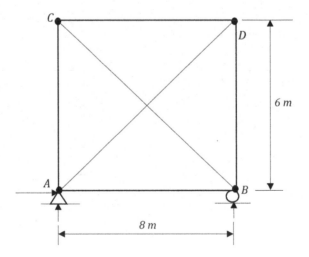

FIGURE 16.9   Original truss as per example problem statement.

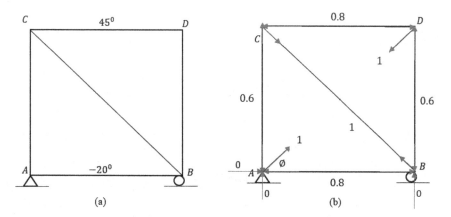

FIGURE 16.10 Primary truss subjected to (a) temperature change and (b) unit tensile force along direction of redundant *AD*.

Next, we apply prescribed temperature changes in the primary truss as shown in Figure 16.10 (a), and separately a unit load applied in the direction of *AD* as shown in Figure 16.10 (b). The virtual work done by any member due to change in temperature will be:

$$\Delta_{ADO} + f_{AD,AD}F_{AD} = 0$$

The above equation is the compatibility equation for the member *AD*.

where Δ_{ADO} is the relative displacement between joints *A* and *D* of the primary structure due to temperature changes. $f_{AD,AD}$ is the flexibility coefficient denoting the relative displacement between the same joints due to a unit value of the redundant F_{AD}. The virtual work expression for Δ_{ADO} can be given as:

$$\Delta_{ADO} = \sum \alpha(\Delta T) L u_{AD}$$

where, u_{AD} is the force in the member due to unit load in the member *AD* as shown in Figure 16.10 (b). Since there is no externally applied load on the truss, applying global equilibrium equations for planner structure, we can determine the support reactions of the truss, which will be all zeros. That is why in Figure 16.10 (b), we have shown the support reactions with 0 values. Now, the value u_{AD} is calculated using standard procedure for truss analysis as we have learned in earlier chapters on the same. As for example, let us calculate the member force *AC* due to unit tensile force in member *AD* in the primary truss as shown in Figure 16.10 (b). Resolving the forces and applying force equilibrium equation in *y* direction, we get,

$$F_{AC} + 1 \times \sin \varnothing = 0$$

or,

$$F_{AC} = -\sin \varnothing = -\frac{6}{10} = -0.6 \text{ kN}$$

Thus,

$$u_{AC} = -0.6 \text{ kN}$$

Following the above procedure of method of joints (or any other methods as one may wish), we can determine all the member forces from the primary truss in Figure 16.10 (b).

The calculation process has been provided below in tabular form for ease of understanding:

Member	L (m)	A (sq. m)	ΔT (°C)	$u_{AD}\left(\dfrac{\text{kN}}{\text{kN}}\right)$ (kN/kN)	$(\Delta T)Lu_{AD}$ (°C m)	$u_{AD}^2 L/A$ (1/m)	$F = u_{AD}F_{AD}$ (kN)
AB	8	0.005	45	−0.8	−288	1024	−32.067
CD	8	0.005	−20	−0.8	128	1024	−32.067
AC	6	0.005	0	−0.6	0	432	−24.05
BD	6	0.005	0	−0.6	0	432	−24.05
AD	10	0.003	0	1.0	0	3333.344	40.085
BC	10	0.003	0	1.0	0	3333.344	40.085
Σ					−160	9578.688	

Now, applying the above mentioned formula,

$$\Delta_{ADO} = \sum \alpha(\Delta T)Lu_{AD} = 1.2\left(10^{-5}\right)(-160)(1.0) = -0.00192 \text{ m} = -1.92 \text{ mm}$$

$$f_{AD,AD} = \frac{1}{E}\sum \frac{u_{AD}^2 L}{A} = \frac{9578.667}{200\left(10^6\right)} = 47.893\frac{\left(10^{-6}\right)\text{m}}{\text{kN}} = 0.0479 \text{ mm/kN}$$

$$F_{AD} = -\frac{\Delta_{ADO}}{f_{AD,AD}} = 40.084 \text{ kN (tension)}$$

With the above obtained values, we can use the compatibility equation to obtain,

$$-1.92 + (0.0479)F_{AD} = 0$$

or,

$$F_{AD} = 40.084 \text{ kN }(T)$$

Since the truss is statically determinate externally, the support reactions will be zero due temperature changes. Also, member forces in other members can be determined from the table and using the relationship, $F = u_{AD}F_{AD}$.

17 Influence Lines for Statically Indeterminate Structures

17.1 INTRODUCTION OF INFLUENCE LINES FOR STATICALLY INDETERMINATE STRUCTURES

In this chapter, we will discuss the methods for constructing influence lines for statically indeterminate structures. It may be recalled from Chapter 11 of Part II of this book, that an influence line is a diagram of a response function of a structure as a function of the position of a downward unit load moving across the structure. The core procedure for constructing influence lines for indeterminate structures is very much the same as that of determinate structures. The steps essentially demand calculating the values of the response function of interest by varying positions of a unit load on the structure and drawing the response function values as ordinates versus the position of the unit load as abscissa to obtain the influence line.

As we have seen, the influence lines for forces and moments of determinate structures consist of straight-line segments. The influence lines are drawn in Chapter 11 by evaluating the ordinates for only a few locations of the unit load and connecting them with straight lines. The influence lines for indeterminate structures, as we will discover soon, are generally curved lines. Thus, the formation of influence lines for indeterminate structures requires calculation of many more ordinates than necessary in the case of determinate structures.

Although any of the methods of analysis of indeterminate structures presented in Part III can be used for computing the ordinates of influence lines, we will use the method of consistent deformations discussed in the previous chapter to learn the method of drawing the influence lines for indeterminate structures.

17.2 INFLUENCE LINES FOR BEAMS

To begin our discussion on influence line diagrams, let us consider the continuous beam with the support conditions as shown in Figure 17.1 (a).

As per Figure 17.1, the unit load is placed (for an instant) at a distance x from the left support. Now our target is to draw the influence line for the support reaction at B (i.e., for B_y). To be able to do that, we need to express B_y in terms of variable x. Since the beam is indeterminate to the first degree, we need to remove one redundant support reaction from the indeterminate beam to make the primary structure. So, the roller support from B has been removed to convert it to primary structure. Once the primary structure is formed, we apply separately the unit load at any arbitrary

DOI: 10.1201/9781003081227-20

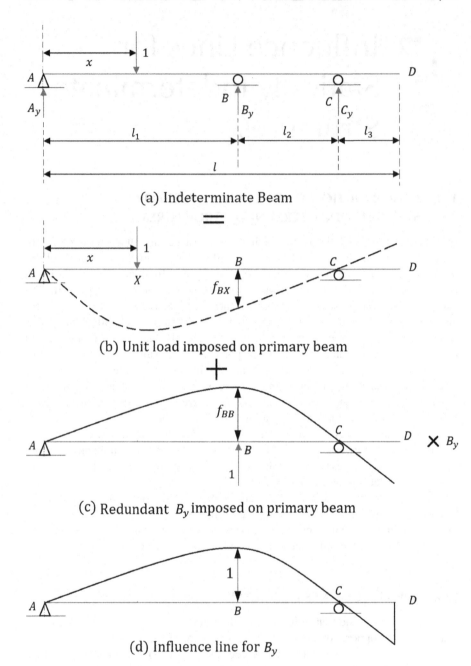

(a) Indeterminate Beam

==

(b) Unit load imposed on primary beam

+

(c) Redundant B_y imposed on primary beam

(d) Influence line for B_y

FIGURE 17.1 Finding influence line for reaction of an indeterminate beam.

point X at a distance x from the left support and redundant reaction B_y as shown in Figure 17.1 (b) and (c) to form the compatibility equation. Since net displacement at support B will be zero, hence, we get:

$$f_{BX} + \widehat{f_{BB}}\left(B_y\right) = 0$$

From which we get:

$$B_y = -\frac{f_{BX}}{\widehat{f_{BB}}}$$

In which, flexibility coefficient, f_{BX} gives the deflection at point B when a unit load is placed at a distance x from left support. And $\widehat{f_{BB}}$ represents the deflection at point B due to a unit redundant force at B. We can apply the above equation easily to get the influence line for the support reaction at B. To be able to do that, we can take help from the Maxwell's law of reciprocal deflection introduced in the earlier chapter. According to this principle, the deflection at B due to a unit load at a distance x is the same as deflection at x due to a unit load at point B. Thus, in summary, we have the following relationship:

$$f_{BX} = f_{XB}$$

With this, the above equation changes to:

$$B_y = -\frac{f_{XB}}{\widehat{f_{BB}}}$$

This is the equation of influence line for reaction at B. Note that the deflections of f_{XB} and $\widehat{f_{BB}}$ are considered positive when in the upward direction. To understand the concept, let us consider the following example problem.

Example 17.1: Draw the influence lines for the reactions at the supports A, B, and C of the indeterminate beam, and shear force and bending moment at point E as shown in Figure 17.2.

SOLUTION: This beam is indeterminate to the first degree. So first, the support at point B is considered as redundant. Now, if we remove this redundant, we will get the primary beam as shown in Figure 17.2 (b).

Influence line for redundant B_y

The value of redundant B_y for an arbitrary position X of the unit load can be determined by solving the compatibility equation:

$$f_{BX} + \widehat{f_{BB}}\left(B_y\right) = 0 \tag{17.1}$$

From which,
$$B_y = -\frac{f_{BX}}{\widehat{f_{BB}}} \tag{17.2}$$

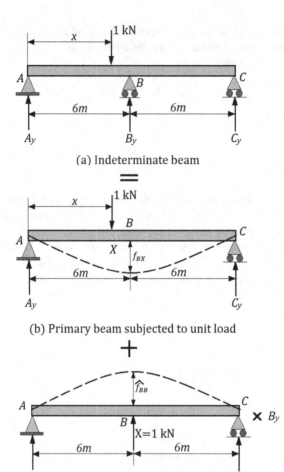

(a) Indeterminate beam

=

(b) Primary beam subjected to unit load

+

(c) Primary beam subjected to redundant B_y

FIGURE 17.2 Example problem on finding the influence line diagram for reactions, shear, and bending moment.

According to Maxwell's law, $f_{BX} = f_{XB}$, we place the unit load at B on the primary beam as shown in Figure 17.3 (a) and compute the deflections A through C by using conjugate beam method. The conjugate beam is shown in Figure 17.3 (b), from which we obtain the following:

$$f_{BA} = f_{AB} = 0$$

$$f_{BD} = f_{DB} = -\left[\left(\frac{9}{EI}\right) \times 3 - \left(\frac{1}{2}\right) \times 3 \times \left(\frac{1.5}{EI}\right) \times \left(\frac{3}{3}\right)\right] = -\frac{24.75}{EI}$$

$$f_{BB} = -\left[\left(\frac{9}{EI}\right) \times 6 - \left(\frac{1}{2}\right) \times 6 \times \left(\frac{3}{EI}\right) \times \left(\frac{6}{3}\right)\right] = -\frac{36}{EI}$$

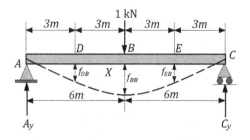

(a) Primary beam subjected to unit load at B

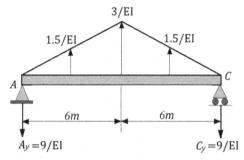

(b) Conjugate beam for unit load at B

FIGURE 17.3   Conjugate beam for unit load at B.

$$f_{BE} = f_{EB} = -\left[\left(\frac{9}{EI}\right) \times 3 - \left(\frac{1}{2}\right) \times 3 \times \left(\frac{1.5}{EI}\right) \times \left(\frac{3}{3}\right)\right] = -\frac{24.75}{EI}$$

$$f_{BC} = f_{CB} = 0$$

The negative signs indicate that these directions occur in the downward directions. The flexibility coefficient $\widehat{f_{BB}}$ in equation (17.1) denotes the upward (positive) deflection of primary beam at B due to the unit value of the redundant B_y as shown in Figure 17.2 (c), whereas the deflection f_{BB} represents the downward (negative) deflection at B due to the external unit load at B.

Thus, we can write, $\widehat{f_{BB}} = -f_{BB} = +\dfrac{36}{EI}$

The ordinates for the influence line for B_y can now be computed by applying equation (17.2) successively for each position of unit load. For example, when the unit load is located at A and C, the value of B_y is given by,

$$B_y = -\frac{f_{BA}}{\widehat{f_{BB}}} = -\frac{f_{BA}}{\widehat{f_{BB}}} = 0$$

When the unit load is located at D, and E, the value of B_y is given by,

$$B_y = -\frac{f_{BD}}{\widehat{f_{BB}}} = -\frac{f_{BE}}{\widehat{f_{BB}}} = \frac{24.75}{36} = 0.6875$$

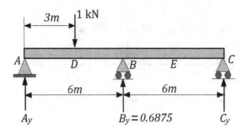

FIGURE 17.4   Finding coordinate of influence line diagram for A_y, when the unit load is at 3 m from support A.

When the unit load is located at B, the value of B_y is given by,

$$B_y = -\frac{f_{BB}}{\hat{f}_{BB}} = \frac{36}{36} = 1$$

Now that the B_y is known, the values of the ordinate of the influence lines for other reactions can be obtained using equation of statics.

Influence line for A_y:

Let us consider the external unit load is located at D, which is 3 m from support A. In this position, the influence line ordinate for B_y is 0.6875. Now, from Figure 17.4, using moment equilibrium about point C, we can easily find out the ordinate of influence line for A_y as below,

$$\circlearrowleft + \sum M_C = 0$$

$$-A_y(12) + 1(9) - 0.6875(6) = 0$$

or, $A_y = 0.41$

Now, we can vary the position of the unit load along the span of the beam and can get the remaining ordinates of influence line diagram.

Similarly, we can find the influence line diagram for C_y as well using equation of statics. The influence line diagrams for A_y, B_y, and C_y are shown in Figure 17.5.

Influence lines for S_E and M_E

The ordinates of the influence lines for the shear and bending moment at E can now be evaluated by placing the unit load successively at points A through C on the indeterminate beam and by using the corresponding values of the reactions computed previously. For example, as shown in Figure 17.6 (a), when the unit load is located at point D, the values of the reactions are $A_y = 0.41$; $B_y = 0.69$; and $C_y = -0.095$.

By considering the equilibrium of the free body of the portion of the beam to the left of E, we obtain,

$$S_E = 0.41 - 1 + 0.69 = 0.1$$

$$M_E = 0.41(9) - 1(6) + 0.69(3) = -0.24$$

The values of remaining ordinates of the influence lines are computed in a similar manner. Figure 17.6 shows the influence line diagrams for S_E and M_E.

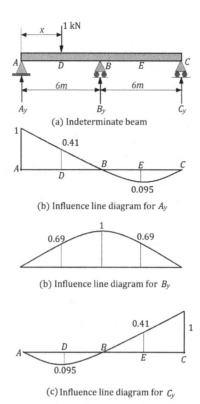

(a) Indeterminate beam

(b) Influence line diagram for A_y

(b) Influence line diagram for B_y

(c) Influence line diagram for C_y

FIGURE 17.5   Influence line diagrams for reaction forces.

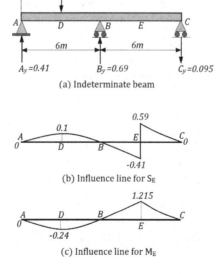

(a) Indeterminate beam

(b) Influence line for S_E

(c) Influence line for M_E

FIGURE 17.6   Influence line diagrams for S_E, and M_E.

17.3 INFLUENCE LINES FOR TRUSSES

Influence lines for trusses can be drawn with the help of flexibility coefficients as that has been considered during the analysis of beams. To understand the concept, let us proceed with Example 17.2.

Example 17.2: Draw the influence line diagrams for member forces of the truss member *BC* and *CE* as shown in Figure 17.7.

SOLUTION: This truss is internally statically indeterminate to the degree one. Now, let us consider the redundant member as the member *CE*. To determine the influence line diagram for F_{CE}, we place a unit load successively to the joints *B* and *C*. For each position of the unit load, we apply the method of consistent deformation to determine the value of member force *CE*, i.e., F_{CE}. First, we remove the member *CE* entirely from the original truss to convert it to determinate primary truss. Under this situation, the primary member with unit load at joint *B* will look something like the diagram as shown in Figure 17.8 (a).

By taking the overall force and moment equilibrium of the determinate truss shown in Figure 17.8, we can determine the support reactions that are as follows:

$$R_A = \frac{2}{3}$$

$$R_B = \frac{1}{3}$$

The above force values will be just interchanged when the unit load is shifted from joint *B* to joint *C* due to symmetry. Under this situation, the force in the member of the truss also needs to be calculated. Since the redundant has been removed, it has become a statically determinate structure and hence, the member forces can be calculated by applying any standard procedure we have learned earlier. The member forces due to unit load at *B* and *C* are given in Figure 17.8 (a) and (b), respectively for quick reference. Also, internal forces on the member when a unit tensile load applied on the redundant member *CE* are also provided in Figure 17.8 (c). Students need to check the correctness of these values on their own.

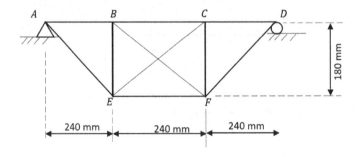

FIGURE 17.7 Example problem for truss member force influence line.

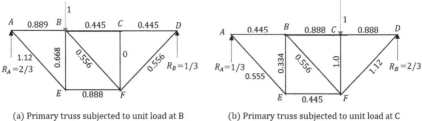

(a) Primary truss subjected to unit load at B
- u_B force

(b) Primary truss subjected to unit load at C
- u_C force

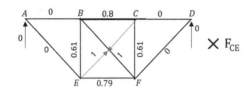

(c) Primary truss subjected to unit tensile force in member CE
- u_{CE} force

FIGURE 17.8   Primary truss with unit load at B, C, and along the redundant member CE.

When the unit load is located at B, the compatibility equation can be written as

$$f_{CE,B} + f_{CE,CE}\ F_{CE} = 0$$

where $f_{CE,B}$ denotes the relative displacement between joints C and E of the primary truss due to unit load at B and $f_{CE,CE}$ relative displacement between the same nodes due to a unit value of redundant force F_{CE}. Once the compatibility formed, we can develop the following table for ease of analysis using compatibility method.

Member	L(mm)	A(mm²)	u_B	u_C	u_{CE}	$\dfrac{u_B u_{CE} L}{A}$	$\dfrac{u_C u_{CE} L}{A}$	$\dfrac{u^2_{CE} L}{A}$
AB	240	6	−0.888	−0.445	0	0	0	0
BC	240	6	−0.445	−0.888	−0.8	15.806	28.416	25.6
CD	240	6	−0.445	−0.888	0	0	0	0
EF	240	6	0.888	−0.445	−0.79	−28.061	−14.062	24.964
BE	180	4	−0.668	−0.334	−0.61	18.337	9.168	16.745
CF	180	4	0	−1.0	−0.61	0	27.45	16.745
AE	300	6	1.12	0.556	0	0	0	0
BF	300	4	−0.556	0.556	1	−41.7	41.7	75
CE	300	4	0	0	1	0	0	75
DF	300	6	0.556	1.12	0	0	0	0
					Σ	−35.618	120.796	234.054

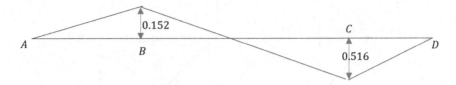

FIGURE 17.9   Influence line diagram for member force F_{CE}.

Now, using virtual work method, we obtain:

$$f_{CE,B} = \frac{1}{E}\sum \frac{u_B u_{CE} L}{A} = -\frac{35.618}{E}$$

$$f_{CE,CE} = \frac{1}{E}\sum \frac{u^2_{CE} L}{A} = \frac{234.054}{E}$$

By substituting these values in the above compatibility equation, we get:

$$F_{CE} = -\frac{f_{CE,B}}{f_{CE,CE}} = 0.152\ (T)$$

Similarly, when the unit load is placed at C., the compatibility equation may be written as:

$$f_{CE,C} + f_{CE,CE}F_{CE} = 0$$

Using the above table, we have:

$$f_{CE,C} = \frac{1}{E}\sum \frac{u_C u_{CE} L}{A} = \frac{120.796}{E}$$

Substituting this value in the above equation we get:

$$F_{CE} = -\frac{f_{CE,C}}{f_{CE,CE}} = -\frac{120.796}{234.054} = -0.516\ (C)$$

Thus, we determine the ordinate at B and C, respectively, from the above results. The influence line diagram for F_{CE} is shown in Figure 17.9.

Following this procedure, the influence lines for all other member forces can be drawn.

17.4   QUALITATIVE INFLUENCE LINES BY THE MÜLLER-BRESLAU'S PRINCIPLE AND INFLUENCE LINE FOR FRAMES

In many practical situations, such as when designing continuous beams or building frames acted upon by uniformly distributed live loads, it is sufficient to draw only the qualitative influence lines to predict where to place the live loads to produce the maximum effect of the response functions. As we have already seen in statically determinate structures, the Müller-Breslau's principle offers a convenient method of forming qualitative influence lines for indeterminate structures.

As we can remember from our determinate structural analysis, this principle states that the influence line for a force (or moment) is given by the deflected shape of the released structure obtained by removing the restraint corresponding to the response function from the original structure and giving the released structure a unit displacement (or rotation) at the location and in the direction of the response function, so that only the response function and the unit load perform external work.

The method of drawing qualitative influence lines for indeterminate structures is the same as that of determinate structures discussed in Chapter 11 of this book. The procedure needs the following steps to be adopted for analysis:

1. Removing from the original structure the restraint corresponding to the response function to obtain the released structure.
2. Imposing a small displacement (or rotation) to the released structure at the location and in the positive direction of the response function.
3. Drawing a deflected shape of the released structure in line with its support and continuity parameters.

The influence lines for indeterminate structures are generally found to be curved lines. Once a qualitative influence line for force or bending moment has been constructed, it can be used to decide where to place the live loads to maximize the value of the forces and/or moments to cause severity. As discussed in earlier sections, the value of shear force or bending moment due to a uniformly distributed live load is maximum positive (or negative) when the load is located in those portions of the structure where the ordinates of the response function influence line are positive (or negative). Because the influence line ordinates tend to diminish rapidly with distance from the point of application of the unit load, live loads placed more than three span lengths away from the location of the unit load generally have a negligible effect on the value of the overall reaction or bending moments. With a known live load pattern, an approximate analysis of the structure can be performed to determine the maximum value of the force or moments that we desire and are required for drawing the influence lines. We shall follow the Example 17.3 that will demonstrate the procedure for qualitative influence line construction using the abovementioned procedure.

Example 17.3: Draw qualitative influence lines for the vertical reactions at supports A and B for the continuous beam shown in Figure 17.10. Also, show the arrangements of a uniformly distributed downward live load w to cause the maximum positive reactions at supports A and B, the maximum negative bending moment at B.

SOLUTION: *Influence line for R_A*

To obtain influence line for R_A, following the Muller-Breslau principle, we first remove the support at A (thus, we remove the redundant reaction at A), and then apply a small displacement in the direction of the redundant force R_A. In this situation, please keep in mind that the other supports to be kept as it is present in the original structure. Thus, due to this small displacement at A, the overall deflected shape of the beam will be the influence line for reaction at A. Pictorially, the qualitative influence line for R_A will be something like as shown in Figure 17.11 (a).

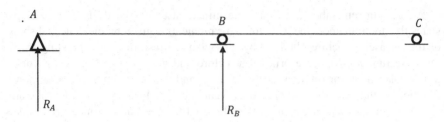

FIGURE 17.10 Example problem for qualitative influence line diagram.

Now to maximize the effect due to live load w, we need to place the same over the span AB of the beam where the ordinate of the influence line diagram is positive.

Influence line for R_B

Following the same procedure as that for R_A, we can form the influence line for R_B by drawing the deflected shape of the beam, maintaining the constraints and direction of redundant support reaction. The same has been drawn in Figure 17.11 (b) for reference.

As can be seen for R_B influence line, the total ordinate is positive near the span and goes near the supports. Hence, for maximum response against the live load, the load should be placed on the beam A to C over the whole span.

Influence line for M_B

To determine the influence line for bending moment at B, we need to remove the support at B and then replace the existing roller support by a hinge and apply small angular displacement at the same location as shown in Figure 17.12. Due to this applied rotational displacement, the deflected shape of the entire beam has also been drawn, and that is the influence line diagram for the bending moment at B.

Thus, to cause maximum negative bending moment, we place the live load in the span AB and BC, where the ordinate of the influence line diagram is negative.

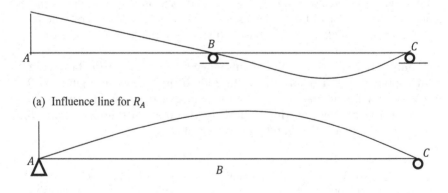

(a) Influence line for R_A

(b) Influence line for R_B

FIGURE 17.11 Qualitative influence line for (a) R_A and (b) R_B diagrams as per example problem.

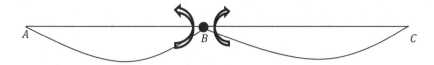

FIGURE 17.12   Qualitative influence line diagrams for bending moment at *B*.

Once we have dealt with the continuous beams, let us focus on the formation of influence lines for statically determinate frames influence lines.

The Müller-Breslau principle provides a quick method for generating the general shape of the influence line for building frames. Once the influence line shape is determined, one can immediately decide on the location of the live loads to form the greatest influence of the functions (reaction, shear, or moment) in the frame.

The shape of the influence line for the positive moment at the center *I* of girder *FG* of the building frame in Figure 17.13 is shown by the broken lines. Thus, uniform loads would be placed only on girders *AB*, *CD*, and *FG* in order to create the largest negative moment at *I*. With the frame in addition to imposed live load at the specified locations in this manner, indeterminate analysis of the frame can then be applied to calculate the critical moment at *I*.

Similarly, for the maximum positive moment at midspan of *BC*, we can place the live load on *EF*, *GH* and *BC* to produce the maximum adverse effect due to live load as that we have told just before for the maximum negative moment at the midspan of *FG*.

So from the example, we find that the Müller-Breslau's principle indeed provides a very handy tool to determine the qualitative influence line diagram of the support reactions and bending moment as required. Thus, by applying this principle, we can easily determine the location and placement of live loads on any number of spans of a continuous beam to obtain the maximum response.

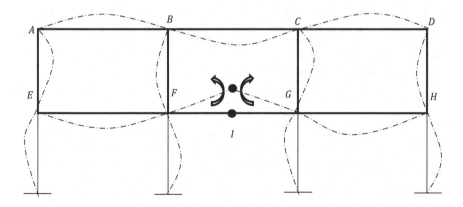

FIGURE 17.13   Building frame qualitative influence lines.

17.5 ALTERNATE APPROACH FOR FINDING INFLUENCE LINE DIAGRAMS FOR INDETERMINATE BEAMS

This approach for finding the influence lines of indeterminate beams is based on a mathematical model derived from the fundamental use of the flexibility method. The mathematical model is based on describing the forces and deformations of the beam as mathematical functions related by the consecutive integration process. This new approach was reported by Dr. Moujalli Hourani [1] of Manhattan College in the proceeding of the American Society of Engineering Education Annual Conference and Exposition in 2002. It was reported that using this new technique, students showed a major improvement in their capabilities to solve problems of influence lines for indeterminate beams. They were capable of developing their computer programs using Excel/Quattro and Mapple/MathCAD to solve problems of influence lines for multi-span beams with various boundary conditions. This approach is not based on new theories or principles but rather on a new methodology to solve a typical structural problem.

Along with the flexibility method, a mathematical model was used as summarized below to evaluate the influence lines:

1. Determine the degree of static indeterminacy.
2. Choose the unknowns/redundants, name them X_1, X_2..........., X_n. The redundant can be external reactions, internal forces or both. Make sure that the structure remains stable.
3. Remove the dedundants, i.e., set the unknowns equal to zero. The structure is now statically determinate, and it is called the released structure.
4. Remove all the loads and apply a unit load corresponding to X_1, at the location of X_1.
5. Repeat the process for all the X's.
6. Find the displacements corresponding to the unknowns in the released structure.
7. Find the flexibility coefficients.
8. Use the principle of superposition and apply the compatibility conditions at the locations of all the redundants.
9. Describe the load as a function Y_L in the X-Y coordinate system, as shown in Figure 17.14. Since the influence line is based on applying a unit concentrated load, then, $Y_L = 0$.
10. The shear is equal to $Y_V = \int Y_L dx = C_1$
11. The bending moment $Y_M = \int Y_V dx = C_1 x + C_2$
12. The slope of the elastic curve is described by,

$$Y_S = \int \left(\frac{1}{EI}\right) Y_M \, dx = \left(\frac{1}{EI}\right) \times \left(C_1 \frac{x^2}{2}\right) + C_2 x + C_3, \text{ assuming constant } EI.$$

13. The transverse deflection of the beam is described by,

$$Y_D = \left(\frac{1}{EI}\right) \times \left(C_1 \frac{x^3}{6}\right) + C_2 \frac{x^2}{2} + C_3 x + C_4$$

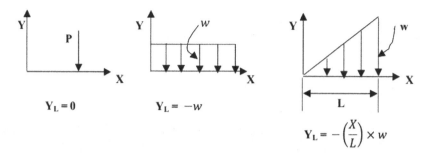

FIGURE 17.14   Description of load function, Y_L.

The constants of integration can be determined by applying the geometric (slope, deflection, and compatibility conditions) and the loading (shear, moment, and joint equilibrium) boundary conditions. We will understand further about this concept with the help of some example problems.

Example 17.4: Draw the influence line for the vertical reaction at B, R_B, for the beam shown in Figure 17.15. The beam is statically indeterminate to the first degree. Choose as a redundant the vertical reaction at B, $B_y = X_1$.

SOLUTION: Let us remove the redundant at B to get the primary structure. Now the beam has become statically determinate as shown in Figure 17.15 (b). Now, remove the applied load and apply a unit vertical load at B. This virtual beam is shown in Figure 17.15 (c). Making use of the principle of superposition, the relationship between these three figures is summarized by:

$$\text{Figure } 17.15\ (a) = \text{Figure } 17.15\ (b) + \text{Figure } 17.15\ (c) \times X_1$$

Applying the compatibility condition for the vertical deflection at B, gives:

$$-\delta BD + F11 \times X1 = 0$$

Using Maxwell's law of reciprocal deflection, we can set $\delta_{BD} = \delta_{DB}$, and the reaction R_B can be found from the above equation to be,

$$R_B = X_1 = \delta_{DB}/F_{11}$$

As it can be seen from the above equation, the only figure needed is Figure 17.15 (c). In the above equations, X_1 = the vertical reaction at B due to a unit vertical load at any point D along the beam; δ_{DB} = the vertical displacement at D due to a unit vertical load at B; F_{11} = the vertical displacement at B due to a unit vertical load at B. The moment diagram of the beam in Figure 17.16 is drawn below:

Due to the discontinuity in the moment diagram, the equations for the two segments AB and BC are needed to describe the moment functions, $x_1 - y_1$ from A to B, and $x_2 - y_2$ from C to B.

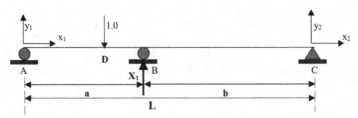

(a) Two span indeterminate continuous beam

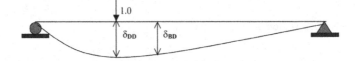

(b) Unit load applied on the the released beam

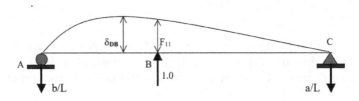

(c) Redundant B$_y$ imposed on released beam

FIGURE 17.15 Finding influence line diagram of vertical reaction for two span indeterminate beam.

From A to B (origin at A)

$$(Y_M)_1 = -\left(\frac{b}{L}\right)x$$

$$(Y_S)_1 = \left(\frac{1}{EI}\right) \times \left[\left(-\frac{b}{L}\right) \times \frac{x^2}{2} + C_3\right]$$

$$(Y_D)_1 = \left(\frac{1}{EI}\right) \times \left[\left(-\frac{b}{L}\right) \times \frac{x^3}{6} + C_3 x + C_4\right]$$

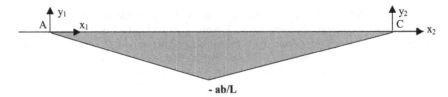

FIGURE 17.16 Moment diagram for the virtual beam.

From C to B (origin at C)

$$(Y_M)_2 = \left(\frac{a}{L}\right) x$$

$$(Y_S)_2 = \left(\frac{1}{EI}\right) \times \left[\left(\frac{a}{L}\right) \times \frac{x^2}{2} + K_3\right]$$

$$(Y_D)_2 = \left(\frac{1}{EI}\right) \times \left[\left(\frac{a}{L}\right) \times \frac{x^3}{6} + K_3 x + K_4\right]$$

Applying the boundary conditions to find the constants of integration.

At $x_1 = 0$, $(Y_D)_1 = 0$, which implies, $C_4 = 0$. At $x_2 = 0$, $(Y_D)_2 = 0$, which implies, $K_4 = 0$.

The compatibility condition for the slope at B, can be written as, $YS1$ at $x1 = a = YS2$ at $x2 = -b$,, which gives,

$$-b\frac{a^2}{2L} + C_3 = a\frac{b^2}{2L} + K_3$$

Simplifying the above equation yields,

$$C_3 - K_3 = \left(\frac{ab}{2L}\right) \times (a + b) = \frac{ab}{2} \tag{17.3}$$

The compatibility condition for the vertical deflection at B, can be written as, $\left[(Y_D)_1\right]_{at\ x1=a} = \left[(Y_D)_2\right]_{at\ x2=-b}$, which gives,

$$\frac{-b \times a^3}{6L} + C_3 a = -\frac{ab^3}{6L} - K_3 b$$

Simplifying the above equation yields,

$$C_3 a + K_3 b = \left(\frac{ab}{6}\right) \times (a - b) \tag{17.4}$$

Solving equations (17.1) and (17.2) for C_3 and K_3, yield

$$C_3 = \left(\frac{ab}{6L}\right) \times (2b + a)$$

$$K_3 = -\left(\frac{ab}{6L}\right) \times (b + 2a)$$

Having solved for all the constants of integration, the equations describing the vertical displacement at any point along the beam due to unit vertical load at B can be written as,

$$\left(Y_D\right)_1 = \left(\frac{1}{6EIL}\right) \times \left[-bx^3 + ab(a+2b)x\right] \text{ for, } 0 \le x \le a$$

$$\left(Y_D\right)_2 = \left(\frac{1}{6EIL}\right) \times \left[ax^3 - ab(2a+b)x\right] \text{ for, } -b \le x \le 0$$

Here, Y_D represents the vertical deflection at any point D due to a unit vertical load at B, δ_{DB}.

$F_{11} = \left(Y_D\right)_1$ at $x_1 = a$. So, we get, $F_{11} = \left(\frac{2}{6EIL}\right)a^2b^2$, and the vertical reaction at B, B_y can be written as,

$$X_1 = \left(Y_D\right)_1 / F_{11} = \left(\frac{1}{6EIL}\right) \times \frac{\left[-bx^3 + xab(a+2b)\right]}{\left[\left(\frac{2}{6EIL}\right) \times a^2b^2\right]} \text{ for } 0 \le x \le a$$

TABLE 17.1

Summary of the Reaction Forces for the Given Indeterminate Beam

Origin At	Valid Range	Equations of Reactions
A	$0 \le x_1 \le a$	$A_y = \dfrac{x^3 - x\left(3a^2 + 2ab\right) + 2a^2L}{2a^2b}$
		$B_y = \dfrac{-x^3 + x\left(a^2 + 2ab\right)}{2abL}$
		$C_y = \dfrac{x\left(x^2 - a^2\right)}{2a^2b}$
	$a \le x \le L$	$A_y = \dfrac{-(x-L)^3(a/b) + ab(x-L)}{2a^2L}$
		$B_y = \dfrac{(x-L)\left(x^2 - 2Lx + a^2\right)}{2b^2a}$
		$C_y = \dfrac{-x^3 + 3Lx^2 + ax(a-4L) + a^2L}{2b^2L}$
C	$-b \le x_2 \le 0$	$A_y = \dfrac{-x^3(a/b) + abx}{2a^2L}$
		$B_y = \dfrac{x^3 - x\left(b^2 + 2ab\right)}{2b^2a}$
		$C_y = -\dfrac{x^3 - x\left(2ab + 3b^2\right) - 2b^2L}{2a^2L}$

Simplifying we get,

$$X_1 = B_y = \frac{-x^3 + x\left(a^2 + 2ab\right)}{2a^2b} \quad \text{for } 0 \le x \le a$$

$$X_1 = B_y = \frac{x^3 + x\left(b^2 + 2ab\right)}{2b^2a} \quad \text{for } -b \le x \le 0$$

Knowing, B_y, the other two reactions can be found by static equilibrium. The equations of the reaction forces are summarized in Table 17.1.

These equations are used to determine the internal resisting forces at any point in the beam. Readers are encouraged to plot the influence line diagrams of various reaction forces using the above equations making one computer program.

18 Slope Deflection Method

18.1 INTRODUCTION

In this chapter, we will discuss the whole idea for analyzing structures using the slope deflection method. Once these concepts are presented and detailed steps are developed, we will form the general slope deflection equations and then use them to analyze statically indeterminate beams and frames. It is to be understood that the slope deflection method is a displacement method or stiffness method, unlike the force method or the method of consistent deformation discussed earlier. Here the unknown displacements are found first, solving the structure's equilibrium equations.

18.2 SLOPE DEFLECTION EQUATIONS AND ANALYSIS OF CONTINUOUS BEAMS

The method of consistent deformation studied in Chapter 16 is called a force method of analysis because it demands developing equations that relate to the unknown forces or moments in a structure. Unfortunately, its use is limited to systems that are not highly indeterminate. This is because tedious work is required to set up the compatibility equations, and moreover, each equation written involves all the unknown forces, making it difficult to solve the resulting set of equations unless computer software is used. In contrast, the slope-deflection method, which is a displacement method of analysis, is not as difficult to handle. As we shall understand soon, it requires less time to write the necessary equations for the solution of a problem and solve these equations for the unknown displacements or deformations and associated internal forces and moments. Also, this process can be easily programmed on a computer and applied to analyze a wide range of indeterminate structures.

The slope-deflection method was originally invented by Heinrich Manderla and Otto Mohr for the purpose of studying secondary stresses in trusses. Later, in 1915, G. A. Maney developed the modern version of this technique and applied this principle to the analysis of indeterminate beams as well as framed structures.

The slope-deflection method is so named since it deals with the unknown slopes and deflections to the applied load on a structure. To develop the general form of the slope-deflection equations, we will consider the typical span AB of a continuous prismatic beam as shown in Figure 18.1, which is subjected to arbitrary loading and has a constant EI. We like to relate the beam's internal end moments with its degrees of freedom, namely, its angular and linear displacements, which could be caused by a relative settlement between the supports, temperature change, due to the presence of external force, etc. or due to any combination of the effects. Since we will be developing a formula, the moments and angular displacements, cord rotations will be considered positive when they act clockwise on the span, as shown in Figure 18.1. Moreover, the linear displacement Δ is considered positive, as shown in the figure,

DOI: 10.1201/9781003081227-21

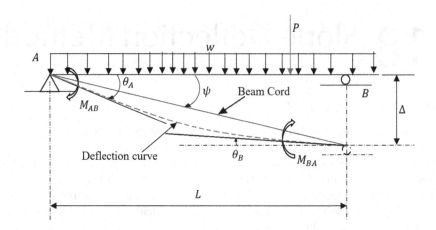

FIGURE 18.1   Two spans of a continuous beam with general loading with deflection diagrams.

since this displacement causes the centerline of the span and the span's cord angle to rotate clockwise. In Figure 18.1, the straight line joining the two nodes A to deflected node B is the cord line of the beam, and the dotted line is the deflection curve of the beam due to the external loading.

The slope-deflection equations can be formulated by using the principle of super-position by taking separately the moments developed at each support due to each of the displacements θ_A, θ_B, Δ and then the external loads.

For angular displacement θ_A only at A, let us consider the following beam with B end fixed as shown in Figure 18.2.

Now, to determine the moment M_{AB} to cause this angular displacement, we seek help from the conjugate beam method. Since the deflection at the two ends of the real beam is zero, corresponding moments at the end of the conjugate beam will also be zero. As θ_A is clockwise, the shear at end A' of the conjugate beam acts downward. Also, support conditions in real beams need to be changed while forming the con-jugate beam. For students, it is advisable to stop here and go through Chapter 9 on conjugate beam analysis for a quick recap. So, applying the appropriate loading and support conditions, the bending moments at the end of the conjugate beam will be:

$$\circlearrowleft + \sum M_{A'} = 0; \quad \frac{1}{2}\left(\frac{M_{AB}}{EI}\right)L\frac{L}{3} - \frac{1}{2}\left(\frac{M_{BA}}{EI}\right)L\frac{2L}{3} = 0$$

$$\circlearrowleft + \sum M_{B'} = 0; \quad \frac{1}{2}\left(\frac{M_{BA}}{EI}\right)L\frac{L}{3} - \frac{1}{2}\left(\frac{M_{AB}}{EI}\right)L\frac{2L}{3} + \theta_A L = 0$$

Solving these two simultaneous equations, we get,

$$M_{AB} = \frac{4EI}{L}\theta_A$$

$$M_{BA} = \frac{2EI}{L}\theta_A$$

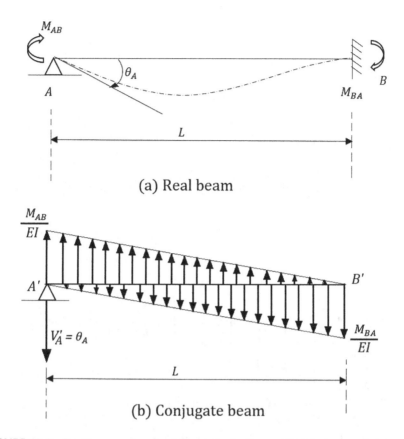

(a) Real beam

(b) Conjugate beam

FIGURE 18.2   Real beam and conjugate beam for finding angular displacement at A.

Similarly, for separate rotational displacement θ_B, we can form the same type of beam with A end fixed. Then applying the conjugate beam method, we can determine the following relationships between angular displacement and moment at the supports:

$$M_{BA} = \frac{4EI}{L}\theta_B$$

$$M_{AB} = \frac{2EI}{L}\theta_B$$

Once the relationships are formed for angular displacements, we can form a similar relationship for the linear displacement Δ that occurred due to the settlement of the support B as per Figure 18.1. For this analysis, we have both ends fixed beam with settlement occurred at the fixed-end B as shown in Figure 18.3.

So, the cord of the beam rotates clockwise, but both ends do not rotate. This results in equal but opposite support moments and shear force at both beams' end, as shown in Figure 18.3. To form the relationship, we again seek the help of the

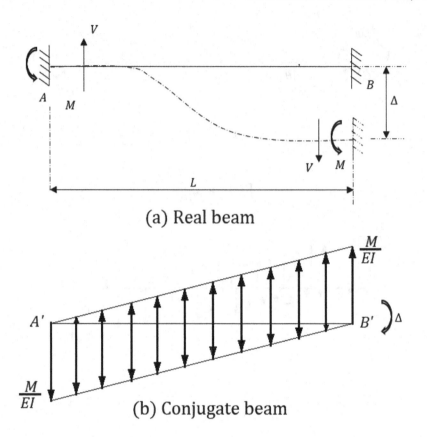

FIGURE 18.3   Fixed-end beam with support settlement at *B*.

conjugate beam theory, and in this case, both ends of the conjugate beam will be the free end as per requirement. However, the displacement at *B* of the real beam and the moment at the end of the conjugate beam at the same point should be the same as Δ.

$$\left[\frac{1}{2}\frac{M}{EI}(L)\left(\frac{2}{3}L\right)\right]-\left[\frac{1}{2}(L)\left(\frac{1}{3}L\right)\right]-\Delta=0$$

$$M_{AB}=M_{BA}=M=-\frac{6EI}{L^2}\Delta$$

By our sign convention, this moment is indeed negative as for equilibrium, and it acts in the anticlockwise direction.

It is to be noted that linear and angular displacements at the nodes occur due to the loading in the span of the beam. Hence, we need to develop a method to transfer the loading on the beam into equivalent force and moment at the nodes to be included in the slope deflection equation. This is simply done by finding reaction moments that each load develops at the nodes. So, the final moment that helps us transfer the span loading of beams at the supports is the fixed-end moment at each support's points or nodes for the given class of loading on the span.

So, in addition to the above, we have the following two moments that need to be included in the slope deflection equation:

$$M_{AB} = (FEM)_{AB}$$

$$M_{BA} = (FEM)_{BA}$$

Let us find out the fixed-end moments for the beam shown in Figure 18.3 under uniformly distributed load (*udl*) throughout its span. The conjugate beam is shown in Figure 18.3 (b). As both ends are fixed, the slope will be zero for both ends. So, from the conjugate beam, we can write,

$$\uparrow + \sum F_y = 0$$

or,

$$\frac{2}{3} L \times \frac{wL^2}{8EI} - 2\left[\frac{1}{2}\left(\frac{M}{EI}\right)L\right]$$

Therefore,

$$M = \frac{wL^2}{12}$$

This is the amount of fixed end moment if *udl* is imposed throughout the span.

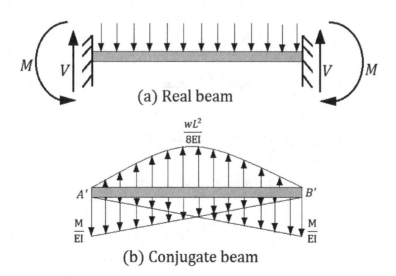

FIGURE 18.4 Fixed-end moment for a uniformly distributed load.

Figures 18.5 and 18.6 show the fixed end moments for various loading and boundary conditions of the beams.

Now we can combine all the terms, i.e., end moments due to each displacement and loading, and can get the following slope deflection equation for the two nodes of a continuous beam:

$$M_{AB} = \frac{2EI}{L}\left[2\theta_A + \theta_B - 3\frac{\Delta}{L}\right] + (FEM)_{AB}$$

$$M_{BA} = \frac{2EI}{L}\left[2\theta_B + \theta_A - 3\frac{\Delta}{L}\right] + (FEM)_{BA}$$

These equations are very neat and very elegant in solving. For each span of a continuous beam, we write these equations, and then for internal nodes, we add the moments at the two sides of the same node, which should add up to zero to maintain the moment equilibrium at those interior nodes. From these, we solve the simultaneous

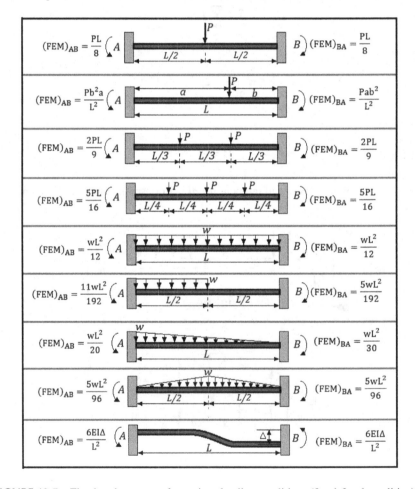

FIGURE 18.5   Fixed-end moments for various loading conditions (fixed-fixed condition).

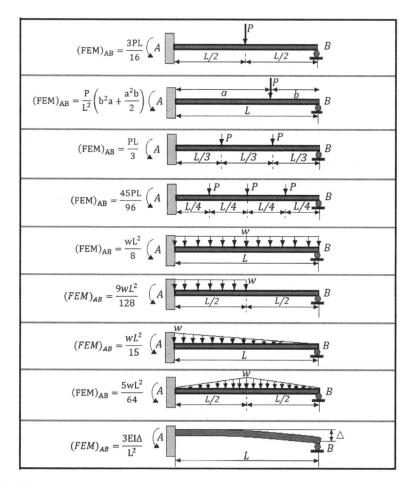

FIGURE 18.6   Fixed-end moments for various loading conditions (fixed-hinged condition).

equations to determine the unknown parameters θ_A and θ_B. These necessary steps will be clear once we will go through Example 18.1.

Example 18.1: Draw the Shear and Bending Moment diagram of the continuous beam shown in Figure 18.7.

SOLUTION: We start the analysis by calculating the fixed-end moments at each node due to each span of the beam.

$$(\text{FEM})_{AB} = -\frac{wl^2}{12} = -96 \text{ kNm}$$

$$(\text{FEM})_{BA} = \frac{wl^2}{12} = 96 \text{ kNm}$$

$$(\text{FEM})_{BC} = -\frac{3PL}{16} = -18 \text{ kNm}$$

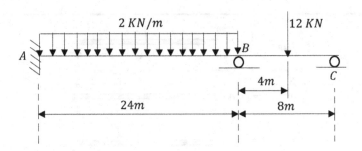

FIGURE 18.7 Example problem.

The last node C of the continuous beam is hinged and $(FEM)_{CB} = 0$. Also, since node A is fixed, $\theta_A = 0$. As well as since the far-end C is pinned, the final moment will also be zero at this location, i.e., $M_{CB} = 0$.

So, with these values in hand, let us write the slope deflection equation as follows:

$$M_{AB} = \frac{2EI}{24}[2 \times 0 + \theta_B - 3 \times 0] - 96$$

or,

$$M_{AB} = 0.0833EI\theta_B - 96$$

$$M_{BA} = \frac{2EI}{L}[2\theta_B + 0 - 3 \times 0] + 96$$

or,

$$M_{BA} = 0.166EI\theta_B + 96$$

Now, for the span BC, we have,

$$M_{BC} = \frac{2EI}{L}\left[2\theta_B + \theta_C - 3\frac{\Delta}{L}\right] + (FEM)_{BC}$$

or,

$$M_{BC} = \frac{2EI}{L}[2\theta_B + \theta_C - 3 \times 0] - 18$$

or,

$$M_{BC} = 0.50EI\theta_B + 0.25EI\theta_C - 18$$

Also, for the far-end C, we have:

$$M_{CB} = \frac{2EI}{L}\left[2\theta_C + \theta_B - 3\frac{\Delta}{L}\right] + (FEM)_{CB}$$

or,

$$0 = \frac{2EI}{L}[2\theta_C + \theta_B - 3 \times 0] + 0$$

From the above equation, we have the following relationship:

$$2\theta_C = -\theta_B$$

or,

$$\theta_C = -\frac{\theta_B}{2}$$

So, substituting these values in other equations for θ_B we get,

$$M_{BC} = 0.5EI\theta_B - 0.25EI \times \frac{\theta_B}{2} - 18$$

or,

$$M_{BC} = 0.375EI\theta_B - 18$$

Now, since there are two moments at two sides of the same node B, these should sum up to zero to maintain the moment equilibrium equation.

Thus, we have, $M_{BA} + M_{BC} = 0$.

or,

$$0.166EI\theta_B + 96 + 0.375EI\theta_B - 18 = 0$$

or,

$$\theta_B = -\frac{144.177}{EI}$$

Substituting these values in the above three equations, we get,

$$M_{AB} = -0.0833EI\frac{144.177}{EI} - 96 = -108 \text{ kN}$$

$$M_{BA} = -0.166EI\frac{144.177}{EI} + 96 = 72.06 \text{ kN}$$

$$M_{BC} = -0.375EI\frac{144.177}{EI} - 18 = -72.06 \text{ kN}$$

Now, to determine the shear force, we need to draw the free-body diagrams as shown in Figure 18.8.

So, from the above-left side free body diagram, we have,

$$V_A \times 24 + 72.06 - 108 - 48 \times 12 = 0$$

or,

$$V_A = 24.41 \text{ kN}$$

$$V_{BA} \times 24 - 72.06 + 108 - 48 \times 12 = 0$$

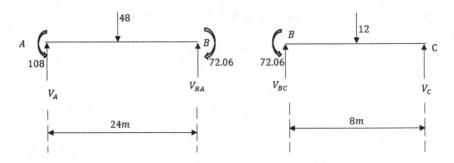

FIGURE 18.8 Free body diagram of the beam span with end moments.

or,

$$V_{BA} = -22.5 \text{ kN}$$

$$V_{BC} \times 8 - 12 \times 4 - 72.06 = 0$$

$$V_{BC} = 15.00 \text{ kN}$$

Also,

$$V_C + 15.00 - 12 = 0$$

or,

$$V_c = -3 \text{ kN}$$

With the above values, the shear and bending moment diagram of the complete beam can be drawn as shown in Figure 18.9.

Bending moment at the mid-span of span AB ($M_{AB/2}$) can be calculated by taking a section at that point and taking a moment of all forces about the same of the left section. So, we have,

$$M_{AB/2} + 12 \times 2 \times 6 - 24.4 \times 12 + 108 = 0$$

or,

$$M_{AB/2} = 40.08 \text{ kNm}$$

Similarly, the moment at mid-span of BC is:

$$M_{BC/2} + 15 \times 4 - 72.06 = 0$$

$$M_{BC/2} = -12.06 \text{ kNm}$$

All above values are placed in the bending and shear force diagrams for ease of understanding.

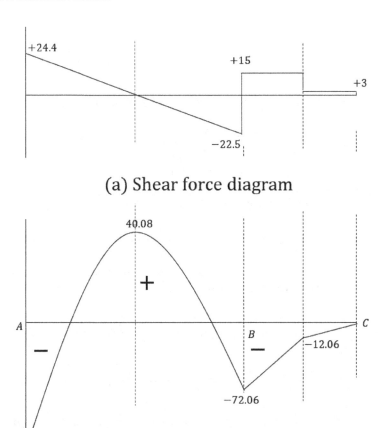

(a) Shear force diagram

(b) Bending moment diagram

FIGURE 18.9   Shear force and bending moment diagram of the example problem.

18.3   MEMBERS WITH FAR END HINGED

The above formulations of slope deflection are derived based on the conditions where the members are considered to be rigidly connected to joints at both sides so that the member end rotations are equal to the rotations of the adjacent joints. But it may so happen that the far end of a member is joined with a hinged connection. In those cases, the end moment will become zero at those far ends, and the slope deflection equations can also be modified accordingly. Let us assume that the end B is hinged on a member AB. Then the slope deflection equations can be written as follows:

$$M_{AB} = \frac{2EI}{L}\left[2\theta_A + \theta_B - 3\frac{\Delta}{L}\right] + (\text{FEM})_{AB}$$

$$M_{BA} = \frac{2EI}{L}\left[2\theta_B + \theta_A - 3\frac{\Delta}{L}\right] + (\text{FEM})_{BA} = 0$$

Solving $M_{BA} = 0$, we get that,

$$\theta_B = -\frac{\theta_A}{2} - \frac{3}{2}\frac{\Delta}{L} - \frac{L}{4EI}(FEM)_{BA}$$

Now, we can eliminate θ_B and rewrite the slope deflection equation as:

$$M_{AB} = \frac{3EI}{L}\left(\theta_A - \frac{\Delta}{L}\right) + \left(FEM_{AB} - \frac{FEM_{BA}}{2}\right)$$

$$M_{BA} = 0$$

Let us go through the following example to understand the concept.

Example 18.2: Figure 18.10 shows a continuous beam *ABCD*. Find the moments at *A*, *B*, *C*, and *D* if the end *A* rotates by 0.004 radians in the clockwise direction and the support *B* sinks by 6 mm. Take $E = 200$ kN/mm^2 and $I = 9 \times 10^7$ mm^4.

SOLUTION: For the data of this problem, $EI = (200 \times 9 \times 10^7)/10^6 = 18000$ kNm^2

Span AB

$$M_{AB} = \frac{2E(2I)}{L}\left(2\theta_A + \theta_B - \frac{3\Delta}{L}\right) = \frac{4EI}{4}\left(2 \times 0.004 + \theta_B - \frac{3 \times 6}{4000}\right) = 63 + 18000\theta_B$$

$$M_{BA} = \frac{2E(2I)}{L}\left(2\theta_B + \theta_A - \frac{3\Delta}{L}\right) = \frac{4EI}{4}\left(2\theta_B + 0.004 - \frac{3 \times 6}{4000}\right) = -9 + 36000\theta_B$$

Here the clockwise movement of the cord is taken as positive.

Span BC

$$M_{BC} = \frac{2E(4I)}{L}\left(2\theta_B + \theta_C + \frac{3\Delta}{L}\right) = \frac{8EI}{8}\left(2\theta_B + \theta_C + \frac{3 \times 6}{8000}\right) = 40.5 + 36,000\theta_B + 18,000\theta_C$$

$$M_{CB} = \frac{2E(4I)}{L}\left(2\theta_C + \theta_B + \frac{3\Delta}{L}\right) = \frac{8EI}{8}\left(2\theta_C + \theta_B + \frac{3 \times 6}{8000}\right) = 40.5 + 18,000\theta_B + 36,000\theta_C$$

Here the anticlockwise movement of the cord is taken as negative.

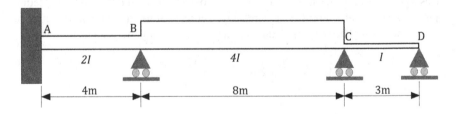

FIGURE 18.10 Problem on a continuous beam with support settlement.

Span CD

$$M_{CD} = \frac{3EI}{L}\left(\theta_C - \frac{\Delta}{L}\right) + \left(FEM_{CD} - \frac{FEM_{DC}}{2}\right)$$

or,

$$M_{CD} = \frac{3EI}{3}\left(\theta_C - 0\right) + \left(0 - \frac{0}{2}\right) = 18,000\theta_C$$

Equilibrium condition at *B*,

$$M_{BA} + M_{BC} = 0$$

or,

$$-9 + 36,000\theta_B + 40.5 + 36,000\theta_B + 18,000\theta_C = 0$$

or,

$$72,000\theta_B + 18,000\theta_C = -31.5 \qquad (18.1)$$

Equilibrium condition at *C*,

$$M_{CB} + M_{CD} = 0$$

or,

$$40.5 + 18,000\theta_B + 36,000\theta_C + 18,000\theta_C = 0$$

or,

$$18,000\theta_B + 54,000\theta_C = -40.5. \qquad (18.2)$$

Solving equations (18.1) and (18.2), we get,

$$\theta_B = -2.73 \times 10^{-4}$$

$$\theta_C = -6.59 \times 10^{-4}$$

Substituting for θ_B and θ_C we get,

$$M_{AB} = 63 + 18,000\theta_B = 63 + 18,000\left(-2.73 \times 10^{-4}\right) = 58.1 \text{ kN}$$

$$M_{BA} = -9 + 36,000\theta_B = -9 + 36,000\left(-2.73 \times 10^{-4}\right) = -18.82 \text{ kN}$$

$$M_{BC} = 40.5 + 36,000\theta_B + 18,000\theta_C$$
$$= 40.5 + 36,000\left(-2.73 \times 10^{-4}\right) + 18,000\left(-6.59 \times 10^{-4}\right) = 18.81 \text{ kN}$$

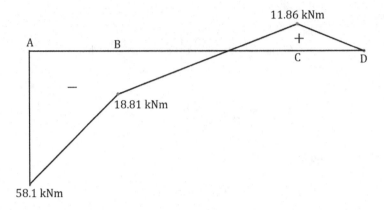

11.86 kNm

+

A   B   C   D

−

18.81 kNm

58.1 kNm

FIGURE 18.11 Bending moment diagram of the example problem.

$$M_{CB} = 40.5 + 18,000\theta_B + 36,000\theta_C$$

$$= 40.5 + 18,000\left(-2.73 \times 10^{-4}\right) + 36,000\left(-6.59 \times 10^{-4}\right) = 11.86 \text{ kN}$$

$$M_{CD} = 18,000\theta_C = 18,000\left(-6.59 \times 10^{-4}\right) = -11.86 \text{ kN}$$

$$M_{DC} = 0$$

The bending moment diagram is shown in Figure 18.11.

18.4 ANALYSIS OF FRAMES WITHOUT ANY SIDESWAY

A frame with the proper restrained condition will not sidesway toward left or right. Moreover, even an improperly restrained frame may not move if it is symmetric in terms of geometry and loading. The number of independent joint translations, i.e., sidesway degrees of freedom for an arbitrary plane frame subjected to general coplanar loading, can be expressed as:

$$ss = 2j - \left[2(f+h) + r + m\right]$$

where j is the number of joints; f is the number of fixed supports; h is the number of hinged supports; r is the number of roller supports; and m is the number of inextensible members. In the coplanar system, the independent joint translation can happen either horizontally or vertically, i.e., in the above expression, $2j$ depicts the total number of translational degrees of freedom available. In the subtracted items, the fixed and hinged support restrain two translations per node. The roller support restrains one translation. Each inextensible member connecting two joints prevents one joint translation in its axial direction.

For analysis of frames without sidesway, we will do that by solving a problem discussed in Example 18.3.

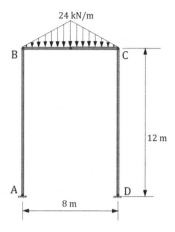

24 kN/m

B

C

12 m

A

D

8 m

FIGURE 18.12 Frame without sway problem.

Example 18.3: Analyze the frame shown in Figure 18.12 and draw the bending moment diagram.

SOLUTION: In this example, the frame consists of four joints ($j = 4$), three members ($m = 3$), two fixed supports ($f = 2$). So, the sidesway degrees of freedom, $ss = 2 \times 4 - [2(2+0)+0+3] = 1$. From this, we can understand that the frame can undergo one independent joint translation. But further, if we observe, we can see this frame is loaded symmetrically with respect to its axis of symmetry. So, this frame is treated without sidesway condition.

In the above frame, all joints are fixed joints. In this situation, we first calculate the fixed-end moments of each beam segment as follows:

$$\left(FEM\right)_{AB} = 0 = \left(FEM\right)_{CD}$$

$$\left(FEM\right)_{BA} = 0 = \left(FEM\right)_{DC}$$

$$\left(FEM\right)_{BC} = -\frac{5wl^2}{96} = -80 \text{ kNm}$$

$$\left(FEM\right)_{CB} = \frac{5wl^2}{96} = 80 \text{ kNm}$$

Note that, θ_A and $\theta_D = 0$, since these two ends are fixed supported. Also, since there will be no sidesway, Δ in general, will be 0.

So,

$$M_{AB} = \frac{2EI}{L}\left[2\theta_A + \theta_B - 3\frac{\Delta}{L}\right] + \left(FEM\right)_{AB}$$

or,

$$M_{AB} = \frac{2EI}{12}[2 \times 0 + \theta_B - 3 \times 0] + 0$$

or,

$$M_{AB} = 0.167EI\theta_B$$

Similarly,

$$M_{BA} = \frac{2EI}{L}\left[2\theta_B + \theta_A - 3\frac{\Delta}{L}\right] + (\text{FEM})_{BA}$$

or,

$$M_{BA} = \frac{2EI}{12}[2\theta_B + 0 - 3 \times 0] + (\text{FEM})_{BA} + 0$$

or,

$$M_{BA} = 0.334EI\theta_B$$

Then,

$$M_{BC} = \frac{2EI}{L}\left[2\theta_B + \theta_C - 3\frac{\Delta}{L}\right] + (\text{FEM})_{BC}$$

or,

$$M_{BC} = 0.5EI\theta_B + 0.25EI\theta_C - 80$$

Similarly,

$$M_{CB} = \frac{2EI}{L}\left[2\theta_C + \theta_B - 3\frac{\Delta}{L}\right] + (\text{FEM})_{CB}$$

or,

$$M_{CB} = 0.5EI\theta_C + 0.25EI\theta_B + 80$$

$$M_{CD} = \frac{2EI}{L}\left[2\theta_C + \theta_D - 3\frac{\Delta}{L}\right] + (\text{FEM})_{CD}$$

or,

$$M_{CD} = \frac{2EI}{8}[2\theta_C + 0 - 3 \times 0] + 0$$

$$M_{CD} = 0.334EI\theta_C$$

$$M_{DC} = \frac{2EI}{L}\left[2\theta_D + \theta_C - 3\frac{\Delta}{L}\right] + (\text{FEM})_{DC}$$

or,

$$M_{DC} = \frac{2EI}{8}\left[2\times0+\theta_C-3\times0\right]+0$$

or,

$$M_{DC} = 0.167EI\theta_C$$

Now, applying the local moment equilibrium equation at joints B and C, respectively, we get,

$$M_{BA} + M_{BC} = 0$$

$$M_{CB} + M_{CD} = 0$$

$$0.834EI\theta_B + 0.25EI\theta_C = 80$$

$$0.834EI\theta_C + 0.25EI\theta_B = -80$$

Solving these two simultaneous equations, we get,

$$\theta_B = \frac{137.27}{EI}$$

$$\theta_C = -\frac{137.27}{EI}$$

which shows that the members BA and CD are deflected by similar angle opposite in direction without any sway. The qualitative deflected shape of the structure without sway is shown in Figure 18.13.

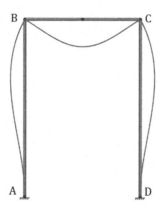

FIGURE 18.13 Example problem of frame qualitative deflected shape without sidesway.

Since the main unknown parameters are calculated, substituting these values in the moment equation, we get,

$$M_{AB} = 0.167EI\theta_B = 22.9 \text{ kNm}$$

$$M_{BA} = 0.334EI\theta_B = 45.76 \text{ kNm}$$

$$M_{BC} = 0.5EI\theta_B + 0.25EI\theta_C - 80 = -45.76 \text{ kNm}$$

$$M_{CB} = 0.5EI\theta_C + 0.25EI\theta_B + 80 = 40.75 \text{ kNm}$$

$$M_{CD} = 0.334EI\theta_C = -22.9 \text{ kNm}$$

Using these end moments, support reaction of the frame can be obtained by using force and moment equilibrium equations alone member-wise. Since this is a simple work, it has been left as an exercise for the readers.

The frame's shear force and bending moment diagram will be something like that shown in Figure 18.14.

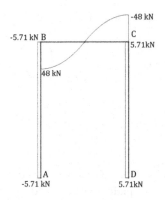

(a) Shear force diagram

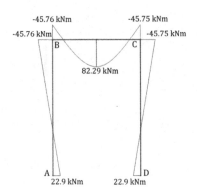

(b) Bending moment diagram

FIGURE 18.14   Shear force and bending moment diagram of example problem frame.

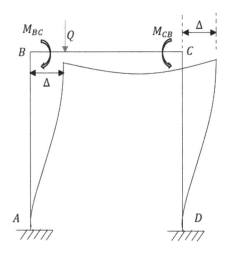

FIGURE 18.15   Frame with sidesway deflected shape.

In the next section, we will discuss the analysis procedure for frames with sidesway.

18.5   ANALYSIS OF FRAMES WITH SIDESWAY

A frame will move or shift sideways when it or the loading acting on it is non-symmetric. To understand this effect, consider the frame shown in Figure 18.15. Here, the loading Q produces unequal moments at the joints B and C. These unequal moments try to move joint B to the right, whereas try to displace joint C to the left. Since M_{BC} is larger than M_{CB}, the net effect is a sidesway or sideway displacement by Δ of both joints B and C to the right, as shown in Figure 18.15. When dealing with a slope-deflection equation to each column of this frame, we must consider the column rotation as unknown in the equation. So, an extra equilibrium equation must be included for the solution.

The techniques to solve this type of problem will be explained with the help of Example 18.4.

Example 18.4: Determine the moments at each joint of the frame shown in Figure 18.16. *EI* is constant.

SOLUTION: Since the frame is nonsymmetrical in nature, applied load at end B will cause the frame to displace by Δ, which remains the same for node C also. Hence, both the members, AB and CD, are rotated with respect to their support point by an angle, say $\alpha_{AB} = \Delta/12$ and $\alpha_{DC} = \Delta/18$, respectively, in the clockwise direction. So, from this relationship, we can deduce that,

$$\alpha_{AB} = \frac{18}{12}\alpha_{DC}$$

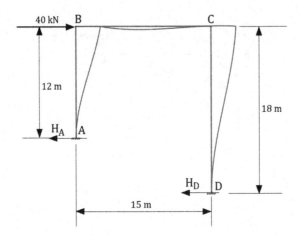

FIGURE 18.16 Frame with sidesway example problem.

So, we need to incorporate it in the slope deflection equation while analyzing the same.

$$M_{AB} = \frac{2EI}{12}\left[2(0) + \theta_B - 3\left(\frac{18}{12}\alpha_{DC}\right)\right] + 0 = EI\left(0.166\theta_B - 0.75\alpha_{DC}\right)$$

$$M_{BA} = \frac{2EI}{12}\left[2\theta_B + 0 - 3\left(\frac{18}{12}\alpha_{DC}\right)\right] + 0 = EI\left(0.333\theta_B - 0.75\alpha_{DC}\right)$$

$$M_{BC} = \frac{2EI}{15}\left[2\theta_B + \theta_C - 3(0)\right] + 0 = EI\left(0.267\theta_B + 0.133\theta_C\right)$$

$$M_{CB} = \frac{2EI}{15}\left[2\theta_C + \theta_B - 3(0)\right] + 0 = EI\left(0.267\theta_C + 0.133\theta_B\right)$$

$$M_{CD} = \frac{2EI}{15}\left[2\theta_C + 0 - 3(\alpha_{DC})\right] + 0 = EI\left(0.222\theta_C - 0.333\alpha_{DC}\right)$$

$$M_{DC} = \frac{2EI}{15}\left[2(0) + \theta_C - 3(\alpha_{DC})\right] + 0 = EI\left(0.111\theta_C - 0.333\alpha_{DC}\right)$$

So, in the above six equations, we have nine unknown forces. Another two equations will be related to moment equilibrium at the joints B and C, namely,

$$M_{AB} + M_{BC} = 0 \tag{18.3}$$

$$M_{CB} + M_{CD} = 0 \tag{18.4}$$

Since horizontal displacement is along the positive x-axis, we can have the following force equilibrium condition for support reactions:

$$40 - H_A - H_D = 0$$

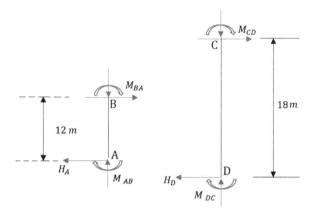

FIGURE 18.17   Column free body diagram for example problem.

The base support reactions or column shears can be related to the internal nodal moments as per Figure 18.17.

So, from the above free body, we get,

$$H_A = -\frac{M_{AB} + M_{BA}}{12}$$

$$H_D = -\frac{M_{DC} + M_{CD}}{18}$$

So, from the horizontal equilibrium equation, we get,

$$40 + \frac{M_{AB} + M_{BA}}{12} + \frac{M_{DC} + M_{CD}}{18} = 0. \tag{18.5}$$

Now, substituting the earlier calculated moment expressions into equations (18.3)–(18.5) and on simplification, we get,

$$0.6\theta_B + 0.133\theta_C - 0.75\alpha_{DC} = 0$$

$$0.133\theta_B + 0.489\theta_C - 0.333\alpha_{DC} = 0$$

$$0.6\theta_B + 0.222\theta_C - 1.944\alpha_{DC} = -\frac{480}{EI}$$

Solving simultaneously, we get from the above equations,

$$EI\theta_B = 438.81$$

$$EI\theta_C = 136.18$$

$$EI\alpha_{DC} = 375.26$$

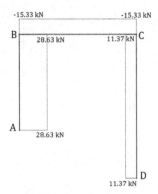

(a) Shear force diagram

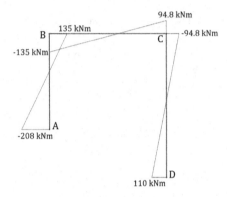

(b) Bending moment diagram

FIGURE 18.18   Shear force and bending moment diagram of nonsymmetrical example frame.

Now, substituting back these above values in the original equations we get,

$$M_{AB} = -208 \text{ kNm}$$

$$M_{BA} = -135 \text{ kNm}$$

$$M_{BC} = 135 \text{ kNm}$$

$$M_{CB} = 94.8 \text{ kNm}$$

$$M_{CD} = -94.8 \text{ kNm}$$

$$M_{DC} = 110 \text{ kNm}$$

Students should pay utmost concentration to the formation of free body diagrams for columns, which in turn provides us additional equations to solve the problem. This procedure of column analysis is very important, and it will be frequently used in subsequent chapters that follow. The shear force and bending moment diagram for this problem is shown in Figure 18.18.

19 Moment Distribution Method

19.1 INTRODUCTION

The moment distribution method is another displacement method of analysis that is a very elegant and convenient approach to apply once few elastic constants have been determined. This is an iterative method in which formulating the equations for unknowns is not even required. In this chapter, we will first state some of the important definitions and points useful for understanding this analysis method. Then we shall apply this method to solve problems related to statically indeterminate beams and frames.

19.2 GENERAL PRINCIPLES AND DEFINITIONS

Prof. Hardy Cross invented the method of analyzing beams and frames using moment distribution in 1930, in an era when computer facilities were not available. That is why this method is also called the Hardy Cross method. By the time this method was published, it had attracted the immediate attention of scholars, engineers, and scientific professionals all over the world. Soon it has been recognized as one of the most notable discoveries in structural analysis during the 20th century.

As will be explained in detail, moment distribution is a method of repetitive approximations that may be carried out to any desired degree of accuracy. The method, at first, assumes that each joint of a structure is fixed. Then, by unfixing and fixing each joint in succession, the internal moments at the joints are 'distributed' and balanced until the joints have rotated to their final or nearly ultimate positions. Readers will find that this method of calculation is both repetitive and simple to apply. Before we jump into the method directly, we will develop some key insights and definitions, which will be found useful toward formulating the complete analysis procedure.

19.2.1 SIGN CONVENTION

We will consider the same sign convention as that has been used for the slope deflection equations: clockwise moments that act on the member will be taken as positive, whereas counterclockwise moments will be taken as negative.

19.2.2 FIXED-END MOMENTS (FEMs)

The moments at the 'walls' or perfectly fixed joints of a loaded member are called fixed-end moments (FEMs). FEMs are already taught in the previous chapter, and we need to remember these formulae for calculating numerical values of the moment

DOI: 10.1201/9781003081227-22

based on the type of load applied to the structure. As for example, if a point load, P, acts on the midspan of a beam of length l, the FEM will be $\pm Pl/8$. In this way, if the beam is subjected to a uniformly distributed load of intensity w, FEMs will be $\pm wl^2/12$. Different FEM due to the general class of loading is provided in Chapter 18, in a tabular form.

19.2.3 MEMBER STIFFNESS FACTORS

Consider a beam with one end pinned and another end fixed, as shown in Figure 19.1. In this situation, a moment has been applied at the pinned end M_{AB}. Due to this moment, the pinned end will be rotated by an angle θ_A as shown in Figure 19.1 (a).

Using slope deflection equation derived in the previous chapter, the moment can be related to the rotation of the cord of the beam by the following equation: here, $\theta_B = \psi = \text{FEM}_{AB} = 0$.

$$M_{AB} = \frac{2EI}{l}(2\theta_A)$$

i.e.,

$$M_{AB} = \frac{4EI}{l}\theta_A$$

The bending stiffness coefficient is defined as the moment that must be applied at the end of a member to cause a unit angular displacement of that end, i.e.,

Setting, $\theta_A = 1$ radian we get,

$$K = \frac{4EI}{l}$$

(a) Beam with far end fixed

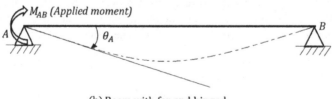

(b) Beam with far end hinged

FIGURE 19.1   Beam with end moments for stiffness coefficient calculation.

However, we will come up with situations where the far-end support may be pinned or rocker supports in many cases. In that case, the stiffness factor will be changed, and the fixed-beam stiffness factor formula cannot be used. Now, if the beam's far end is also hinged as shown in Figure 19.1 (b), we can write,

$$M_{AB} = \frac{3EI}{l}\theta_A$$

In this case, the bending stiffness will be

$$K = \frac{3EI}{l}$$

Here we can observe that the bending stiffness of the beam is reduced by 25% when the fixed support at B is replaced by hinged support.

19.2.4 JOINT STIFFNESS FACTOR

If several members are connected by a fixed joint at some location, following the principle of superposition, the total stiffness factor at the joint will be the algebraic sum of each and every member's stiffness factors. This total sum of stiffness factors indicates the moment required at a joint to rotate the joint by a unit angular displacement. This is further explained in the following.

Suppose members AB, CB, DB, and EB are rigidly connected to joint B, as shown in Figure 19.2. Now, a moment M is applied at joint B, causing it to rotate by an angle θ. This applied moment, M, will now be resisted by all the four members connected at joint B. If we consider the moment equilibrium of joint B as shown in Figure 19.3, we can write,

$$M + M_{BA} + M_{BC} + M_{BD} + M_{BE} = 0$$

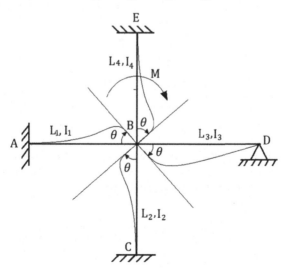

FIGURE 19.2 Joint stiffness and distribution factor.

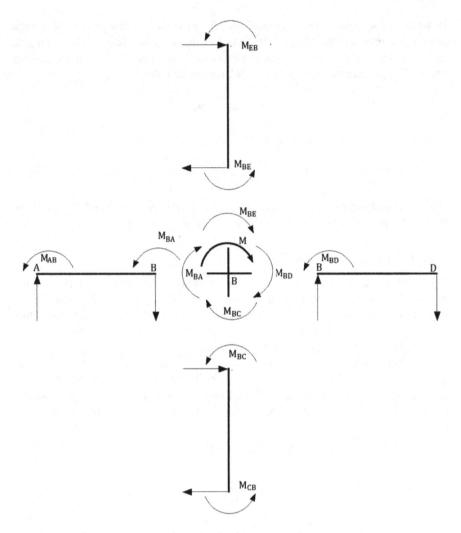

FIGURE 19.3   Free-body diagram of joint B and four members.

or,

$$M = -(M_{BA} + M_{BC} + M_{BD} + M_{BE})$$

As all the members are rigidly connected at joint B, the rotations of the ends B of these members are the same as that of the joint. Now, the moments at B of the members can be expressed in terms of joint rotation θ.

$$M_{BA} = \left(\frac{4EI_1}{l_1}\right)\theta = K_{BA}\theta$$

$$M_{BC} = \left(\frac{4EI_2}{l_2}\right)\theta = K_{BC}\theta$$

$$M_{BD} = \left(\frac{3EI_3}{l_3}\right)\theta = K_{BD}\theta$$

$$M_{BE} = \left(\frac{4EI_4}{l_4}\right)\theta = K_{BE}\theta$$

Now we can rewrite the earlier expression as,

$$M = -\left(\frac{4EI_1}{l_1} + \frac{4EI_2}{l_2} + \frac{3EI_3}{l_3} + \frac{4EI_4}{l_4}\right)\theta$$

or,

$$M = (K_{BA} + K_{BC} + K_{BD} + K_{BE})\theta$$

or,

$$M = -\left(\sum K_B\right)\theta$$

Here $\sum K_B$ is the joint stiffness factor or sum of the bending stiffness of all members connected to joint B.

19.2.5 DISTRIBUTION FACTOR (DF)

If a moment M is applied to a fixed connected joint, the connecting members will each supply a portion of the resisting moment required to maintain the moment equilibrium at that joint. That portion of the total resisting moment supplied by each member is known as the distribution factor (DF). To calculate its value, let us consider the joint shown in Figure 19.2. The rotational stiffness of a joint is defined as the moment required to cause a unit rotation of the joint.

or,

$$\frac{M}{\theta} = -\left(\sum K_B\right)$$

So,

$$\theta = -\frac{M}{\sum K_B}$$

After replacing the value θ in the earlier expressions of the member end moments, we can write,

$$M_{BA} = -\left(\frac{K_{BA}}{\sum K_B}\right)M$$

$$M_{BC} = -\left(\frac{K_{BC}}{\Sigma K_B}\right)M$$

$$M_{BD} = -\left(\frac{K_{BD}}{\Sigma K_B}\right)M$$

$$M_{BA} = -\left(\frac{K_{BE}}{\Sigma K_B}\right)M$$

From this, we can understand that the applied moment M is distributed to the four members in proportion to their bending stiffness. The ratio $\left(K/\Sigma K_B\right)$ is called the DF for that member at end B. It represents the fraction of applied moment M that is distributed to end B of a member.

If an applied moment M causes the joint to rotate an amount θ, each member j rotates by this same amount. If the stiffness factor of jth member is K_j, the moment contributed by that member to maintain equilibrium at the joint will be $M_j = K_j\theta$. Since the total stiffness factor at the joint is given by $K = \Sigma_{j=1}^{n} K_j$, hence,

$$\text{DF}_i = \frac{M_i}{M} = \frac{\theta K_j}{\theta \Sigma_{j=1}^{n} K_j} = \frac{K_j}{\Sigma K_j}$$

So, by taking all the individual member stiffness coefficients, we can use the above formula to calculate the DF for each member by simply substituting the values in the above equation.

19.2.6 MEMBER RELATIVE STIFFNESS FACTOR

For continuous beams, we frequently found that elastic constant E remains the same throughout the entire beam. In this case, we can remove the common factor $4E$, as in the DF equation, this common term gets canceled from numerator and denominator. Hence, it is convenient just to determine the member's relative stiffness factor as:

$$K_j = \frac{I_j}{L_j}$$

when the far end of the beam is fixed and used for computation of DFs. If the far end of a member is hinged, the relative stiffness factor will be,

$$K_j = \frac{3}{4}\frac{I_j}{L_j}$$

19.2.7 CARRY OVER FACTOR

When we apply a moment M_{AB} at end A, as shown in Figure 19.1 (a), a moment M_{BA} is developed at the fixed-end B. This developed moment at the fixed-end B due to the applied moment at end A is called the carry-over moment.

We know that $M_{AB} = (4EI/l)\theta_A$ and $M_{BA} = (2EI/l)\theta_A$. So, equating these two relationships, we get:

$$M_{BA} = \frac{1}{2}M_{AB}$$

or,

$$M' = \frac{1}{2}M$$

This indicates that moment M at pin end has been transferred to the wall or fixed support as $M/2$. Hence, in the case of a beam with a far-end fixed, the carry-over factor is +1/2. The plus sign indicates that both moments are acting in the same direction (clockwise).

19.3 BASIC CONCEPT OF MOMENT DISTRIBUTION METHOD

To understand the basic concept of moment distribution, it is better to do it with the help of an example. Refer to Figure 19.4 that indicates the typical arrangement of a beam with loading conditions. We will apply the moment distribution method to analyze the beam in a stepwise detailed manner, so that student gains a good basic understanding of the approach.

The stiffness factor on either side of the node B is:

$$K_{BA} = \frac{4E(300)}{15} = 4E(20)\ \text{mm}^4/\text{m}$$

$$K_{BC} = \frac{4E(600)}{20} = 4E(30)\ \text{mm}^4/\text{m}$$

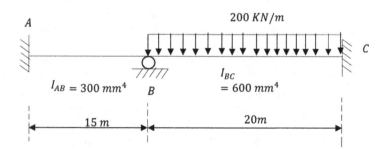

FIGURE 19.4 Example problem of moment distribution.

So, the DFs will be:

$$DF_{BA} = \frac{4E(20)}{4E(20)+4E(30)} = 0.4$$

$$DF_{BC} = \frac{4E(30)}{4E(20)+4E(30)} = 0.6$$

At the wall or fixed support end, the stiffness is infinite. Hence, the DFs in this respect will be:

$$DF_{AB} = \frac{4E(20)}{\infty+4E(30)} = 0$$

$$DF_{CB} = \frac{4E(30)}{\infty+4E(30)} = 0$$

Once the DFs are calculated, we will now calculate the FEMs, which is done as follows:

$$(FEM)_{BC} = -\frac{\omega l^2}{12} = -6666.67 \text{ kNm}$$

$$(FEM)_{CB} = \frac{\omega l^2}{12} = 6666.67 \text{ kNm}$$

Now, to maintain equilibrium at joint B, we need to apply an equal but opposite moment at the same joint B. Hence, the applied moment will be = +6666.67 kNm. Now, this additional moment applied at B will be distributed in the span BA and BC as per the DFs. So, BA will carry = +6666.67 kNm × 0.4 = +2666.67 kNm and BC will carry +6666.67 kNm × 0.6 = +4000.00 kNm. These two moments are located at joint B. Hence, as per the carry-over factor, the half of these moments will be carried over to the farthest end of the beams. So, carried over moment at joint A will be = 1/2 × 2666.67 = 1333.34 kNm and at C will be = 1/2 × 4000 = 2000.00 kNm.

All the above methods are carried over by a tabular form, which helps us to keep track of everything in a sequential manner. Table 19.1 is shown below for a ready reference.

So, as we can see from Table 19.1, at joint B, the moment equilibrium is ensured (since +2666.67 − 2666.67 = 0).

As long as we get this joint equilibrium ensured, we keep on this process, and the iteration continues till, at each joint where a number of members meet, moment equilibrium is satisfied. This tabular form is the elegance of this method that we have told at the beginning of this chapter. The final bending moment diagram is left as an exercise for the reader.

TABLE 19.1
Moment Distribution Process for this Problem

Joint	Fixed A	B		Fixed C	
Member	AB	BA	BC		CB
DF	0	0.4	0.6		0
FEM			−6666.67		6666.67
Balance B		+2666.67	+4000.0		
Carry over	$1333.34 \leftarrow CO = +\frac{1}{2}$			$CO = +\frac{1}{2} \rightarrow$	2000.0
Final moments	*+1333.34*	*+2666.67*	*−2666.67*		*+8666.67*

19.4   STIFFNESS FACTOR MODIFICATIONS

We have already seen that for a beam with the far end fixed, the stiffness factor of the beam is $4EI/l$ and in section 19.2.3, we have seen if a beam's far end is hinged, the stiffness factor gets reduced by 25% and becomes $3EI/l$. In this section, again, we would like to develop the stiffness factor when the far end is hinged along with few special cases by the conjugate beam method.

19.4.1   MEMBER PIN SUPPORTED AT FAR END

Let us consider the following situation in which the rear end of the beam is pin supported, as shown in Figure 19.5.

Now taking a moment of all forces about support B, we get from conjugate beam:

$$V_A \times l - \frac{1}{2}\left(\frac{M}{EI}\right)\left(\frac{2}{3}\right)l = 0$$

or,

$$V_A = \theta = \frac{Ml}{3EI}$$

or,

$$M = \frac{3EI}{l}\theta$$

Thus, for a beam with far end pinned or rocker support, the stiffness factor will be:

$$K = \frac{3EI}{l}$$

Also, the carry-over factor will be zero (0) since the pin end does not support any moment. Hence, the moment will be zero at that support point.

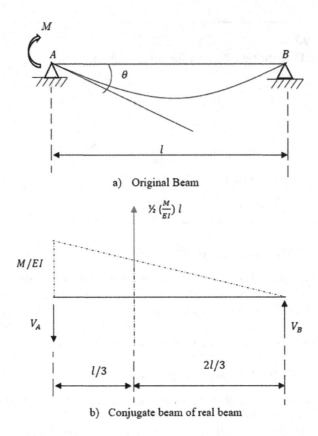

FIGURE 19.5 Pin or rocker supported far end and stiffness coefficient: (a) original beam and (b) conjugate beam of real beam.

19.4.2 SYMMETRIC BEAM AND LOADING

If the beam is symmetric with respect to both loading and geometry, the bending moment diagram of the beam will also be symmetric. As a result, modification of stiffness factor can be modified based on the center span bending moment, as will be clear from Figure 19.6.

Due to symmetry, the internal moment for joints B and C will be the same. Let us consider this moment to be M, as shown in Figure 19.6 (b) for span BC conjugate beam. We can write the following equation by taking a moment about joint C of the conjugate beam forces:

$$-V_B'l + \frac{M}{EI}l\left(\frac{l}{2}\right) = 0$$

So,

$$V_B' = \theta = \frac{Ml}{2EI}$$

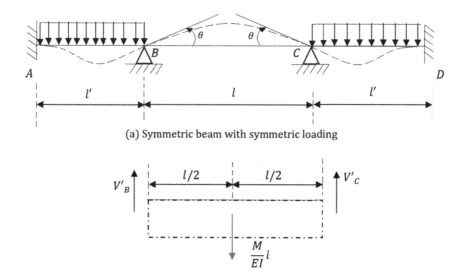

(a) Symmetric beam with symmetric loading

(b) Conjugate beam for span BC

FIGURE 19.6   Symmetric beam in geometry and loading.

or,

$$M = \frac{2EI}{l}\theta$$

So, for symmetric loading, the stiffness factor will be:

$$K = \frac{2EI}{l}$$

Thus, moments for the half span of the beam can be distributed by using the above stiffness of the factor measured at the center of the span of the beam.

19.4.3   SYMMETRIC BEAM WITH UNSYMMETRIC LOADING

In the case of a symmetric beam with antisymmetric loading, we can, like the previous case, analyze the beam concerning the center span of the beam, provided the stiffness factor takes a different form than the usual one.

For the previous case, if the beam is loaded only for the portion AB and there is no load in the BC span, the conjugate beam for the span AB of the beam will be as shown in Figure 19.7.

As carried out in the earlier section, taking a moment of all forces about joint C of the conjugate beam, as shown in Figure 19.7, we get:

$$-V_B' l + \frac{1}{2}\left(\frac{M}{EI}\right)\left(\frac{l}{2}\right)\left(\frac{5l}{6}\right) - \frac{1}{2}\left(\frac{M}{EI}\right)\left(\frac{l}{2}\right)\left(\frac{l}{6}\right) = 0$$

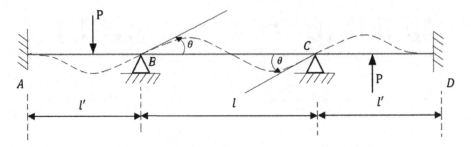

(a) Symmetric beam with antisymmetric loading

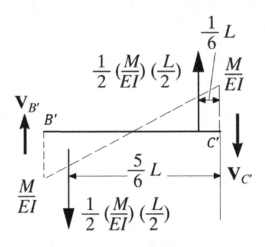

(b) Conjugate beam for span BC

FIGURE 19.7   Symmetric beam in geometry and antisymmetric loading.

or,

$$V_B' = \theta = \frac{Ml}{6EI}$$

or,

$$M = \frac{6EI}{l}\theta$$

So, the stiffness at the center of the span will be:

$$K = \frac{6EI}{l}$$

19.5   ANALYSIS OF CONTINUOUS BEAMS

Primary discussion on a continuous beam has been provided in Section 19.3. In this section, we will make a detailed analysis and draw the shear force and bending moment diagrams of a continuous beam as shown in Figure 19.8, following all the previous concepts given in earlier sections.

At the very first, the DFs at each joint first need to be calculated:

$$K_{AB} = \frac{4EI}{12}, \; K_{BC} = \frac{4EI}{12}, \; K_{CD} = \frac{4EI}{8}, \; K_{AB} = K_{DC} = \infty$$

So,

$$DF_{AB} = DF_{DC} = 0$$

$$DF_{BA} = DF_{BC} = \frac{(4EI/12)}{4EI/12 + 4EI/12} = 0.5, \; DF_{CB} = \frac{4EI/12}{4EI/12 + 4EI/8} = 0.4,$$

$$DF_{CD} = \frac{4EI/8}{4EI/12 + 4EI/8} = 0.6$$

Next, we need to find out the FEMs of each span by considering imaginary clamps at both ends of each span.

$$FEM_{BC} = -\frac{wl^2}{12} = -180 \text{ kNm}, \; FEM_{CB} = 180 \text{ kNm},$$

$$FEM_{CD} = -\frac{Pl}{8} = -200 \text{ kNm}, \; FEM_{DC} = \frac{Pl}{8} = 200 \text{ kNm}$$

Now, equipped with the above inputs, let us form the moment distribution table to complete the analysis as done before.

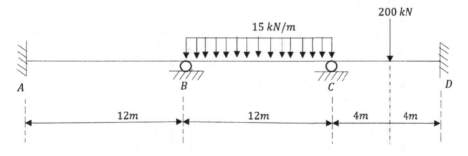

FIGURE 19.8   A continuous beam arrangement and loading details.

Joint	A	B		C		D
Member	AB	BA	BC	CB	CD	DC
DF	0	0.5	0.5	0.4	0.6	0
FEM	0	0	−180	180	−200	200
Balancing Moment			+180	$(200 - 180) = +20$		
Dist. Mom.	0	$180 \times 0.5 = 90$	$180 \times 0.5 = 90$	$20 \times 0.4 = 8$	$20 \times 0.6 = 12$	0
Carry over Mom.	$90/2 = 45$	0	$8/2 = 4$	$90/2 = 45$	0	$12/2 = 6$
Dist. Mom.	0	$-4 \times 0.5 = -2$	$-4 \times 0.5 = -2$	$-45 \times 0.4 = -18$	$-45 \times 0.6 = -27$	0
Carry over Mom.	$-2/2 = -1$	0	$-18/2 = -9$	$-2/2 = -1$	0	$-27/2 = -13.5$
Dist. Mom.	0	$9 \times 0.5 = 4.5$	$9 \times 0.5 = 4.5$	$1 \times 0.4 = 0.4$	$1 \times 0.6 = 0.6$	0
Carry over Mom.	$4.5/2 = 2.25$	0	$0.4/2 = 0.2$	$4.5/2 = 2.25$	0	$0.6/2 = 0.3$
Dist. Mom.	0	$-0.2 \times 0.5 = -0.1$	$-0.2 \times 0.5 = -0.1$	$-2.25 \times 0.4 = -0.9$	$-2.25 \times 0.6 = -1.35$	0
ΣM (FEM + Dist. Mom + carry over Mom.)	46.25	92.40	−92.40	215.75	−215.75	192.8

So, the joint equilibrium of moments is ensured, as seen from the final summation row of the above table. We have provided the calculation stepwise and detailedly so that readers gain sufficient confidence in solving other problems after studying this. Detailed understanding of various steps involved is a must for all students who want to gain complete control of this beautiful and elegant analysis process. The overall steps involved in the moment distribution method is summarized as below:

1. Calculate the DFs for members rigidly connected to the joints. The sum of all the DFs at a joint must be equal to one.
2. Assuming all the free joints are clamped against rotation, calculate the FEMs due to the external loads or support settlements (if any) for each span. The clockwise moment is considered positive.
3. Calculate and distribute the unbalanced moment to the members connected to a particular joint as per their DFs.
4. Carry over one-half of each distributed moment to the far end of the member.
5. Repeat steps (3) and (4) until either all free joints are balanced or the unbalanced moments at these joints become negligibly small.

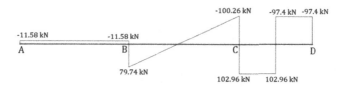

(a) Shear force diagram

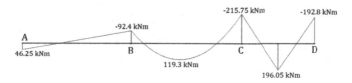

(b) Bending moment diagram

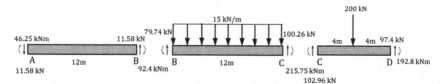

(c) Equilibrium of each span

FIGURE 19.9 Example problem of (a) shear force, (b) bending moment, and (c) free-body diagram of various beam spans.

6. Find out the final moments by algebraically summing up the fixed end, distributed, and carry-over moments at each member end.
7. By considering the equilibrium of each span and joint, find out the end shears and support reactions, respectively.
8. Draw the shear force and bending moment diagram.

The beam's shear force and bending moment diagram is given in Figure 19.9 (a) and (b), respectively. Upon drawing the free-body diagram of respective beam elements as shown in Figure 19.9 (c), one can determine the shear force and support reactions by applying equilibrium equations. The calculation part has been left as an exercise for the students.

19.6 ANALYSIS OF FRAMES WITHOUT SIDESWAY

For frames without any sidesway, the moment distribution method can be conveniently applied as has been done for continuous beams. Following the same tabular form, we will analyze a frame to determine the unknown moments at its joints. Moreover, in special cases, the steps involved can be reduced by adopting modified stiffness methods as and when required.

Example 19.1: Determine the internal moments at the joints of the below frame. All the supports are fixed – except *E, D* which are pinned, and El is cinstant for the entire frame.

SOLUTION:

Joint	A	B		C			D	E
Member	AB	BA	BC	CB	CD	CE	DC	EC
DF	0	0.545	0.455	0.330	0.298	0.372	1	1
FEM	0	0	–216	+216	0	0	0	0
Balancing Mom.			216	–216				
Dist. Mom	0	117.72	96.12	–71.28	–64.368	–80.352	0	0
Carry over Mom.	117.72/2 = 58.86	0	–71.28/2 = –35.64	96.12/2 = 48.06	0	0	0	0
Dist. Mom		19.42	16.22	–15.86	–14.32	–17.88	0	0
Carry over Mom.	9.71	0	–15.86/2 = –7.93	16.22/2 = 8.11	0	0	0	0
Dist. Mom	0	4.321	3.529	–2.676	–2.417	–3.017	0	0
Carry over Mom.	2.16	0	–1.338	1.765	0	0	0	0
Dist. Mom	0	0.729	0.595	–0.582	–0.526	–0.6576	0	0
Carry over Mom.	0.365	0	–0.291	0.298	0	0	0	0
Dist. Mom	71.095	142.19	142.19	183.835	–81.631	102.91	0	0

Since *E* and *D* are pinned, the stiffness factors for *CD* and *CE* can be computed by $K = 3EI/l$, i.e., $K_{CD} = 3EI/15$ and $K_{CE} = 3EI/12$.

$$\text{So, DF}_{AB} = 0,$$

$$DF_{BA} = \frac{4EI/15}{4EI/15 + 4EI/18} = 0.545$$

$$DF_{BC} = 1 - 0.545 = 0.455$$

$$DF_{CB} = \frac{4EI/18}{4EI/18 + 3EI/15 + 3EI/12} = 0.330$$

$$DF_{CD} = \frac{3EI/15}{4EI/18 + 3EI/15 + 3EI/12} = 0.298$$

$$DF_{CE} = 1 - 0.330 - 0.298 = 0.372$$

$$DF_{DC} = 1$$

$$DF_{EC} = 1$$

We used these values in the tabular form on the previous page to analyze the frame using the moment distribution method. The shear force and bending moment diagram of this frame is shown in Figure 19.11 (a) and (b), respectively.

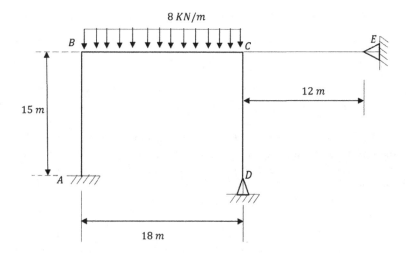

FIGURE 19.10 Example problem of frame without sidesway.

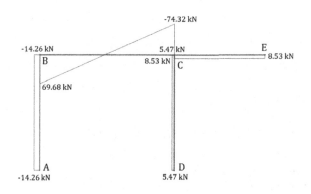

(a) Shear force diagram

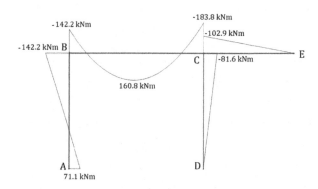

(b) Bending moment diagram

FIGURE 19.11 (a) Shear force and (b) bending moment diagram of the frame without sidesway.

19.7   ANALYSIS OF FRAMES WITH SIDESWAY

In Chapter 18, on the slope deflection method, we have explained that frames that are unsymmetrical in geometry or applied load or both are subject to sway. A situation for a frame with unsymmetrical loading acting on it is given in Figure 19.12. In this frame, the applied force tends to sway the frame to the right by a deflection Δ. Due to this deflection, there will be unequal moments distributed in the frames. We can determine the final moments in the frame due to sidesway by the application of the principle of superposition. First, we assume that there is a virtual support at C, which makes the frame nonsway type. Then we calculate the support reaction R at this support C by completing the moment distribution of this nonsway frame. After that, we apply the same but opposite restraining force R in the frame and calculate the internal moments once again. One method for doing this last step is to assume a value for any one of the internal moments, say, M'_{BA}. Using moment distribution and statics, we can then determine the deflection Δ' and R' against the assumed value of M'_{BA}. Since linear deformation occurs within the elastic limit, actual force R will be proportional to this. So that we can write,

$$\frac{M_{BA}}{R} = \frac{M'_{BA}}{R'}$$

From which we get,

$$\therefore M_{BA} = M'_{BA}\left(R/R'\right)$$

To understand the abovementioned process in detail, we will go through a problem discussed in Example 19.2, which will clear all our doubts and enable us to gain confidence in the analysis procedure.

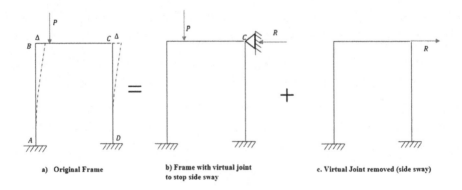

a) Original Frame   b) Frame with virtual joint   c. Virtual Joint removed (side sway)
                            to stop side sway

FIGURE 19.12   Frame with sidesway – preliminary analysis method.

Example 19.2: Determine the moments at each joint of the frame as shown in Figure 19.13.

SOLUTION: We first make the frame prevent from sidesway as was done in Figure 19.13 (b), i.e., with virtual support at C.

Now, with the virtual supports, we first calculate the FEMs against the applied loading:

$$FEM_{BC} = -20\frac{4^2(1)}{5^2} = -12.8 \text{ kNm}$$

$$FEM_{CB} = 20\frac{1^2(4)}{5^2} = 3.2 \text{ kNm}$$

Then, the DFs are calculated:

$$DF_{AB} = 0, DF_{BC} = \frac{4EI/5}{4EI/5 + 4EI/5} = 0.5, DF_{BA} = 1 - DF_{BC} = 0.5$$

$$DF_{CB} = \frac{4EI/5}{4EI/5 + 4EI/5} = 0.5, DF_{CD} = \frac{4EI/5}{4EI/5 + 4EI/5} = 0.5, DF_{DC} = 0$$

Now, we are in a position to solve the first stage of this problem by usual moment distribution method as given in the tabular form:

Joint	A	B		C		D
Member	AB	BA	BC	CB	CD	DC
DF	0	0.5	0.5	0.5	0.5	0
FEM	0	0	−12.8	3.2	0	0
Balancing Mom.		+12.8		−3.2		
Dist. Mom.		6.4	6.4	−1.6	−1.6	
Carry over Mom.	3.2	0	−0.8	3.2	0	−0.8
Dist. Mom.	0	0.40	0.40	−1.6	−1.6	0
Carry over Mom.	0.20	0	−0.80	0.20	0	−0.80
Dist. Mom.	0	0.40	0.40	−0.10	−0.10	0
Carry over Mom.	0.20	0	−0.05	0.20	0	−0.05
Dist. Mom.	0	0.025	0.025	−0.1	−0.1	0
ΣM (FEM + Dist. Mom. + carry over Mom.)	3.6	7.225	−7.225	3.4	−3.4	−1.65

Once done with this, we need to draw free-body diagram of the columns to determine end shear force acting in the column (Figure 19.14).

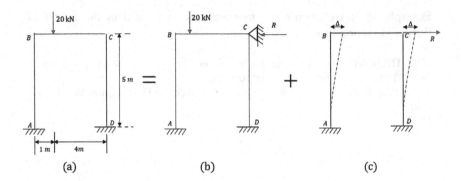

FIGURE 19.13 Example problem with sidesway and virtual support.

Taking moment about joint B in Column BA, we get:

$$R_A(5) - 3.6 - 7.225 = 0$$

or,

$$R_A = 2.165 \text{ kN}$$

Similarly, by taking moment about joint C of column CD, we get:

$$R_D(5) - 3.4 - 1.65 = 0$$

or,

$$R_D = 1.01 \text{ kN}$$

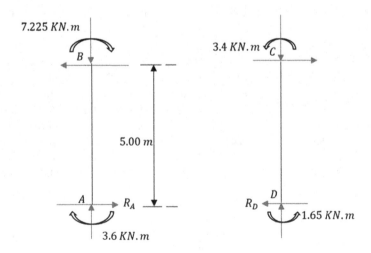

FIGURE 19.14 Column free-body diagram for force analysis.

So, from the overall frame, applying global equilibrium force equation in horizontal direction, we get:

$$R - 2.165 + 1.01 = 0$$

or,

$$R = 1.155 \text{ kN}$$

Now, we must apply an equal but opposite force of R on the frame at point C and calculate the internal moments accordingly. This situation is shown in Figure 19.13 (c) with the deflection diagram for better understanding of the reader.

In the above case, the joints B and C are assumed to be restrained against rotation. Under this situation, the deflection Δ induces moments in the columns, which can be determined by using the concept of slope deflection equations. Consider a member AB fixed at both ends and support B settles by an amount Δ. Also let us consider that there is no load acting on the beam. So, applying slope deflection equation we get:

$$M_{AB} = \frac{2EI}{L}\left[2\theta_A + \theta_B - 3\frac{\Delta}{L}\right] + \left(\text{FEM}\right)_{AB}$$

Here, $\theta_A = \theta_B = 0$, $\left(\text{FEM}\right)_{AB} = 0$, and the cord members AB and DC rotate clockwise ($\psi = \Delta/L = +$ve). So, from the above equation we get:

$$M_{AB} = -\frac{6EI\Delta}{L^2}$$

Hence, for our columns DC and AB, we have induced moment due to the deflection:

$$-\frac{6EI\Delta}{5^2}$$

Also, since both ends, B and C of the frame in Figure 19.15, are happened to be displaced by the same amount Δ', and AB, DC have the same E, I, and length, FEM in AB and DC will be the same. So, let us arbitrarily choose values of FEMs as follows:

$$\text{FEM}_{AB} = \text{FEM}_{BA} = \text{FEM}_{CD} = \text{FEM}_{DC} = -100 \text{ kNm}$$

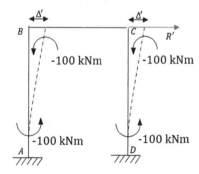

FIGURE 19.15 Arbitrarily assumed fixed-end moments.

Negative sign is due to anticlockwise FEM induced. With these values in hand, let us carry out the moment distribution once again for the entire frame in the tabular form as given below:

Joint	A	B		C		D
Member	AB	BA	BC	CB	CD	DC
DF	0	0.5	0.5	0.5	0.5	0
FEM	−100	−100	0	0	−100	−100
Balancing Mom.		100		100		
Dist. Mom	0	50	50	50	50	0
Carry over Mom.	25	0	25	25	0	25
Dist. Mom	0	−12.5	−12.5	−12.5	−12.5	0
Carry over Mom.	−6.25	0	−6.25	−6.25	0	−6.25
Dist. Mom	0	3.125	3.125	3.125	3.125	0
Carry over Mom.	1.56	0	1.56	1.56	0	1.56
Dist. Mom	0	−0.78	−0.78	−0.78	−0.78	0
Carry over Mom.	−0.39	0	−0.39	−0.39	0	−0.39
Dist. Mom	0	0.195	0.195	0.195	0.195	0
ΣM (FEM + Dist. Mom + carry over Mom.)	−80.08	−59.96	59.96	59.96	−59.96	−80.08

Once the table is ready, we can draw the same column free-body diagrams as done in Figure 19.14 and analyze the case for column horizontal support reactions. In this case, the support reactions in horizontal direction will be:

$$R_A = 28.01 \text{ kN}$$

$$R_D = 28.01 \text{ kN}$$

Thus, for the entire frame, applying the global equilibrium of force condition in horizontal direction, we get:

$$R' - 28.01 - 28.01 = 0$$

or,

$$R' = 56.02 \text{ kN}$$

So, the end moments at the joint of this frame will be:

$$M_{AB} = 3.6 + \left(\frac{1.155}{56.02} \right)(-80.08) = 1.95 \text{ kNm}$$

$$M_{BA} = 7.225 + \left(\frac{1.155}{56.02} \right)(-59.96) = 5.99 \text{ kNm}$$

$$M_{BC} = -7.225 + \left(\frac{1.155}{56.02} \right)(59.96) = -5.99 \text{ kNm}$$

$$M_{CB} = 3.4 + \left(\frac{1.155}{56.02}\right)(59.96) = 4.64 \text{ kNm}$$

$$M_{CD} = -3.4 + \left(\frac{1.155}{56.02}\right)(-59.96) = -4.64 \text{ kNm}$$

$$M_{DC} = -1.65 + \left(\frac{1.155}{56.02}\right)(-80.08) = -3.30 \text{ kNm}$$

Now the analysis is complete. The final shear force and bending moment diagram of this frame is shown in Figure 19.16.

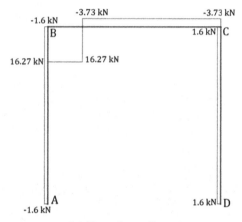

(a) Shear force diagram

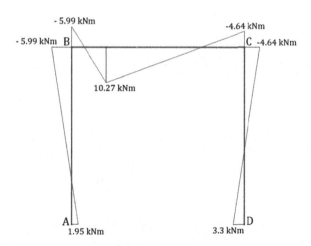

(b) Bending moment diagram

FIGURE 19.16   (a) Shear force and (b) bending moment diagram of the given frame.

19.7.1 MULTISTOREY FRAMES

For multistorey frames, we analyze them in the same manner as that of a single-storied frame explained in detail above. To elaborate the process, let us consider a multistorey frame as shown in Figure 19.17. Under action of an unsymmetrical loading, the frame will sway as shown in the same diagram. To do the analysis, we first apply two supports at two suitable nodes of the frame to make it into nonsway-type frame as shown in Figure 19.17 (b). After that, we carry out the moment distribution of the frame with these virtual supports and calculate the virtual support reactions R_1 and R_2 from the global equilibrium equations.

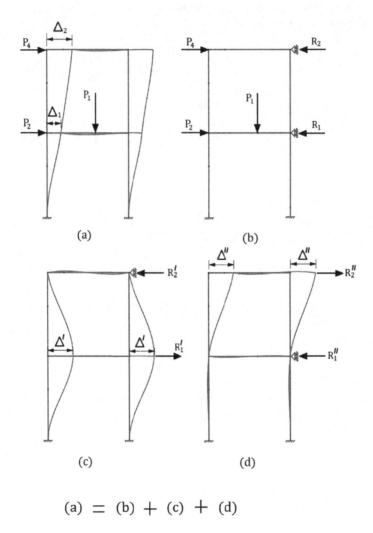

$$(a) = (b) + (c) + (d)$$

FIGURE 19.17 Multistorey frame with sway: (a) original frame, (b) frame with virtual supports to make nonsway-type (c), and (d) multistorey frame with support reactions for second stage analysis.

Once the first part of analysis is complete, we apply the horizontal virtual displacements one by one in two nodes keeping the other node as restrained and carry out the moment distributions separately to calculate the virtual support reactions R_1', R_2', and R_1'', R_2'', respectively. These displacements cause FEMs in the frame and which can be assigned specific numerical values in the frame. By distributing these FEMs and with the help of equations of equilibrium, we can find out R_1', R_2', and R_1'', R_2''. See Figure 19.17 (c) and (d) for understanding the process. Thus, at this stage, we have completed the second-stage analysis process of the multistorey frame.

Since in the last two steps as per Figure 19.17 (c) and (d), we have to assume internal FEM values twice; hence, there must be some correction factors that need to be considered while forming the equilibrium equations, which are given below:

$$R_2 = -C'R_2' + C''R_2''$$

$$R_1 = +C'R_1' - C''R_1''$$

By solving the above two simultaneous equations, we will determine the values of the correction factors C' and C''. These correction factors are then multiplied with the moments we get after completing analysis in stage one and two analysis. The resultant moments are then found by adding these corrected moments to those obtained for frame, shown in Figure 19.17 (b).

As this process of analysis for multistorey frame for large buildings will become very much difficult to handle manually, computer programming or different structural analysis software is called for analysis of large multistorey frame. Also, multistorey frames involve several joint displacements independently, which is also very difficult to analyze manually without some assumptions. The more assumptions we take, more erroneous the result becomes. Matrix analysis procedures may also be adopted to solve this type of problems much efficiently than moment distribution method. In the later chapter, we will provide introduction of matrix structural analysis and provide some insight about the analysis process of multistorey frames.

20 Kani's Method or Rotation Contribution Method

20.1 INTRODUCTION

This method was introduced by Gasper Kani of Germany in 1947. This iterative method is suitable for approximate analysis of multistory frames, beams, or other statically indeterminate structures in their entirety, i.e., approximate values of the internal forces for the whole frame, beams, or other indeterminate structures can be determined by applying this method. This method is seldom used today due to the advent of advanced structural analysis software programs. Still, for interested readers of this subject, it is important to learn this method to have a quick analysis to get approximate results of frames and beams, which can be supplied as input toward preliminary design drawing preparations.

20.2 BASIC CONCEPT

In this section, we will discuss the basic concept of Kani's method or rotation contribution method. First, we will develop the idea without any lateral displacement at the ends of a member. After that, we will formulate the same, considering the lateral displacements at the ends.

20.2.1 Members without Relative Lateral Displacement

Let us consider AB, as shown in Figure 20.1 (a), one of the spans of a framed or a continuous structure. The deformed shape of the member AB is shown in Figure 20.1 (b). Let us assume that no lateral displacement occurs at the ends of member AB. Let the end A rotates by an angle θ_A under the influence of end moment M_{AB}, and the end B rotates by an angle θ_B under the influence of end moment M_{BA}. Let us assume that clockwise end moments and clockwise rotations at the ends are positive. Now the final end moments M_{AB} and M_{BA} can be split up into the components as shown in Figure 20.1 (c)–(e).

First of all, the ends A and B of the member are considered as fixed. Corresponding to this condition, the fixed end moments at A and B are determined as $\bar{M}_{AB}$ and $\bar{M}_{BA}$, respectively. Now, maintaining the fixity at end B, let the end A is rotated through an angle θ_A by applying a moment $2M'_{AB}$. Due to $2M'_{AB}$ at end A, a moment of magnitude M'_{AB} in the same direction will be induced at end B. This moment is called as rotation contribution of end A. Similarly at this stage, maintaining the fixity at end A, let the end B is rotated through an angle θ_B by applying a moment $2M'_{BA}$. Due to $2M'_{BA}$ at

DOI: 10.1201/9781003081227-23

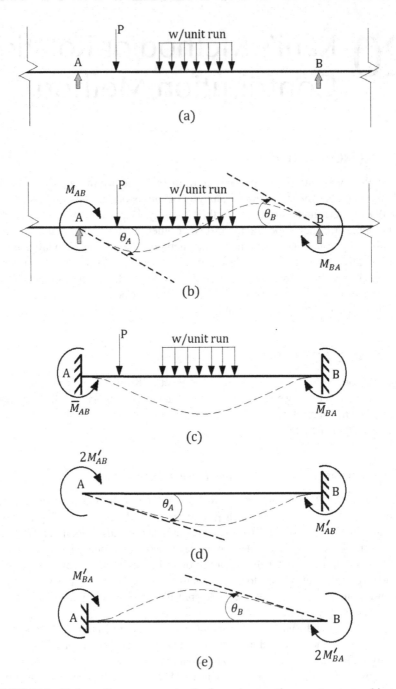

FIGURE 20.1 Deformation components of a framed or continuous structure subjected to external loading.

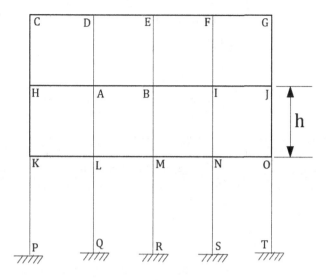

FIGURE 20.2 A multistorey frame subjected to vertical loading only.

end B, a moment of magnitude M'_{BA} in the same direction will be induced at end A. This moment is called as rotation contribution of end B. So, the final moments M_{AB} and M_{BA} can be expressed as follows:

$$M_{AB} = \bar{M}_{AB} + 2M'_{AB} + M'_{BA}$$

$$M_{BA} = \bar{M}_{BA} + 2M'_{BA} + M'_{AB}$$

The expression for the final moment at the near end of a member can be expressed as the algebraic sum of:

1. The fixed end moment at the near end.
2. Twice the rotation contribution of the near end.
3. Rotation contribution of the far end.

Now let us consider a multistorey frame as shown in Figure 20.2. Suppose no lateral joint displacement is occurring for any of the members of this frame. Let us consider joint A, where HA, LA, DA, and BA members meet together. The end moments meeting at this joint A can be written as:

$$M_{AB} = \bar{M}_{AB} + 2M'_{AB} + M'_{BA}$$

$$M_{AL} = \bar{M}_{AL} + 2M'_{AL} + M'_{LA}$$

$$M_{AH} = \bar{M}_{AH} + 2M'_{AH} + M'_{HA}$$

$$M_{AD} = \bar{M}_{AD} + 2M'_{AD} + M'_{DA}$$

For the equilibrium of joint A, the sum of all the end moments at A must be zero.
i.e.,

$$\sum M_A = 0$$

$$\sum M_{Aj} = \sum \bar{M}_{Aj} + 2\sum M'_{Aj} + \sum M'_{jA} = 0 \qquad (20.1)$$

where $\sum \bar{M}_{Aj}$ is the algebraic sum of the fixed end moments at A for all members meeting at A. $\sum M'_{Aj}$ is the algebraic sum of the rotation contributions at A for all members meeting at A. $\sum M'_{jA}$ is the algebraic sum of the rotation contributions at the far end joints with respect to joint A.

From equation (20.1), we can write,

$$\sum M'_{Aj} = \left(-\frac{1}{2}\right)\left\{\sum \bar{M}_{Aj} + \sum M'_{jA}\right\} \qquad (20.2)$$

We know for the member AB, with the end B fixed as shown in Figure 20.1 (d), the moment required at A to rotate the joint A by θ_A can be expressed as:

$$2M'_{AB} = \frac{4EI_{AB}}{L_{AB}}\theta_A = 4EK_{AB}\theta_A$$

where $I_{AB}/L_{AB} = K_{AB}$, $E = $ Young's modulus.
i.e.,

$$M'_{AB} = 2EK_{AB}\theta_A \qquad (20.3)$$

Now, as all the members meeting at joint A are rigidly connected, they will undergo the same amount of rotation θ_A. Assuming Young's modulus are the same for all the members, now we can write:

$$\sum M'_{Aj} = 2E\theta_A \sum K_{Aj} \qquad (20.4)$$

Now, dividing equation (20.3) by equation (20.4), we get,

$$\frac{M'_{AB}}{\sum M'_{Aj}} = \frac{2EK_{AB}\theta_A}{2E\theta_A \sum K_{Aj}}$$

or,

$$M'_{AB} = \left\{\frac{K_{AB}}{\sum K_{Aj}}\right\}\sum M'_{Aj} \qquad (20.5)$$

Now, from equations (20.2) and (20.5), we get,

$$M'_{AB} = \left(-\frac{1}{2}\frac{K_{AB}}{\sum K_{Aj}}\right)\left\{\sum \bar{M}_{Aj} + \sum M'_{jA}\right\} \qquad (20.6)$$

The ratio $\left(-(1/2)\left(K_{AB}/\sum K_{Aj}\right)\right)$ is called the rotation factor for the member AB at joint A. Let us consider, $\mu_{AB} = \left(-(1/2)\left(K_{AB}/\sum K_{Aj}\right)\right)$, we can rewrite equation (20.6) as,

$$M'_{AB} = \mu_{AB}\left\{\sum \bar{M}_{Aj} + \sum M'_{jA}\right\} \tag{20.7}$$

From equation (20.7), the summation of fixed end moments, $\sum \bar{M}_{Aj}$, can be computed easily and is a known quantity. The summation of far end rotation contributions, $\sum M'_{jA}$, is also a known quantity, as in the first trial, we take the far end rotation moment as zero to start with. Thus from equation (20.7), we can easily find the near-end rotation contributions through some cycles until more accurate values are obtained. If the end of a member is fixed, the rotation of that end is zero, and the rotation contribution of that end is zero as well. If the end of a member is hinged or pinned, it is convenient to consider the end as fixed and to take the relative stiffness as $\frac{3}{4}\frac{I}{L}$.

We will give the details of this analysis procedure by carrying out the detailed calculations of Example 20.1, which follow this section.

Example 20.1: Solve the continuous beam shown in Figure 20.3 by rotation contribution method.

SOLUTION: The fixed end moments are calculated as next, considering clockwise moments as positive:

$$\bar{M}_{AB} = -\frac{100 \times 1 \times 2^2}{3^2} = -44.44 \text{ kNm}$$

$$\bar{M}_{BA} = \frac{100 \times 2 \times 1^2}{3^2} = 22.22 \text{ kNm}$$

$$\bar{M}_{BC} = -\frac{50 \times 4^2}{12} = -66.67 \text{ kNm}$$

$$\bar{M}_{CB} = \frac{50 \times 4^2}{12} = 66.67 \text{ kNm}$$

$$\bar{M}_{CD} = -\frac{60 \times 4}{8} = -30 \text{ kNm}$$

$$\bar{M}_{DC} = \frac{60 \times 4}{8} = 30 \text{ kNm}$$

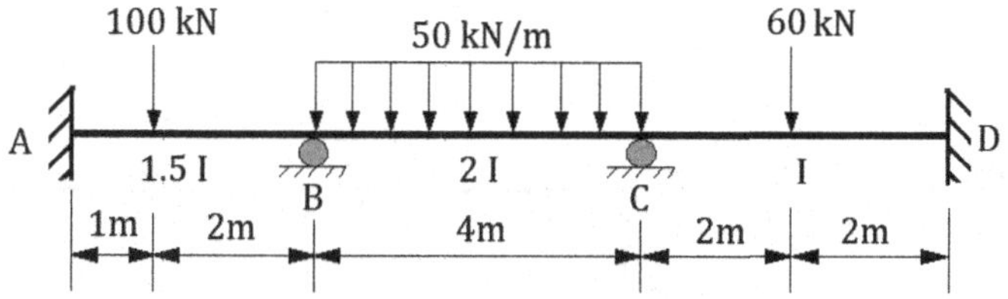

FIGURE 20.3 Example problem of a continuous beam by Kani's method.

TABLE 20.1

Calculating Rotation Factors for the Example Problem

Joint	Members	Relative Stiffness, K	Total Relative Stiffness, ΣK	Rotation Factor, $\mu = \left(-\dfrac{1}{2}\dfrac{K}{\Sigma K}\right)$
B	BA	$\dfrac{1.5I}{3}$	I	$-\dfrac{1}{4}$
	BC	$\dfrac{2I}{4}$		$-\dfrac{1}{4}$
C	CB	$\dfrac{2I}{4}$	$\dfrac{3I}{4}$	$-\dfrac{1}{3}$
	CD	$\dfrac{I}{4}$		$-\dfrac{1}{6}$

Next, we will evaluate the rotation factors at joints B and C using Table 20.1. The sum of the fixed end moments at joint B,

$$\sum \bar{M}_B = \bar{M}_{BA} + \bar{M}_{BC}$$

or,

$$\sum \bar{M}_B = 22.22 - 66.67 = -44.45 \text{ kNm}$$

The sum of the fixed end moments at joint C,

$$\sum \bar{M}_C = \bar{M}_{CB} + \bar{M}_{CD}$$

or,

$$\sum \bar{M}_C = 66.67 - 30 = 36.67 \text{ kNm}$$

The scheme for proceeding with the method of rotation contribution is shown in Figure 20.4, where the joints B and C are shown as two square boxes. The sum of the fixed end moments at B and C is written in the smaller square boxes for B and C, respectively. The rotation factors $-1/4$ for members BA and BC at B, and the rotation factors $-1/3$ and $-1/6$ for members CB and CD at C are also shown in the figure. The fixed end moments are entered above the horizontal line outside the square boxes.

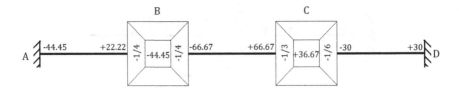

FIGURE 20.4 The scheme for proceeding with the method of rotation contribution.

The rotation contribution can now be determined by a successive iteration process as explained next:

CYCLE 1

Consider joint B:
Applying equation (20.7) to this joint, we get,

$$M'_{BA} = \mu_{BA} \left\{ \sum \bar{M}_B + \sum M'_{jB} \right\}_{j=A,C}$$

and

$$M'_{BC} = \mu_{BC} \left\{ \sum \bar{M}_B + \sum M'_{jB} \right\}_{j=A,C}$$

where $\sum \bar{M}_B$ is the sum of fixed end moments at $B = 22.22 - 66.67 = -44.45$ kNm, $\sum M'_{AB} = 0$, since end A is fixed end, $\sum M'_{CB} = 0$, assumed to start with. Substituting these values, we get,

$$M'_{BA} = \left(-\frac{1}{4} \right)\{-44.45 + 0 + 0\} = 11.11 \text{ kNm}$$

$$M'_{BC} = \left(-\frac{1}{4} \right)\{-44.45 + 0 + 0\} = 11.11 \text{ kNm}$$

Consider joint C:
Rotation moments M'_{CB} and M'_{CD} will be determined as follows:

$$M'_{CB} = \mu_{CB} \left\{ \sum \bar{M}_C + \sum M'_{jC} \right\}_{j=B,D}$$

or,

$$M'_{CB} = \left(-\frac{1}{3} \right)\{36.67 + 11.11 + 0\} = -15.93 \text{ kNm}$$

$$M'_{CD} = \mu_{CD} \left\{ \sum \bar{M}_C + \sum M'_{jC} \right\}_{j=B,D}$$

or,

$$M'_{CD} = \left(-\frac{1}{6} \right)\{36.67 + 11.11 + 0\} = -7.96 \text{ kNm}$$

CYCLE 2

Consider joint B:

$$M'_{BA} = \mu_{BA} \left\{ \sum \bar{M}_B + \sum M'_{jB} \right\}_{j=A,C}$$

or,

$$M'_{BA} = \left(-\frac{1}{4} \right)\{-44.45 + 0 - 15.93\} = 15.095 \text{ kNm}$$

$$M'_{BC} = \mu_{BC} \left\{ \sum \bar{M}_B + \sum M'_{jB} \right\}_{j=A,C}$$

or,

$$M'_{BC} = \left(-\frac{1}{4}\right)\{-44.45 + 0 - 15.93\} = 15.095 \text{ kNm}$$

Consider joint C:

$$M'_{CB} = \mu_{CB}\left\{\sum \bar{M}_C + \sum M'_{jC}\right\}_{j=B,D}$$

or,

$$M'_{CB} = \left(-\frac{1}{3}\right)\{36.67 + 15.095 + 0\} = -17.255 \text{ kNm}$$

$$M'_{CD} = \mu_{CD}\left\{\sum \bar{M}_C + \sum M'_{jC}\right\}_{j=B,D}$$

or,

$$M'_{CD} = \left(-\frac{1}{6}\right)\{36.67 + 15.095 + 0\} = -8.63 \text{ kNm}$$

CYCLE 3

Consider joint B:

$$M'_{BA} = \mu_{BA}\left\{\sum \bar{M}_B + \sum M'_{jB}\right\}_{j=A,C}$$

or,

$$M'_{BA} = \left(-\frac{1}{4}\right)\{-44.45 + 0 - 17.255\} = 15.43 \text{ kNm}$$

$$M'_{BC} = \mu_{BC}\left\{\sum \bar{M}_B + \sum M'_{jB}\right\}_{j=A,C}$$

or,

$$M'_{BC} = \left(-\frac{1}{4}\right)\{-44.45 + 0 - 17.255\} = 15.43 \text{ kNm}$$

Consider joint C:

$$M'_{CB} = \mu_{CB}\left\{\sum \bar{M}_C + \sum M'_{jC}\right\}_{j=B,D}$$

or,

$$M'_{CB} = \left(-\frac{1}{3}\right)\{36.67 + 15.43 + 0\} = -17.37 \text{ kNm}$$

$$M'_{CD} = \mu_{CD}\left\{\sum \bar{M}_C + \sum M'_{jC}\right\}_{j=B,D}$$

or,

$$M'_{CD} = \left(-\frac{1}{6}\right)\{36.67 + 15.43 + 0\} = -8.68 \text{ kNm}$$

CYCLE 4

Consider joint B:

$$M'_{BA} = \mu_{BA} \left\{ \sum \bar{M}_B + \sum M'_{jB} \right\}_{j=A,C}$$

or,

$$M'_{BA} = \left(-\frac{1}{4} \right) \{ -44.45 + 0 - 17.37 \} = 15.455 \text{ kNm}$$

$$M'_{BC} = \mu_{BC} \left\{ \sum \bar{M}_B + \sum M'_{jB} \right\}_{j=A,C}$$

or,

$$M'_{BC} = \left(-\frac{1}{4} \right) \{ -44.45 + 0 - 17.37 \} = 15.455 \text{ kNm}$$

Consider joint C:

$$M'_{CB} = \mu_{CB} \left\{ \sum \bar{M}_C + \sum M'_{jC} \right\}_{j=B,D}$$

or,

$$M'_{CB} = \left(-\frac{1}{3} \right) \{ 36.67 + 15.455 + 0 \} = -17.375 \text{ kNm}$$

$$M'_{CD} = \mu_{CD} \left\{ \sum \bar{M}_C + \sum M'_{jC} \right\}_{j=B,D}$$

or,

$$M'_{CD} = \left(-\frac{1}{6} \right) \{ 36.67 + 15.455 + 0 \} = -8.6875 \text{ kNm}$$

The scheme for proceeding with the rotation contribution method after Cycle 4 is shown in Figure 20.5.

After the fourth cycle, we may stop our iteration process as the difference between two successive cycles, as can be seen from Figure 20.5, is becoming negligible. Now we can proceed to calculate the end moments as follows with the outcome of this fourth cycle.

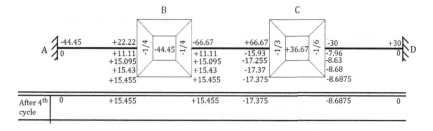

FIGURE 20.5 The scheme for proceeding with the method of rotation contribution after Cycle 4.

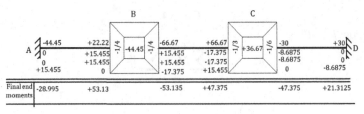

(a) Determining end moments by Kani's method

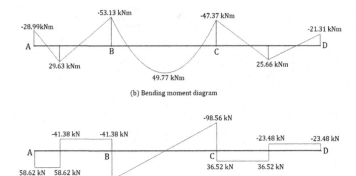

(b) Bending moment diagram

(c) Shear force diagram

FIGURE 20.6 (a) Determining the end moments by Kani's method, (b) bending moment diagram, and (c) shear force diagram.

For any span and from equation (20.1), we can write,

Final end moment at the near end = fixed end moment at the near end + twice the rotation contribution of the near end + rotation contribution of the far end.

The calculation for the final end moment is shown in Figure 20.6 (a). The corresponding bending moment and shear force diagram for this continuous beam is shown in Figure 20.6 (b) and (c), respectively.

20.2.2 MEMBERS WITH RELATIVE LATERAL DISPLACEMENT

In this section, we will reformulate the rotation contribution method considering the members whose ends have undergone lateral displacement of Δ amount as shown in Figure 20.7.

From the slope deflection equations, we can calculate the end moments due to this lateral displacement Δ as follows:

$$M_{AB} = \frac{2EI}{L}\left(2\theta_A + \theta_B - 3\psi_{AB}\right) + \bar{M}_{AB}$$

where $\theta_A = \theta_B = 0$, as both ends are fixed; $\psi_{AB} = -\frac{\Delta}{L}$, negative as the cord rotation is anticlockwise; $\bar{M}_{AB} = 0$, as no external load is present on the member AB. After substituting all these values in the abovementioned equation, we can get,

$$M_{AB} = \frac{2EI}{L}\left(+\frac{3\Delta}{L}\right) = +\frac{6EI\Delta}{L^2} = M_{BA}$$

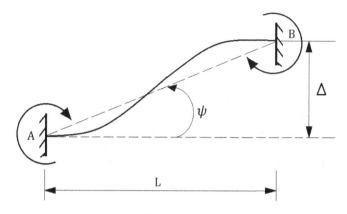

FIGURE 20.7 Member with relative lateral displacement Δ.

We have considered the clockwise moments as positive. And here also, we have obtained the end moments as positive, so they should be clockwise in nature due to anticlockwise cord rotation, as shown in the figure.

Let us rename, $M_{AB} = M_{BA} = M''_{AB} = M''_{BA} = + \frac{6EI\Delta}{L^2}$

When this kind of lateral displacements occur for a member, the final moments at A and B are given by,

$$M_{AB} = \bar{M}_{AB} + 2M'_{AB} + M'_{BA} + M''_{AB}$$

$$M_{BA} = \bar{M}_{BA} + 2M'_{BA} + M'_{AB} + M''_{BA}$$

The quantity $M''_{AB} = M''_{BA}$ is called the displacement contribution of the member AB.

The expression for the final moment at the near end of a member can be expressed as the algebraic sum of:

1. The fixed end moment at the near end.
2. Twice the rotation contribution of the near end.
3. Rotation contribution of the far end.
4. The displacement contributions of the near end.

If several members meet at joint A, we can similarly derive the following expression like as equation (20.7).

$$M'_{AB} = \mu_{AB} \left\{ \sum \bar{M}_{Aj} + \sum M'_{jA} + \sum M''_{Aj} \right\}$$

Now, using the abovementioned equation, we can easily find out the rotation contributions iteratively, including the effect of lateral displacement, and finally can obtain the end moments.

If the lateral displacement Δ is known, the displacement contribution, i.e., the fixed end moment generated due to lateral displacement, can also be found. In that case, the net fixed end moment is calculated at each end. If Δ is not known, then additional equations are to be used to find out the results.

Let us go through Example 20.2 to understand the effect of lateral displacement.

Example 20.2: Find the member end moments of the three-span continuous beam due to the support settlement of 3.5 mm at B, 9 mm at C, and the loading condition shown in Figure 20.8 (a). For all members, take $I = 3.5 \times 10^7$ mm^4 and $E = 200$ kN/mm^2.

SOLUTION: Let us consider clockwise moment and clockwise cord rotation as positive.

Fixed end moments

Span AB
Resultant fixed end moment at support

$$A = -\frac{60 \times 6^2}{12} - \frac{6 \times 200 \times 3.5 \times 10^7 \times 3.5}{6^2 \times 10^9} = -180 - 4.083 = -184.083 \text{ kNm}$$

Resultant fixed end moment at support $B = +180 - 4.083 = 175.917$ kNm

Span BC
Resultant fixed end moment at support

$$B = -\frac{150 \times 5}{8} - \frac{6 \times 200 \times 3.5 \times 10^7 \times (9 - 3.5)}{5^2 \times 10^9} = -93.75 - 9.24 = -102.99 \text{ kNm}$$

Resultant fixed end moment at support $C = 93.75 - 9.24 = 84.51$ kNm

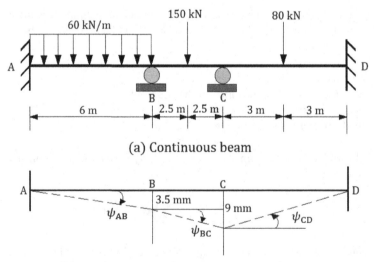

(a) Continuous beam

(b) Cord rotations due to support settlements

FIGURE 20.8 Example problem on Kani's method considering support settlement. (a) Continuous beam and (b) cord rotations due to support settlements.

TABLE 20.2
Calculating Rotation Factors for the Example Problem

Joint	Members	Relative Stiffness, K	Total Relative Stiffness, ΣK	Rotation Factor, $\mu = \left(-\dfrac{1}{2}\dfrac{K}{\Sigma K} \right)$
B	BA	$\dfrac{I}{6}$	$\dfrac{11I}{30}$	$-\dfrac{5}{22}$
	BC	$\dfrac{I}{5}$		$-\dfrac{6}{22}$
C	CB	$\dfrac{I}{5}$	$\dfrac{11I}{30}$	$-\dfrac{6}{22}$
	CD	$\dfrac{I}{6}$		$-\dfrac{5}{22}$

Span CD
Resultant fixed end moment at support

$$C = -\frac{80 \times 6}{8} + \frac{6 \times 200 \times 3.5 \times 10^7 \times 9}{6^2 \times 10^9} = -60 + 10.5 = -49.5 \text{ kNm}$$

Resultant fixed end moment at support $D = +60 + 10.5 = 70.5$ kNm
Next, we will evaluate the rotation factors at joints B and C using Table 20.2.
The scheme for proceeding with the rotation contribution method after Cycle 4 is shown in Figure 20.9.
The calculation for the final end moment is shown in Figure 20.10 (a). The corresponding bending moment and shear force diagram for this continuous beam are shown in Figure 20.10 (b) and (c), respectively.

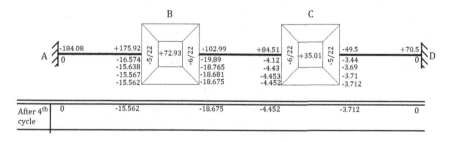

FIGURE 20.9 The scheme for proceeding with Kani's method after Cycle 4.

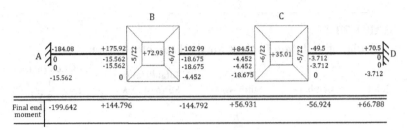

(a) Determining end moments by Kani's method

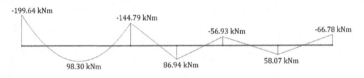

(b) Bending moment diagram

(c) Shear force diagram

FIGURE 20.10 (a) Determining the end moments by Kani's method for support settlement problem, (b) bending moment diagram, and (c) shear force diagram.

20.3 ANALYSIS OF FRAMES WITH SIDESWAY WITH VERTICAL LOADINGS

Let us consider the multistory frame shown in Figure 20.2 again and let HK represent a vertical member of the frame. Let M_{HK} and M_{KH} be the end moments in the end H and K, respectively, and the horizontal forces exerted by the frame on the member HK at H and K be F_H as shown in Figure 20.11 under equilibrium condition.

As the member HK is under equilibrium condition, we can write,

$$M_{HK} + M_{KH} + F_H h = 0$$

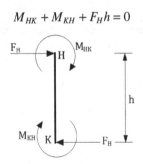

FIGURE 20.11 Equilibrium of a vertical column of a multistory frame.

or,

$$F_H = -\frac{M_{HK} + M_{KH}}{h}$$

F_H represents the shear force at any section of the member *HK*. Let *HK, AL, BM, IN, JO* be the vertical members of a story. Applying a similar concept, we can finally find out the sum of shear forces in all the columns of a particular story as follows:

$$\sum F_H = -\frac{\sum M_{HK} + \sum M_{KH}}{h}$$

where $\sum M_{HK}$ is the sum of the end moments at the upper ends of all the columns of a particular story. $\sum M_{KH}$ is the sum of the end moments at the lower ends of all the columns of a particular story. $\sum F_H$ is the story shear for a particular story. *h* is the height of the columns of a particular story.

Obviously, the story shear for a particular story should be equal to the sum of all horizontal external loads above that particular story. But in this case, the multistory frame is only subjected to external vertical loads. So, for each story, the story shear is equal to zero. So, we can write, $\sum M_{HK} + \sum M_{KH} = 0$, for a particular story. From the earlier discussions, we can write the general expressions for the end moments for column *HK* as next,

$$M_{HK} = \bar{M}_{HK} + 2M'_{HK} + M'_{KH} + M''_{HK}$$

$$M_{KH} = \bar{M}_{KH} + 2M'_{KH} + M'_{HK} + M''_{KH}$$

The terminologies used in the abovementioned two equations are already explained. Since the loading on the frame is vertical only, $\bar{M}_{HK} = \bar{M}_{KH} = 0$. Now, if we add the abovementioned two equations, we will get,

$$M_{HK} + M_{KH} = 3M'_{HK} + 3M'_{KH} + 2M''_{HK}$$

But,

$$\sum M_{HK} + \sum M_{KH} = 0$$

$$\therefore 3\sum M'_{HK} + 3\sum M'_{KH} + 2\sum M''_{HK} = 0$$

Therefore,

$$\sum M''_{HK} = -\frac{3}{2}\left[\sum M'_{HK} + \sum M'_{KH}\right] \tag{20.8}$$

The abovementioned equation establishes a relationship between the rotation contributions and the displacement contributions. We already know for any member the displacement contribution is,

$$= \frac{6EI\Delta}{L^2}$$

Now, the relative lateral displacement Δ is the same for all the columns of a story. Suppose the length, L and Young's modulus, E is assumed to be the same for all the columns. In that case, we can say the displacement contribution for a column of a story is proportional to the moment of inertia of the section of the column.

i.e.,

$$M''_{HK} \propto I$$

But we know the relative stiffness, $K_{HK} = I_{HK}/L_{HK}$. So, $M''_{HK} \propto K_{HK}$, if $L_{HK} = L$, for all the columns of a story.

Therefore,

$$\frac{M''_{HK}}{\Sigma M''_{HK}} = \frac{K_{HK}}{\Sigma K_{HK}}$$

$$M''_{HK} = \frac{K_{HK}}{\Sigma K_{HK}} \sum M''_{HK} \tag{20.9}$$

From Eqs. (20.8) and (20.9), we get,

$$M''_{HK} = \frac{K_{HK}}{\Sigma K_{HK}} \left(-\frac{3}{2}\right) \left[\sum M'_{HK} + \sum M'_{KH} \right] \tag{20.10}$$

The quantity $\left(K_{HK}/\Sigma K_{HK} \right)(-3/2)$ is called the displacement factor for member HK.

In equation (20.10), $\Sigma M'_{HK} + \Sigma M'_{KH}$ represents the sum of rotation contributions of the top and bottom ends of all the columns of a particular story. ΣK_{HK} represents the sum of the relative stiffness of all the columns of a story concerned.

Let us consider, $\gamma_{HK} = \left(-\frac{3}{2}\frac{K_{HK}}{\Sigma K_{HK}}\right)$, we can rewrite equation (20.10) as,

$$M''_{HK} = \gamma_{HK} \left[\sum M'_{HK} + \sum M'_{KH} \right] \tag{20.11}$$

Let us discuss Example 20.3 to understand the procedure just discussed.

Example 20.3: Find the member end moments for the portal frame shown in Figure 20.12.

SOLUTION: *Fixed end moments*

$$\bar{M}_{AB} = \bar{M}_{BA} = \bar{M}_{CD} = \bar{M}_{DC} = 0$$

$$\bar{M}_{BC} = -\frac{190 \times 4 \times 8^2}{12^2} = -337.78 \text{ kNm}$$

$$\bar{M}_{CB} = \frac{190 \times 8 \times 4^2}{12^2} = 168.89 \text{ kNm}$$

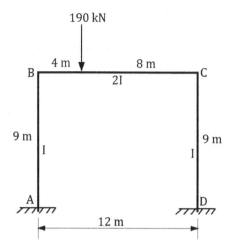

FIGURE 20.12 Example problem on Kani's method considering a portal frame under vertical unsymmetrical load.

Rotation factors

Rotation factors are calculated using Table 20.3.

Displacement factors

Displacement factors are calculated using Table 20.4.

The sum of the fixed end moments at joint B,

$$\sum \bar{M}_B = \bar{M}_{BA} + \bar{M}_{BC}$$

or,

$$\sum \bar{M}_B = 0 - 337.78 = -337.78 \text{ kNm}$$

TABLE 20.3

Calculating Rotation Factors for the Given Portal Frame

Joint	Member	Relative Stiffness, K	Total Relative Stiffness, ΣK	Rotation Factor, $\mu = \left(-\dfrac{1}{2} \dfrac{K}{\Sigma K} \right)$
B	BA	$\dfrac{I}{9}$	$\dfrac{5I}{18}$	$-\dfrac{1}{5}$
	BC	$\dfrac{2I}{12}$		$-\dfrac{3}{10}$
C	CB	$\dfrac{2I}{12}$	$\dfrac{5I}{18}$	$-\dfrac{3}{10}$
	CD	$\dfrac{I}{9}$		$-\dfrac{1}{5}$

TABLE 20.4

Calculating Displacement Factors for the Given Portal Frame

Vertical Member	Relative Stiffness, K	Total Relative Stiffness, ΣK	Displacement Factor, $\gamma = \left(-\dfrac{3}{2}\dfrac{K}{\Sigma K} \right)$
AB	$\dfrac{I}{9}$		$-\dfrac{3}{4}$
		$\dfrac{2I}{9}$	
DC	$\dfrac{I}{9}$		$-\dfrac{3}{4}$

The sum of the fixed end moments at joint C,

$$\sum \bar{M}_C = \bar{M}_{CB} + \bar{M}_{CD}$$

or,

$$\sum \bar{M}_C = 168.89 + 0 = 168.89 \text{ kNm}$$

The scheme for proceeding with the method of rotation contribution is shown in Figure 20.13, where the joints B and C are shown as two square boxes. The sum of the fixed end moments at B and C is written in the smaller square boxes for B and C, respectively. The rotation factors $-1/5$ for member BA and $-3/10$ for BC at B and the rotation factors $-1/5$ for member CD and $-3/10$ for members CB at C are also shown in the figure. The displacement factors are entered by the side of each column. The fixed end moments are entered above the horizontal line outside the square boxes.

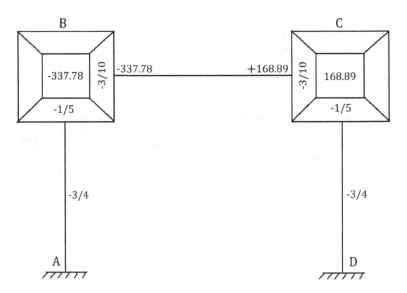

FIGURE 20.13   The scheme for proceeding with Kani's method.

Iteration process: Cycle 1

Joint B

Summation of the fixed end moments	$= -337.78$
Far end rotation contributions	
At A	$= 0$
At C (assumed)	$= 0$
Summation of displacement contributions	
BC	$= 0$
BA (assumed)	$= 0$

$$= -337.78$$

$$M'_{BA} = \left(-\frac{1}{5}\right)(-337.78)$$

$$= 67.556 \text{ kNm}$$

$$M'_{BC} = \left(-\frac{3}{10}\right)(-337.78)$$

$$= 101.334 \text{ kNm}$$

Joint C

Summation of the fixed end moments	$= 168.89$
Far end rotation contributions	
At B	$= 101.334$
At D	$= 0$
Summation of displacement contributions	
CB	$= 0$
CD (assumed)	$= 0$

$$= 270.224$$

$$M'_{CB} = \left(-\frac{3}{10}\right)(270.224)$$

$$= -81.067 \text{ kNm}$$

$$M'_{CD} = \left(-\frac{1}{5}\right)(270.224)$$

$$= -54.045 \text{ kNm}$$

Story one (there is only one story)

Rotation contribution at top of column BA	$= 67.556$
Rotation contribution at top of column CD	$= -54.045$
Rotation contribution at bottom of columns	$= 0$

$$= 13.511$$

$$M''_{AB} = \left(-\frac{3}{4}\right)(13.511)$$

$$= -10.133$$

$$M''_{CD} = \left(-\frac{3}{4}\right)(13.511)$$

$$= -10.133$$

Iteration process: Cycle 2

Joint B

Summation of the fixed end moments	$=-337.78$
Far end rotation contributions	
At A	$=0$
At C	$=-81.067$
Summation of displacement contributions	
BC	$=0$
BA	$=-10.133$

$$= -428.98$$

$$M'_{BA} = \left(-\frac{1}{5}\right)(-428.98)$$

$$= 85.796 \text{ kNm}$$

$$M'_{BC} = \left(-\frac{3}{10}\right)(-428.98)$$

$$= 128.694 \text{ kNm}$$

Joint C

Summation of the fixed end moments	$=168.89$
Far end rotation contributions	
At B	$=128.694$
At D	$=0$
Summation of displacement contributions	
CB	$=0$
CD	$=-10.133$

$$= 287.451$$

$$M'_{CB} = \left(-\frac{3}{10}\right)(287.451)$$

$$= -86.235 \text{ kNm}$$

$$M'_{CD} = \left(-\frac{1}{5}\right)(287.451)$$

$$= -57.49 \text{ kNm}$$

Story one (there is only one story)

Rotation contribution at top of column BA	$=85.796$
Rotation contribution at top of column CD	$=-57.49$
Rotation contribution at bottom of columns	$=0$

$$= 28.306$$

$$M''_{AB} = \left(-\frac{3}{4}\right)(28.306)$$

$$= -21.229$$

$$M''_{CD} = \left(-\frac{3}{4}\right)(28.306)$$

$$= -21.229$$

Iteration process: Cycle 3
Joint B

Summation of the fixed end moments	$= -337.78$
Far end rotation contributions	
At A	$= 0$
At C	$= -86.235$
Summation of displacement contributions	
BC	$= 0$
BA	$= -21.229$
	$= -445.244$

$$M'_{BA} = \left(-\frac{1}{5}\right)(-445.244)$$
$$= 89.05 \text{ kNm}$$

$$M'_{BC} = \left(-\frac{3}{10}\right)(-445.244)$$
$$= 133.57 \text{ kNm}$$

Joint C

Summation of the fixed end moments	$= 168.89$
Far end rotation contributions	
At B	$= 133.57$
At D	$= 0$
Summation of displacement contributions	
CB	$= 0$
CD	$= -21.229$
	$= 281.231$

$$M'_{CB} = \left(-\frac{3}{10}\right)(281.231)$$
$$= -84.37 \text{ kNm}$$

$$M'_{CD} = \left(-\frac{1}{5}\right)(281.231)$$
$$= -56.24 \text{ kNm}$$

Story one (there is only one story)

Rotation contribution at top of column BA	$= 89.05$
Rotation contribution at top of column CD	$= -56.24$
Rotation contribution at bottom of columns	$= 0$
	$= 32.81$

$$M''_{AB} = \left(-\frac{3}{4}\right)(32.81)$$
$$= -24.61$$

$$M''_{CD} = \left(-\frac{3}{4}\right)(32.81)$$
$$= -24.61$$

Iteration process: Cycle 4
Joint B

Summation of the fixed end moments	$= -337.78$
Far end rotation contributions	
At A	$= 0$
At C	$= -84.37$
Summation of displacement contributions	
BC	$= 0$
BA	$= -24.61$

$$= -446.76$$

$$M'_{BA} = \left(-\frac{1}{5}\right)(-446.76)$$

$$= 89.352 \text{ kNm}$$

$$M'_{BC} = \left(-\frac{3}{10}\right)(-446.76)$$

$$= 134.03 \text{ kNm}$$

Joint C

Summation of the fixed end moments	$= 168.89$
Far end rotation contributions	
At B	$= 134.03$
At D	$= 0$
Summation of displacement contributions	
CB	$= 0$
CD	$= -24.61$

$$= 278.31$$

$$M'_{CB} = \left(-\frac{3}{10}\right)(278.31)$$

$$= -83.49 \text{ kNm}$$

$$M'_{CD} = \left(-\frac{1}{5}\right)(278.31)$$

$$= -55.66 \text{ kNm}$$

Story one (there is only one story)

Rotation contribution at top of column BA	$= 89.352$
Rotation contribution at top of column CD	$= -55.66$
Rotation contribution at bottom of columns	$= 0$

$$= 33.692$$

$$M''_{AB} = \left(-\frac{3}{4}\right)(33.692)$$

$$= -25.27$$

$$M''_{CD} = \left(-\frac{3}{4}\right)(33.692)$$

$$= -25.27$$

Iteration process: Cycle 5
Joint B

Summation of the fixed end moments	$= -337.78$
Far end rotation contributions	
At A	$= 0$
At C	$= -83.49$
Summation of displacement contributions	
BC	$= 0$
BA	$= -25.27$

$$= -446.54$$

$$M'_{BA} = \left(-\frac{1}{5}\right)(-446.54)$$
$$= 89.308 \text{ kNm}$$

$$M'_{BC} = \left(-\frac{3}{10}\right)(-446.54)$$
$$= 133.96 \text{ kNm}$$

Joint C

Summation of the fixed end moments	$= 168.89$
Far end rotation contributions	
At B	$= 133.96$
At D	$= 0$
Summation of displacement contributions	
CB	$= 0$
CD	$= -25.27$

$$= 277.58$$

$$M'_{CB} = \left(-\frac{3}{10}\right)(277.58)$$
$$= -83.274 \text{ kNm}$$

$$M'_{CD} = \left(-\frac{1}{5}\right)(277.58)$$
$$= -55.52 \text{ kNm}$$

Story one (there is only one story)

Rotation contribution at top of column BA	$= 89.308$
Rotation contribution at top of column CD	$= -55.52$
Rotation contribution at bottom of columns	$= 0$

$$= 33.788$$

$$M''_{AB} = \left(-\frac{3}{4}\right)(33.788)$$
$$= -25.34$$

$$M''_{CD} = \left(-\frac{3}{4}\right)(33.788)$$
$$= -25.34$$

The iteration process has been carried out up to the fifth cycle, as shown in Figure 20.14 (a). After that, from the values of the fifth cycle, the final end moments have been found, as shown in Figure 20.14 (b).

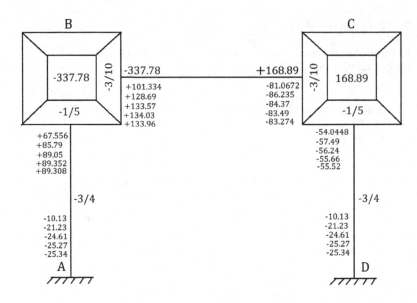

(a) The scheme for proceeding with Kani's method after cycle 5

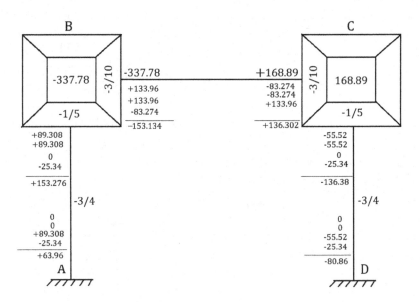

(b) Determining end moments by Kani's method

FIGURE 20.14 Portal frame analysis by Kani's method under vertical load. (a) The scheme for proceeding with Kani's method after Cycle 5 and (b) determining end moments by Kani's method.

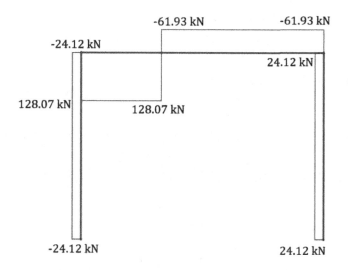

(a) Shear force diagram

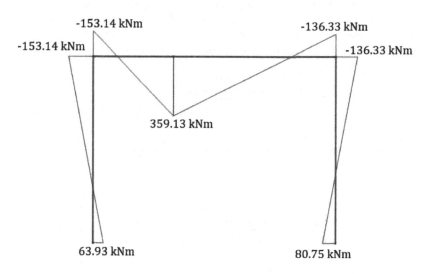

(b) Bending moment diagram

FIGURE 20.15 (a) Shear force diagram and (b) bending moment diagram of the portal frame.

    The shear force and bending moment diagrams have been given in Figure 20.15 (a) and (b), respectively. Note that there is a slight deviation in the values of bending moments as shown in the figure and with the values obtained by Kani's method after the fifth cycle. As we increase our number of cycles, we will get closer to the values shown in the figure.

20.4 ANALYSIS OF FRAMES WITH SIDESWAY WITH VERTICAL LOADING AND HORIZONTAL LOADING AT NODAL POINTS

Let us consider a multistoried frame subjected to horizontal loads applied to nodal points along with the vertical loading, as shown in Figure 20.16. As was discussed in Section 20.3, the sum of the shear forces in all the columns of a particular story can be expressed as,

$$\sum F_H = -\frac{\sum M_{HK} + \sum M_{KH}}{h}$$

Let us consider fth story of a multistory frame. We can write, $S_f = \sum F_H$, where S_f is the story shear of the fth story. Now we can rewrite the abovementioned expression as,

$$S_f = -\frac{\left(\sum M_{HK} + \sum M_{KH}\right) \text{ for the } f\text{th story}}{h_f}$$

where h_f is the height of columns of the fth story, $\sum M_{HK}$ is the sum of the end moments at the upper ends of all the columns of the fth story, $\sum M_{KH}$ is the sum of the end moments at the lower ends of all the columns of the fth story.

Rearranging the abovementioned equation, we can write,

$$\left(\sum M_{HK} + \sum M_{KH}\right) = -S_f h_f$$

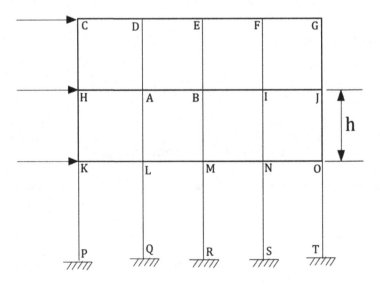

FIGURE 20.16 Multistory frame subjected to horizontal loading at nodal points.

The general expression of the end moments for a particular vertical member HK can be written as:

$$M_{HK} = \bar{M}_{HK} + 2M'_{HK} + M'_{KH} + M''_{HK} = 2M'_{HK} + M'_{KH} + M''_{HK}$$

$$M_{KH} = \bar{M}_{KH} + 2M'_{KH} + M'_{HK} + M''_{KH} = 2M'_{KH} + M'_{HK} + M''_{KH}$$

Since $\bar{M}_{HK} = \bar{M}_{KH} = 0$, because no external loads are acting in between the joints of column HK.

Therefore,

$$M_{HK} + M_{KH} = 3M'_{HK} + 3M'_{KH} + 2M''_{HK}$$

i.e.,

$$\sum M_{HK} + \sum M_{KH} = 3\sum M'_{HK} + 3\sum M'_{KH} + 2\sum M''_{HK} = -S_f h_f$$

or,

$$2\sum M''_{HK} = -S_f h_f - 3\sum M'_{HK} - 3\sum M'_{KH}$$

or,

$$2\sum M''_{HK} = -3\left(\frac{S_f h_f}{3} + \sum M'_{HK} + \sum M'_{KH}\right)$$

$$\therefore \sum M''_{HK} = -\frac{3}{2}\left(\frac{S_f h_f}{3} + \sum M'_{HK} + \sum M'_{KH}\right) \qquad (20.12)$$

For a particular story, the relative lateral displacement is the same for all the columns. Let us assume the height of all the columns and Young's modulus are also the same.

The displacement contribution for a column HK of the fth story,

$$M''_{HK} = \frac{6EI\Delta}{h_f^2} = \frac{6E\Delta}{h_f} \times \frac{I}{h_f} = \frac{6E\Delta}{h_f} K_{HK}$$

i.e.,

$$M''_{HK} \propto K_{HK}$$

So, we can say the displacement contribution of a particular column is proportional to its relative stiffness.

Therefore,

$$\frac{M''_{HK}}{\sum M''_{HK}} = \frac{K_{HK}}{\sum K_{HK}}$$

$$M''_{HK} = \frac{K_{HK}}{\sum K_{HK}} \sum M''_{HK}$$

Now, with the help of this expression, and using equation (20.12), we get,

$$M''_{HK} = \frac{K_{HK}}{\sum K_{HK}}\left(-\frac{3}{2}\right)\left(\frac{S_f h_f}{3} + \sum M'_{HK} + \sum M'_{KH}\right) \qquad (20.13)$$

$$M''_{HK} = \gamma_{HK}\left(\frac{S_f h_f}{3} + \sum M'_{HK} + \sum M'_{KH}\right) \qquad (20.14)$$

where $\gamma_{HK} = (K_{HK}/\sum K_{HK})(-3/2) =$ displacement factor for HK, and the quantity $S_f h_f/3$ is called the story moment. Let us discuss Example 20.4 to clarify the concept just discussed.

Example 20.4: Analyze the portal frame under the influence of horizontal load applied at joint B as shown in Figure 20.17. Assume all the members have the same flexural rigidity.

SOLUTION: *Fixed end moments*

$$\bar{M}_{AB} = \bar{M}_{BA} = \bar{M}_{CD} = \bar{M}_{DC} = \bar{M}_{BC} = \bar{M}_{CB} = 0$$

Rotation factors
Rotation factors are calculated using Table 20.5.
Storey shear

$$S_f = 15 \text{ kN}$$

Storey moment

$$\frac{S_f h_f}{3} = \frac{15 \times 7}{3} = 35 \text{ kNm}$$

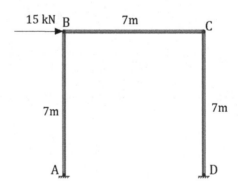

FIGURE 20.17 Example problem on the portal frame under the influence of horizontal nodal load.

TABLE 20.5
Calculating Rotation Factors for the Given Portal Frame under Horizontal Nodal Load

Joint	Member	Relative Stiffness, K	Total Relative Stiffness, ΣK	Rotation Factor, $\mu = \left(-\dfrac{1}{2}\dfrac{K}{\Sigma K}\right)$
B	BA	$\dfrac{I}{7}$	$\dfrac{2I}{7}$	$-\dfrac{1}{4}$
	BC	$\dfrac{I}{7}$		$-\dfrac{1}{4}$
C	CB	$\dfrac{I}{7}$	$\dfrac{2I}{7}$	$-\dfrac{1}{4}$
	CD	$\dfrac{I}{7}$		$-\dfrac{1}{4}$

TABLE 20.6
Calculating Displacement Factors for the Given Portal Frame

Vertical Member	Relative Stiffness, K	Total Relative Stiffness, ΣK	Displacement Factor, $\gamma = \left(-\dfrac{3}{2}\dfrac{K}{\Sigma K}\right)$
AB	$\dfrac{I}{7}$	$\dfrac{2I}{7}$	$-\dfrac{3}{4}$
DC	$\dfrac{I}{7}$		$-\dfrac{3}{4}$

Displacement factors
Displacement factors are calculated using Table 20.6.
Iteration: Cycle 1

	Rotation Contributions		
Joint B		**Joint C**	
Summation of the fixed end moments	0	Summation of the fixed end moments	0
Far end rotation contributions		Far end rotation contributions	
At A (M'_{AB})	0	At B (M'_{BC})	0 (assumed)
At C (M'_{CB})	0 (assumed)	At D (M'_{DC})	0
Displacement contributions		Displacement contributions	
BC (M''_{BC})	0	CB (M''_{CB})	0
BA (M''_{BA})	0 (assumed)	CD (M''_{CD})	0 (assumed)
Total	**0**	**Total**	**0**

Rotation Contributions

Joint B		Joint C	

M'_{BA}

$= \mu_{BA} \left\{ \sum \bar{M}_B \right.$

$+ \sum M'_{jB}$

$\left. + \sum M''_{Bj} \right\}_{j=A,C}$

$M'_{BA} = \left(-\dfrac{1}{4}\right)(0)$

$= 0 = M'_{BC}$

M'_{CB}

$= \mu_{CB} \left\{ \sum \bar{M}_C \right.$

$+ \sum M'_{jC}$

$\left. + \sum M''_{Cj} \right\}_{j=B,D}$

$M'_{CB} = \left(-\dfrac{1}{4}\right)(0)$

$= 0 = M'_{CD}$

Displacement Contributions

Rotation contributions at the top of columns	
For AB (M'_{BA})	0
For DC (M'_{CD})	0
Rotation contributions at the bottom of columns	
For AB (M'_{AB})	0
For DC (M'_{DC})	0
Story moment $\left(\frac{S_f h_f}{3}\right)$	35
Total	**35**

$M''_{BA} = \gamma_{BA} \left(\dfrac{S_f h_f}{3} + \sum M'_{BA} + \sum M'_{AB} \right)$

$M''_{BA} = \left(-\dfrac{3}{4}\right)(35) = -26.25 = M''_{CD}$

Iteration: Cycle 2

Rotation Contributions

Joint B		Joint C	
Summation of the fixed end moments	0	Summation of the fixed end moments	0
Far end rotation contributions		Far end rotation contributions	
At A (M'_{AB})	0	At B (M'_{BC})	+6.56
At C (M'_{CB})	0	At D (M'_{DC})	0
Displacement contributions		Displacement contributions	
BC (M''_{BC})	0	CB (M''_{CB})	0
BA (M''_{BA})	−26.25	CD (M''_{CD})	−26.25
Total	**−26.25**	**Total**	**−19.69**

M'_{BA}

$= \mu_{BA} \left\{ \sum \bar{M}_B \right.$

$+ \sum M'_{jB}$

$\left. + \sum M''_{Bj} \right\}_{j=A,C}$

M'_{BA}

$= \left(-\dfrac{1}{4}\right)(-26.25)$

$= +6.56 = M'_{BC}$

M'_{CB}

$= \mu_{CB} \left\{ \sum \bar{M}_C \right.$

$+ \sum M'_{jC}$

$\left. + \sum M''_{Cj} \right\}_{j=B,D}$

M'_{CB}

$= \left(-\dfrac{1}{4}\right)(-19.69)$

$= +4.92 = M'_{CD}$

Displacement Contributions

Rotation contributions at the top of columns

For AB (M'_{BA})	+6.56
For DC (M'_{CD})	+4.92

Rotation contributions at the bottom of columns

For AB (M'_{AB})	0
For DC (M'_{DC})	0
Story moment $\left(\frac{S_f h_f}{3}\right)$	35
Total	**46.48**

$$M''_{BA} = \gamma_{BA}\left(\frac{S_f h_f}{3} + \sum M'_{BA} + \sum M'_{AB}\right) \qquad M''_{BA} = \left(-\frac{3}{4}\right)(46.48) = -34.86 = M''_{CD}$$

Iteration: Cycle 3

Rotation Contributions

Joint B		Joint C	
Summation of the fixed end moments	0	Summation of the fixed end moments	0
Far end rotation contributions		Far end rotation contributions	
At A (M'_{AB})	0	At B (M'_{BC})	+7.485
At C (M'_{CB})	+4.92	At D (M'_{DC})	0
Displacement contributions		Displacement contributions	
BC (M''_{BC})	0	CB (M''_{CB})	0
BA (M''_{BA})	−34.86	CD (M''_{CD})	−34.86
Total	**−29.94**	**Total**	**−27.375**

M'_{BA}	M'_{BA}	M'_{CB}	M'_{CB}
$= \mu_{BA}\left\{\sum \bar{M}_B\right.$	$=\left(-\frac{1}{4}\right)(-29.94)$	$= \mu_{CB}\left\{\sum \bar{M}_C\right.$	$=\left(-\frac{1}{4}\right)(-27.375)$
$+ \sum M'_{jB}$	$= +7.485 = M'_{BC}$	$+ \sum M'_{jC}$	$= +6.84 = M'_{CD}$
$\left. + \sum M''_{Bj}\right\}_{j=A,C}$		$\left. + \sum M''_{Cj}\right\}_{j=B,D}$	

Displacement Contributions

Rotation contributions at the top of columns

For AB (M'_{BA})	+7.485
For DC (M'_{CD})	+6.84

Rotation contributions at the bottom of columns

For AB (M'_{AB})	0
For DC (M'_{DC})	0
Story moment $\left(\frac{S_f h_f}{3}\right)$	35
Total	**49.33**

$$M''_{BA} = \gamma_{BA}\left(\frac{S_f h_f}{3} + \sum M'_{BA} + \sum M'_{AB}\right) \qquad M''_{BA} = \left(-\frac{3}{4}\right)(49.33) = -36.99 = M''_{CD}$$

Iteration: Cycle 4

Rotation Contributions

	Joint B		Joint C	
Summation of the fixed end moments	0		Summation of the fixed end moments	0
Far end rotation contributions			Far end rotation contributions	
At A (M'_{AB})	0		At B (M'_{BC})	+7.54
At C (M'_{CB})	+6.84		At D (M'_{DC})	0
Displacement contributions			Displacement contributions	
BC (M''_{BC})	0		CB (M''_{CB})	0
BA (M''_{BA})	−36.99		CD (M''_{CD})	−36.99
Total	**−30.15**		**Total**	**−29.45**
M'_{BA}	M'_{BA}		M'_{CB}	M'_{CB}

$$M'_{BA} = \mu_{BA}\left\{\sum \bar{M}_B\right\} + \sum M'_{jB} + \sum M''_{Bj}\Bigg\}_{j=A,C}$$

$$= \left(-\frac{1}{4}\right)(-30.15)$$
$$= +7.54 = M'_{BC}$$

$$M'_{CB} = \mu_{CB}\left\{\sum \bar{M}_C\right\} + \sum M'_{jC} + \sum M''_{Cj}\Bigg\}_{j=B,D}$$

$$= \left(-\frac{1}{4}\right)(-29.45)$$
$$= +7.36 = M'_{CD}$$

Displacement Contributions

Rotation contributions at the top of columns	
For AB (M'_{BA})	+7.54
For DC (M'_{CD})	+7.36
Rotation contributions at the bottom of columns	
For AB (M'_{AB})	0
For DC (M'_{DC})	0
Story moment $\left(\frac{S_f h_f}{3}\right)$	35
Total	**49.9**

$$M''_{BA} = \gamma_{BA}\left(\frac{S_f h_f}{3} + \sum M''_{BA} + \sum M'_{AB}\right)$$

$$M''_{BA} = \left(-\frac{3}{4}\right)(49.9) = -37.425 = M''_{CD}$$

Iteration: Cycle 5

Rotation Contributions

	Joint B		Joint C	
Summation of the fixed end moments	0		Summation of the fixed end moments	0
Far end rotation contributions			Far end rotation contributions	
At A (M'_{AB})	0		At B (M'_{BC})	+7.52
At C (M'_{CB})	+7.36		At D (M'_{DC})	0

Rotation Contributions			
Joint B		**Joint C**	
Displacement contributions		*Displacement contributions*	
BC (M''_{BC})	0	CB (M''_{CB})	0
BA (M''_{BA})	−37.425	CD (M''_{CD})	−37.425
Total	**−30.1**	**Total**	**−29.91**
M'_{BA}	M'_{BA}	M'_{CB}	M'_{CB}

$$= \mu_{BA} \left\{ \sum \bar{M}_B \right.$$
$$+ \sum M'_{jB}$$
$$\left. + \sum M''_{Bj} \right\}_{j=A,C}$$

$$= \left(-\frac{1}{4}\right)(-30.1)$$
$$= +7.52 = M'_{BC}$$

$$= \mu_{CB} \left\{ \sum \bar{M}_C \right.$$
$$+ \sum M'_{jC}$$
$$\left. + \sum M''_{Cj} \right\}_{j=B,D}$$

$$= \left(-\frac{1}{4}\right)(-29.91)$$
$$= +7.48 = M'_{CD}$$

Displacement Contributions	
Rotation contributions at the top of columns	
For AB (M'_{BA})	+7.52
For DC (M'_{CD})	+7.48
Rotation contributions at the bottom of columns	
For AB (M'_{AB})	0
For DC (M'_{DC})	0
Story moment $\left(\frac{S_f h_f}{3}\right)$	35
Total	**50**

$$M''_{BA} = \gamma_{BA}\left(\frac{S_f h_f}{3} + \sum M'_{BA} + \sum M'_{AB}\right) \qquad M''_{BA} = \left(-\frac{3}{4}\right)(50) = -37.5 = M''_{CD}$$

After performing the fifth cycle, it was found that the rotation and displacement contribution values are getting converged, i.e., no more cycle is required. So, the values obtained in the fifth cycle are used to calculate the final end moments. The obtained values from all the iterations for rotation and displacement contribution are given in Table 20.7.

The end moments have been calculated as shown in Figure 20.18.

The shear force and bending moment diagrams for the given portal frame are shown in Figure 20.19 (a) and (b), respectively.

TABLE 20.7
Values of Rotation and Displacement Contributions for All the Iterations

Cycle	M'_{BA}	M'_{BC}	M'_{CB}	M'_{CD}	M''_{BA}	M''_{CD}
First cycle	0	0	0	0	−26.25	−26.25
Second cycle	+6.56	+6.56	+4.92	+4.92	−34.86	−34.86
Third cycle	+7.485	+7.485	+6.84	+6.84	−36.99	−36.99
Fourth cycle	+7.54	+7.54	+7.36	+7.36	−37.425	−37.425
Fifth cycle	+7.52	+7.52	+7.48	+7.48	−37.5	−37.5

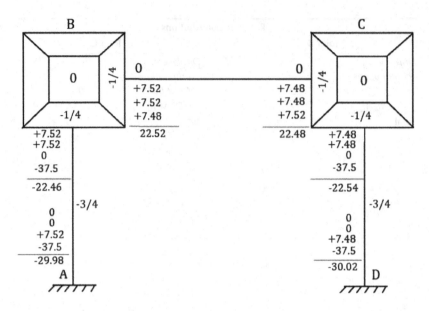

FIGURE 20.18 Determining end moments by Kani's method.

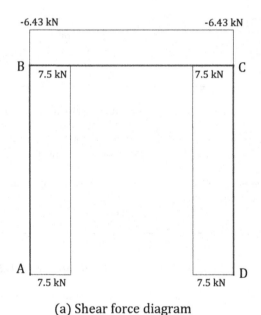

(a) Shear force diagram

FIGURE 20.19 (a) Shear force diagram and (b) bending moment diagram of the portal frame. *(Continued)*

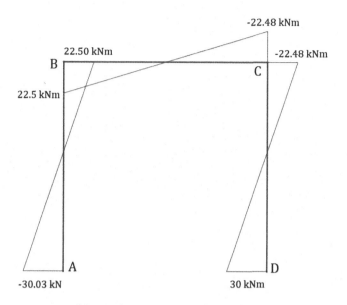

(b) Bending moment diagram

FIGURE 20.19   *(Continued)*

20.5   ANALYSIS OF FRAMES WITH COLUMNS WITH UNEQUAL HEIGHT

Let us consider the frame with unequal columns as shown in Figure 20.20 with a horizontal load P acting at node A.

As the column is unsymmetrical both in geometry and loading conditions, it will sway toward the right in the direction of the applied force. Due to equilibrium, base shear forces drawn near each column's supports will be developed, which will counteract the effect of the horizontal external load P. Under this situation, the base shear in each column can be written as:

$$H_{AB} = \frac{(M_{AB} + M_{BA})}{h_1}$$

Likewise,

$$H_{CD} = \frac{(M_{CD} + M_{DC})}{h_2}$$

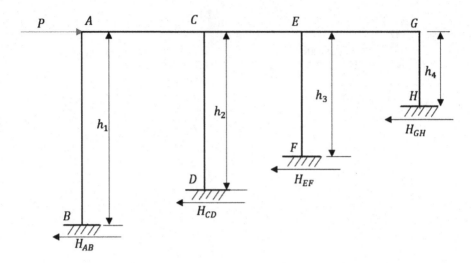

FIGURE 20.20 Frame with sidesway and unequal columns.

$$H_{EF} = \frac{(M_{EF} + M_{FE})}{h_3}$$

$$H_{GH} = \frac{(M_{GH} + M_{HG})}{h_4}$$

Following horizontal force equilibrium conditions, we can write:

$$P + \sum H = 0$$

where $\sum H$ is the summation of all horizontal support reactions, i.e., total horizontal shear in all columns of a particular story, as shown in Figure 20.20.

Thus,

$$P + \sum \frac{M}{h} = 0$$

where $\sum M$ is the sum of the end moments of all the individual columns in a particular story as written above, and h is the corresponding length of those columns.

Now, if we take any reference height of a column, say h_r, and which is the maximum height as well of all the columns, we can modify the abovementioned equations as follows:

$$H_{AB} = \frac{(M_{AB} + M_{BA})}{h_r} \times \frac{h_r}{h_1} = \frac{(M_{AB} + M_{BA})}{h_r} C_{AB}$$

$$\vdots$$

$$H_{GH} = \frac{(M_{GH} + M_{HG})}{h_r} \times \frac{h_r}{h_4} = \frac{(M_{GH} + M_{HG})}{h_r} C_{GH}$$

Now we can substitute these values back in the force equilibrium equations for member AB to get:

$$P + \frac{M_{AB} + M_{BA}}{h_r} C_{AB} = 0$$

$$P + \frac{1}{h_r}\left(\bar{M}_{AB} + 2M'_{AB} + M'_{BA} + M''_{AB} + \bar{M}_{BA} + 2M'_{BA} + M'_{AB} + M''_{BA}\right)C_{AB} = 0$$

or,

$$P + \frac{1}{h_r}\left(2M'_{AB} + M'_{BA} + M''_{AB} + 2M'_{BA} + M'_{AB} + M''_{BA}\right)C_{AB} = 0$$

Since $\bar{M}_{AB} = \bar{M}_{BA} = 0$, as no external loads are applied in between the joints.

$$P + \frac{1}{h_r}\left(3M'_{AB} + 3M'_{BA} + 2M''_{AB}\right)C_{AB} = 0$$

or,

$$M''_{AB} C_{AB} = -\frac{3}{2}\left[\frac{Ph_r}{3} - \left(M'_{AB}C_{AB} + M'_{BA}C_{AB}\right)\right]$$

For all columns, we get:

$$\sum M''_{AB} C_{AB} = -\frac{3}{2}\left[\frac{Ph_r}{3} - \sum\left(M'_{AB}C_{AB} + M'_{BA}C_{AB}\right)\right] \qquad (20.15)$$

Now, moment generated at the ends of AB column due to sway of δ amount:

$$M''_{AB} = \frac{6EI_{AB}\delta}{h_{AB}^2} = \frac{6E\delta}{h_{AB}}\frac{I_{AB}}{h_{AB}} = C\frac{K_{AB}}{h_{AB}} \qquad (20.16)$$

where $C = 6E\delta$. Furthermore, we can expand the abovementioned equation as follows:

$$M''_{AB}\frac{h_r}{h_{AB}} = C \times \frac{K_{AB}}{h_{AB}} \times \frac{h_r}{h_{AB}}$$

or,

$$M''_{AB}C_{AB} = \frac{C}{h_r}K_{AB}\frac{h_r^2}{h_{AB}^2}$$

or,

$$M''_{AB}C_{AB} = \frac{C}{h_r}K_{AB}C_{AB}^2$$

So, summing for all the columns, we get:

$$\sum M''_{AB}C_{AB} = \frac{C}{h_r}\sum K_{AB}C_{AB}^2$$

or,

$$C = h_r\frac{\sum M''_{AB}C_{AB}}{\sum K_{AB}C_{AB}^2} \tag{20.17}$$

Again writing equation (20.16),

$$M''_{AB} = C\frac{K_{AB}}{h_{AB}}$$

$$= h_r\frac{\sum M''_{AB}C_{AB}}{\sum K_{AB}C_{AB}^2} \times \frac{K_{AB}}{h_{AB}}$$

$$= K_{AB}\frac{h_r}{h_{AB}} \times \frac{\sum M''_{AB}C_{AB}}{\sum K_{AB}C_{AB}^2}$$

$$\therefore M''_{AB} = \frac{K_{AB}C_{AB}}{\sum K_{AB}C_{AB}^2} \times \sum M''_{AB}C_{AB} \tag{20.18}$$

Substituting, the value of $\sum M''_{AB} C_{AB}$ from equation (20.15) into equation (20.18),

$$\therefore M''_{AB} = -\frac{K_{AB} C_{AB}}{\sum K_{AB} C^2_{AB}} \times \frac{3}{2}\left[\frac{Ph_r}{3} - \sum\left(M'_{AB} C_{AB} + M'_{BA} C_{AB}\right)\right]$$

Thus, the displacement coefficient is given by:

$$\gamma_{AB} = -\frac{3}{2}\frac{K_{AB} C_{AB}}{\sum K_{AB} C^2_{AB}}$$

And story moment is given by:

$$\frac{Ph_r}{3}$$

If there is no horizontal force acting on the frame system,

$$M''_{AB} = -\frac{K_{AB} C_{AB}}{\sum K_{AB} C^2_{AB}} \times \frac{3}{2}\left[\sum\left(M'_{AB} C_{AB} + M'_{BA} C_{AB}\right)\right]$$

If heights of all columns are identical, then:

$$C_{AB} = C_{CD} = C_{EF} = C_{GH} = 1$$

In such case, the abovementioned equation becomes:

$$M''_{AB} = -\frac{3}{2}\frac{K_{AB}}{\sum K_{AB}}\sum\left(M'_{AB} + M'_{BA}\right)$$

Let us discuss Example 20.5 to clarify the concept just discussed.

Example 20.5: Analyze the frame shown in Figure 20.21 using Kani's method.

SOLUTION: *Fixed end moments:*

$$\bar{M}_{BC} = -\frac{wl^2}{12} = -3 \text{ kNm}$$

$$\bar{M}_{CB} = \frac{wl^2}{12} = 3 \text{ kNm}$$

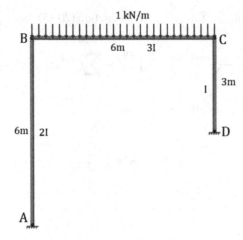

FIGURE 20.21 Example problem of frame system with loading.

Rotation contributions at different joints are as follows:

$$\mu_{BA} = -\frac{1}{2}\frac{2I/6}{2I/6+3I/6} = -0.2$$

$$\mu_{BC} = -\frac{1}{2}\frac{3I/6}{2I/6+3I/6} = -0.3$$

$$\mu_{CB} = -\frac{1}{2}\frac{3I/6}{2I/6+3I/6} = -0.3$$

$$\mu_{CD} = -\frac{1}{2}\frac{I/3}{I/3+3I/6} = -0.2$$

Now, let us consider the height, $h_r = 6$ m. Then, we get:

$$C_{AB} = \frac{h_r}{h_{AB}} = 1$$

$$C_{CD} = \frac{h_r}{h_{CD}} = 2$$

Displacement factors are given by:

$$\gamma_{AB} = -\frac{3}{2}\frac{K_{AB}C_{AB}}{\Sigma K_{AB}C_{AB}^2} = -\frac{3}{2}\frac{2I/6\times1}{2I/6\times1^2+I/3\times0.5^2} = -0.3$$

$$\gamma_{CD} = -\frac{3}{2}\frac{I/3\times0.5}{2I/6\times1^2+I/3\times0.5^2} = -0.6$$

Iteration: Cycle 1

Rotation Contributions			
Joint B		**Joint C**	
Summation of the fixed end moments	−3	Summation of the fixed end moments	3
Far end rotation contributions		Far end rotation contributions	
At A (M'_{AB})	0	At B (M'_{BC})	0.9
At C (M'_{CB})	0 (assumed)	At D (M'_{DC})	0
Displacement contributions		Displacement contributions	
BC (M''_{BC})	0	CB (M''_{CB})	0
BA (M''_{BA})	0 (assumed)	CD (M''_{CD})	0 (assumed)
Total	**−3**	**Total**	**3.9**

$$M'_{BA}$$
$$= \mu_{BA}\left\{\sum \bar{M}_B \right.$$
$$+ \sum M'_{jB}$$
$$\left. + \sum M''_{Bj}\right\}_{j=A,C}$$

$$M'_{BA} = (-0.2)(-3)$$
$$= 0.6$$

$$M'_{CB}$$
$$= \mu_{CB}\left\{\sum \bar{M}_C \right.$$
$$+ \sum M'_{jC}$$
$$\left. + \sum M''_{Cj}\right\}_{j=B,D}$$

$$M'_{CB} = (-0.3)(3.9)$$
$$= -1.17$$

$$M'_{BC}$$
$$= \mu_{BC}\left\{\sum \bar{M}_B \right.$$
$$+ \sum M'_{jB}$$
$$\left. + \sum M''_{Bj}\right\}_{j=A,C}$$

$$M'_{BC} = (-0.3)(-3)$$
$$= 0.9$$

$$M'_{CD}$$
$$= \mu_{CD}\left\{\sum \bar{M}_C \right.$$
$$+ \sum M'_{jC}$$
$$\left. + \sum M''_{Cj}\right\}_{j=B,D}$$

$$M'_{CD} = (-0.2)(3.9)$$
$$= -0.78$$

Displacement Contributions	
Rotation contributions at the top of columns	
For AB (M'_{BA})	0.6
For DC (M'_{CD})	−0.78
Rotation contributions at the bottom of columns	
For AB (M'_{AB})	0
For DC (M'_{DC})	0
Story moment $\left(\frac{Ph_r}{3}\right)$	0

$$M''_{BA} = \gamma_{BA}\left(\frac{Ph_r}{3} + \sum (M'_{BA} + M'_{AB})C_{AB}\right)$$

$$M''_{BA} = (-0.3)\left[\{(0.6+0)\times 1\} + \{(-0.78+0)\times 2\}\right] = 0.288$$

$$M''_{CD} = \gamma_{CD}\left(\frac{Ph_r}{3} + \sum (M'_{CD} + M'_{DC})C_{CD}\right)$$

$$M''_{CD} = (-0.6)\left[\{(0.6+0)\times 1\} + \{(-0.78+0)\times 2\}\right] = 0.576$$

Iteration: Cycle 2

		Rotation Contributions		
	Joint B		**Joint C**	
Summation of the fixed end moments	-3		Summation of the fixed end moments	3
Far end rotation contributions			Far end rotation contributions	
At A (M'_{AB})	0		At B (M'_{BC})	1.165
At C (M'_{CB})	-1.17		At D (M'_{DC})	0
Displacement contributions			Displacement contributions	
BC (M''_{BC})	0		CB (M''_{CB})	0
BA (M''_{BA})	0.288		CD (M''_{CD})	0.576
Total	**-3.882**		**Total**	**4.741**

$$M'_{BA}$$
$$= \mu_{BA}\left\{\sum \bar{M}_B\right\}$$
$$+ \sum M'_{jB}$$
$$+ \sum M''_{Bj}\bigg\}_{j=A,C}$$

$$M'_{BA}$$
$$= (-0.2)(-3.882)$$
$$= 0.7764$$

$$M'_{CB}$$
$$= \mu_{CB}\left\{\sum \bar{M}_C\right\}$$
$$+ \sum M'_{jC}$$
$$+ \sum M''_{Cj}\bigg\}_{j=B,D}$$

$$M'_{CB}$$
$$= (-0.3)(4.741)$$
$$= -1.4222$$

$$M'_{BC}$$
$$= \mu_{BC}\left\{\sum \bar{M}_B\right\}$$
$$+ \sum M'_{jB}$$
$$+ \sum M''_{Bj}\bigg\}_{j=A,C}$$

$$M'_{BC}$$
$$= (-0.3)(-3.882)$$
$$= 1.165$$

$$M'_{CD}$$
$$= \mu_{CD}\left\{\sum \bar{M}_C\right\}$$
$$+ \sum M'_{jC}$$
$$+ \sum M''_{Cj}\bigg\}_{j=B,D}$$

$$M'_{CD}$$
$$= (-0.2)(4.741)$$
$$= -0.9482$$

	Displacement Contributions	
Rotation contributions at the top of columns		
For AB (M'_{BA})	0.7764	
For DC (M'_{CD})	−0.9482	
Rotation contributions at the bottom of columns		
For AB (M'_{AB})	0	
For DC (M'_{DC})	0	
Story moment $\left(\frac{Ph_r}{3}\right)$	0	

$$M''_{BA} = \gamma_{BA}\left(\frac{Ph_r}{3} + \sum (M'_{BA} + M'_{AB})C_{AB}\right)$$

$$M''_{BA} = (-0.3)\Big[\{(0.7764+0)\times 1\} + \{(-0.9482+0)\times 2\}\Big] = 0.336$$

$$M''_{CD} = \gamma_{CD}\left(\frac{Ph_r}{3} + \sum (M'_{CD} + M'_{DC})C_{CD}\right)$$

$$M''_{CD} = (-0.6)\Big[\{(0.7764+0)\times 1\} + \{(-0.9482+0)\times 2\}\Big] = 0.672$$

Iteration: Cycle 3

Rotation Contributions			
Joint B		**Joint C**	
Summation of the fixed end moments	−3	Summation of the fixed end moments	3
Far end rotation contributions		Far end rotation contributions	
At A (M'_{AB})	0	At B (M'_{BC})	1.226
At C ((M'_{CB})	−1.4222	At D (M'_{DC})	0
Displacement contributions		Displacement contributions	
BC (M''_{BC})	0	CB (M''_{CB})	0
BA ((M''_{BA})	0.336	CD (M''_{CD})	0.672
Total	**−4.0862**	**Total**	**4.898**

$$M'_{BA}$$
$$= \mu_{BA}\left\{\sum \bar{M}_B \right.$$
$$+ \sum M'_{jB}$$
$$\left. + \sum M''_{Bj}\right\}_{j=A,C}$$

M'_{BA}
$= (-0.2)(-4.0862)$
$= 0.81724$

$$M'_{CB}$$
$$= \mu_{CB}\left\{\sum \bar{M}_C \right.$$
$$+ \sum M'_{jC}$$
$$\left. + \sum M''_{Cj}\right\}_{j=B,D}$$

M'_{CB}
$= (-0.3)(4.898)$
$= -1.4694$

$$M'_{BC}$$
$$= \mu_{BC}\left\{\sum \bar{M}_B \right.$$
$$+ \sum M'_{jB}$$
$$\left. + \sum M''_{Bj}\right\}_{j=A,C}$$

M'_{BC}
$= (-0.3)(-4.0862)$
$= 1.226$

$$M'_{CD}$$
$$= \mu_{CD}\left\{\sum \bar{M}_C \right.$$
$$+ \sum M'_{jC}$$
$$\left. + \sum M''_{Cj}\right\}_{j=B,D}$$

M'_{CD}
$= (-0.2)(4.898)$
$= -0.9796$

Displacement Contributions	
Rotation contributions at the top of columns	
For AB (M'_{BA})	0.81724
For DC (M'_{CD})	−0.9796
Rotation contributions at the bottom of columns	
For AB (M'_{AB})	0
For DC (M'_{DC})	0
Story moment $\left(\frac{Ph_r}{3}\right)$	0

$$M''_{BA} = \gamma_{BA}\left(\frac{Ph_r}{3} + \sum (M'_{BA} + M'_{AB})C_{AB}\right)$$

$$M''_{BA} = (-0.3)\left[\{(0.81724+0)\times 1\} + \{(-0.9796+0)\times 2\}\right] = 0.3426$$

$$M''_{CD} = \gamma_{CD}\left(\frac{Ph_r}{3} + \sum (M'_{CD} + M'_{DC})C_{CD}\right)$$

$$M''_{CD} = (-0.6)\left[\{(0.81724+0)\times 1\} + \{(-0.9796+0)\times 2\}\right] = 0.6852$$

Iteration: Cycle 4

	Rotation Contributions		
Joint B		**Joint C**	
Summation of the fixed end moments	−3	*Summation of the fixed end moments*	3
Far end rotation contributions		*Far end rotation contributions*	
At A (M'_{AB})	0	At B (M'_{BC})	1.238
At C (M'_{CB})	−1.4694	At D (M'_{DC})	0
Displacement contributions		*Displacement contributions*	
BC (M''_{BC})	0	CB (M''_{CB})	0
BA ((M''_{BA})	0.3426	CD ((M''_{CD})	0.6852
Total	**−4.127**	**Total**	**4.9232**

M'_{BA}

$= \mu_{BA} \left\{ \sum \bar{M}_B \right\}$

$+ \sum M'_{jB}$

$+ \sum M''_{Bj} \Big\}_{j=A,C}$

M'_{BA}

$= (-0.2)(-4.127)$

$= 0.8254$

M'_{CB}

$= \mu_{CB} \left\{ \sum \bar{M}_C \right\}$

$+ \sum M'_{jC}$

$+ \sum M''_{Cj} \Big\}_{j=B,D}$

M'_{CB}

$= (-0.3)(4.9232)$

$= -1.477$

M'_{BC}

$= \mu_{BC} \left\{ \sum \bar{M}_B \right\}$

$+ \sum M'_{jB}$

$+ \sum M''_{Bj} \Big\}_{j=A,C}$

M'_{BC}

$= (-0.3)(-4.127)$

$= 1.238$

M'_{CD}

$= \mu_{CD} \left\{ \sum \bar{M}_C \right\}$

$+ \sum M'_{jC}$

$+ \sum M''_{Cj} \Big\}_{j=B,D}$

M'_{CD}

$= (-0.2)(4.9232)$

$= -0.9846$

Displacement Contributions	
Rotation contributions at the top of columns	
For AB (M'_{BA})	0.8254
For DC (M'_{CD})	−0.9846
Rotation contributions at the bottom of columns	
For AB (M'_{AB})	0
For DC (M'_{DC})	0
Story moment $\left(\frac{Ph_r}{3} \right)$	0
Total	**0.00**

$$M''_{BA} = \gamma_{BA} \left(\frac{Ph_r}{3} + \sum (M'_{BA} + M'_{AB})C_{AB} \right)$$

$$M''_{BA} = (-0.3)\left[\{(0.8254 + 0) \times 1\} + \{(-0.9846 + 0) \times 2\} \right] = 0.3431$$

$$M''_{CD} = \gamma_{CD} \left(\frac{Ph_r}{3} + \sum (M'_{CD} + M'_{DC})C_{CD} \right)$$

$$M''_{CD} = (-0.6)\left[\{(0.81724 + 0) \times 1\} + \{(-0.9796 + 0) \times 2\} \right] = 0.6863$$

TABLE 20.8
Values of Rotation and Displacement Contributions for All the Iterations

Cycle	M'_{BA}	M'_{BC}	M'_{CB}	M'_{CD}	M''_{BA}	M''_{CD}
First cycle	0.6	0.9	−1.17	−0.78	0.288	0.576
Second cycle	0.7764	1.165	−1.4222	−0.9482	0.336	0.672
Third cycle	0.81724	1.226	−1.4694	−0.9796	0.3426	0.6852
Fourth cycle	0.8254	1.238	−1.477	−0.9846	0.3431	0.6863

After performing the fourth cycle, it was found that the rotation and displacement contribution values are getting converged, i.e., no more cycle is required. So, the values obtained in the fourth cycle are used to calculate the final end moments. The obtained values from all the iterations for rotation and displacement contribution are given in Table 20.8.

The end moments have been calculated as shown in Figure 20.22.

The shear force and bending moment diagrams are shown in Figure 20.23 (a) and (b), respectively.

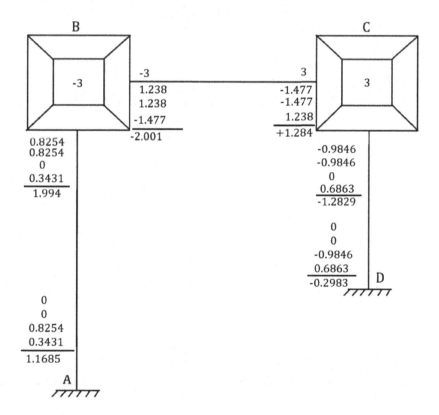

FIGURE 20.22 Determining end moments by Kani's method.

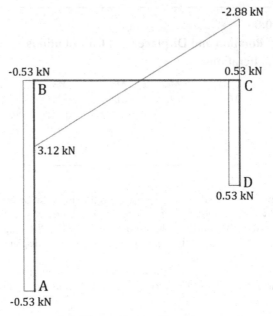

(a) Shear force diagram

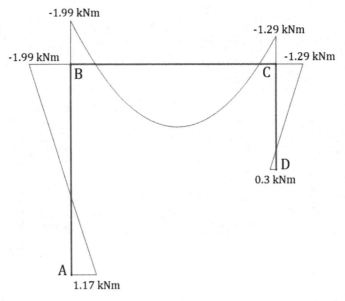

(b) Bending moment diagram

FIGURE 20.23 (a) Shear force diagram and (b) bending moment diagram of the portal frame.

21 Column Analogy Method

21.1 INTRODUCTION

The basic concept of the column analogy method circles around analyzing the statically indeterminate beams of single span, portal frames, closed box frames with some assumptions related to the given properties of that structural element. We adopt this method to determine the redundant force or moments of statically indeterminate structural elements more easily than all other methods we have learned so far. This method provides the exact value of the redundant force and moment in contrast to other approximate methods like moment distribution. This method is based on the similarity between the moments induced in a statically indeterminate structure, and the stresses produced in an eccentrically loaded short column.

21.2 BASIC CONCEPT

We begin with the concept of stress for eccentrically loaded short columns. For such columns, stress at any point is linearly proportional to the distance of the point from the point of application of load (Figure 21.1).

As shown in Figure 21.1, stress (f) at any point A with coordinates (x, y), can be written as:

$$f = a + bx + cy$$

where a, b, c are constants.

So, for any small elemental area δA near point A, the force acting in that area will be:

$$\delta P = f \delta A$$

So, the total force acting on the column section can be expressed as:

$$P = \int \delta P = \int f \delta A = \int f dA$$

or,

$$P = \int (a + bx + cy) dA$$

$$P = a \int dA + b \int x dA + c \int y dA$$

DOI: 10.1201/9781003081227-24

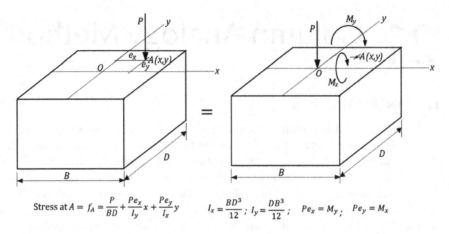

Stress at $A = f_A = \dfrac{P}{BD} + \dfrac{Pe_x}{I_y}x + \dfrac{Pe_y}{I_x}y$ $\quad I_x = \dfrac{BD^3}{12}\, ;\, I_y = \dfrac{DB^3}{12}\, ;\quad Pe_x = M_y\, ;\quad Pe_y = M_x$

FIGURE 21.1 Short column section with eccentric loading.

If the reference axis passes through the center of gravity of the section, the point O coincides with center of gravity.

$$\int x\,dA = 0 = \int y\,dA$$

So, in that case, we are left with:

$$P = a\int dA = aA$$

where A is the total area of cross section of the column.

So,

$$a = \frac{P}{A}$$

The moment of the elementary force δP about x axis is given by:

$$\delta M_x = y\delta P$$

So, the total moment:

$$M_x = \int y\,dP = \int yf\,dA = \int y\left(a + bx + cy\right)dA$$

or,

$$M_x = a\int y\,dA + b\int xy\,dA + c\int y^2\,dA = 0 + bI_{xy} + cI_x \qquad (21.1)$$

where I_{xy} is the cross moment of inertia and I_x is the moment of inertia about x axis.

Similarly, for moment about y axis, we get:

$$M_y = \int x\,dP = \int xf\,dA = \int x\left(a + bx + cy\right)dA$$

or,

$$M_y = 0 + bI_y + cI_{xy} \tag{21.2}$$

So, solving equations (21.1) and (21.2), we get:

$$b = \frac{M_y I_x - M_x I_{xy}}{I_y I_x - I_{xy}^2}$$

$$c = \frac{M_x I_y - M_y I_{xy}}{I_y I_x - I_{xy}^2}$$

Hence, by substituting these values in the original stress equation, we can determine the full equation for stress at any point of the column. Moreover, it is customary to note that if the reference axis is the principal axis, $I_{xy} = 0$.

$$f = \frac{P}{A} + \frac{M_y}{I_y} x + \frac{M_x}{I_x} y$$

Now we set our attention toward the bending of a curved beam, CB as shown in Figure 21.2.

In the following curved beam *CB*, the *CB* portion has been rotated through an angle $\delta\theta$ at point *C*. Due to this rotation, point *B* has been shifted to new location *B'*. Also, the point *T* assumes the new location *T'*. The angle *TCT'* is $\delta\theta$. Let δV and δH are vertical and horizontal displacements of the point *B'* with respect to *B*. Let angle *BTC* is φ. So, angle *B' T' C* will also be φ. Draw a line parallel to *B' T'* at *B*. Let the angle between this line and *BT* be $\delta\varphi$. $\delta\varphi$ will be equal to $\delta\theta$.

Since all the angles are infinitesimally small,

$$BB' = BC \; \delta\theta$$

So,

$$\delta V = BB' \cos\theta$$

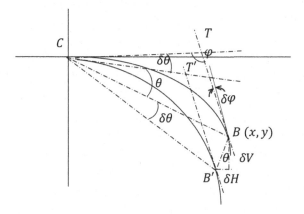

FIGURE 21.2 Bending of a curved beam.

or,

$$\delta V = BC \cos\theta \; \delta\theta$$

Implies,

$$\delta V = x\delta\theta$$

Similarly,

$$\delta H = BB' \sin\theta = BC \sin\theta \; \delta\theta = y\delta\theta$$

Considering the beam as a rigid body excepting a small length δs at C, the change in slope at C will be equal to $M\delta s/EI$, where M is the bending moment of the beam at point C.

The change in slope at B will be $\dfrac{M\delta s}{EI}$.

Thus, we get the following relationships:

$$\delta\varphi = \frac{M\delta s}{EI}$$

$$\delta V = \frac{Mx\delta s}{EI}$$

$$\delta H = \frac{My\delta s}{EI}$$

Now we are ready to learn the column analogy method that will be developed in the following section.

21.3   DEVELOPMENT OF THE COLUMN ANALOGY METHOD

Indeterminate structure consists of two types of bending moments, one moment (M_i) considering the structure as indeterminate one, and the other will be the moment (M_s) in the primary structure by removing the redundancy from the structure. Hence, the total moment can be written as:

$$M = M_s + M_i$$

Let θ be the relative rotation of the ends of the structure, H be the relative horizontal displacement of the ends, and V be the relative vertical displacement of the ends.

$$\theta = \int \frac{Mds}{EI}$$

$$H = \int \frac{Myds}{EI}$$

$$V = \int \frac{Mxds}{EI}$$

In case the relative rotation and displacements at the ends are zero:

$$\theta = 0 = \int \frac{Mds}{EI}$$

or,

$$\theta = 0 = \int \frac{(M_s + M_i)ds}{EI}$$

So,

$$\int \frac{(M_i)ds}{EI} = -\int \frac{(M_s)ds}{EI}$$

Similarly,

$$H = 0 = \int \frac{Myds}{EI} \Rightarrow \int \frac{(M_i y)ds}{EI} = -\int \frac{(M_s)ds}{EI} y$$

Also,

$$V = 0 = \int \frac{Mxds}{EI} \Rightarrow \int \frac{(M_i x)ds}{EI} = -\int \frac{(M_s)ds}{EI} x$$

Now, let us consider a short column of width $1/EI$, and the load intensity is $-M_s$.

So, the total load, $P = -\int \frac{M_s ds}{EI}$
So,

$$\int \frac{M_i ds}{EI} = P$$

or,

$$dP = \frac{M_i ds}{EI}$$

or,

$$\int \frac{M_i yds}{EI} = \int ydP$$

$$\int \frac{M_i xds}{EI} = \int xdP$$

Now, comparing these values with the stress values calculated for column with eccentric load, we get:

$$\int fdA = P$$

$$\int fydA = M_x = \int ydP$$

$$\int fxdA = M_y = \int xdP$$

Analogously, for the imagined column, ds/EI will be the elementary area, f will be the stress at any point due to load intensity $-M_s$ or the total load $-\int M_s ds/EI$.

So, we have got:

$$M_i = \frac{P}{A} + \frac{M_y I_x - M_x I_{xy}}{I_y I_x - I_{xy}^2} \times x + \frac{M_x I_y - M_y I_{xy}}{I_y I_x - I_{xy}^2} \times y$$

or,

$$M_i = \frac{P}{A} + \frac{M_y - M_x \left(I_{xy}/I_x\right)}{I_y \left[1 - \left(I_{xy}^2/I_x I_y\right)\right]} \times x + \frac{M_x - M_y \left(I_{xy}/I_y\right)}{I_x \left[1 - \left(I_{xy}^2/I_y I_x\right)\right]} \times y$$

Putting,

$$M_y' = M_y - M_x \frac{I_{xy}}{I_x}$$

And,

$$M_x' = M_x - M_y \frac{I_{xy}}{I_y}$$

$$I_y' = I_y \left(1 - \frac{I_{xy}^2}{I_x I_y}\right)$$

$$I_x' = I_x \left(1 - \frac{I_{xy}^2}{I_y I_x}\right)$$

Hence, the final bending moment equation becomes:

$$M_i = \frac{P}{A} + \frac{M_y'}{I_y'} \times x + \frac{M_x'}{I_x'} \times y$$

M_s will be positive if it induces tension in the inside fibers and it will be negative if it induces tension in the outer fibers. For positive M_s, P will be tensile and negative, and for negative M_s, P will be compressive and positive. M_i is positive if f is compressive. Also, for any structure, if the support is hinged, it does not offer any resistance to rotation and can take any rotation. The flexural rigidity EI at a simply supported end is zero, and hence, $1/EI$ is taken as infinite, and for a fixed support, there will be no rotations and the flexural rigidity EI is infinite; hence, $1/EI$ is taken as zero. The thickness of the load diagram is the same as the analogous column section. So, if the beam is of uniform flexural rigidity, the thickness of the analogous column and that of the load diagram on this column may be taken as unity, since the quantity $1/EI$ cancels out.

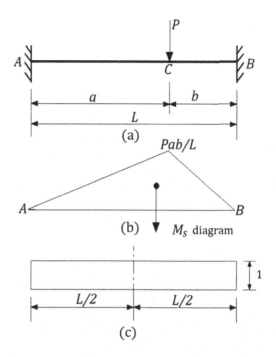

FIGURE 21.3 Example problem on column analogy method for fixed beam.

Example 21.1: **A fixed beam of span** L **carries a point load** P **eccentrically on the span at a distance** a **from the left end and** b **from the right end as shown in Figure 21.3(a). Find the fixed end moments at the ends of the beam.**

SOLUTION: Let M_{AB} and M_{BA} be the fixed end moments at A and B, respectively, and these are redundants. If we remove these two redundants, we will get the basic determinate simply supported beam. For this simply supported beam, the M_s diagram is shown in Figure 21.3 (b). The height of the M_s diagram is $\frac{Pab}{L}$.

The beam is having uniform flexural rigidity throughout its length. So, the thickness of the M_s diagram and the thickness of the analogous column are taken as unity.

Total load on the analogous column $= \frac{1}{2} L \frac{Pab}{L} \times 1 = \frac{Pab}{2}$

C.G. of the M_s diagram from A, $= \dfrac{\frac{1}{2} \times a \times \frac{Pab}{L} \times \frac{2a}{3} + \frac{1}{2} \times b \times \frac{Pab}{L} \times \left(a + \frac{b}{3}\right)}{\frac{1}{2} \times L \times \frac{Pab}{L}} = \frac{L+a}{3}$

$\therefore$ Eccentricity of the load $= \left(\frac{L+a}{3}\right) - \frac{L}{2} = \frac{2a-L}{6}$

Area of the analogous column section $= L \times 1 = L$
Moment of inertia of the analogous column section $= \frac{1 \times L^3}{12} = \frac{L^3}{12}$

Stress at any point of the column section, $M_i = \dfrac{\frac{Pab}{2}}{L} \pm \dfrac{\frac{Pab}{2} \times \frac{2a-L}{6}}{\frac{L^3}{12}} \times x$

or, $M_i = \frac{Pab}{2L} \pm \frac{Pab(2a-L)}{L^3} x$

Stress at A, $M_{iA} = \dfrac{Pab}{2L} - \dfrac{Pab(2a-L)}{L^3}\dfrac{L}{2} = \dfrac{Pab}{L^2}(L-a) = \dfrac{Pab^2}{L^2}$

Stress at B, $M_{iB} = \dfrac{Pab}{2L} + \dfrac{Pab(2a-L)}{L^3}\dfrac{L}{2} = \dfrac{Pab}{2L^2}(L+2a-L) = \dfrac{Pa^2b}{L^2}$

Fixed end moment at A, $M_{AB} = M_{sA} - M_{iA} = 0 - \dfrac{Pab^2}{L^2} = -\dfrac{Pab^2}{L^2}$

Fixed end moment at B, $M_{BA} = M_{sB} - M_{iB} = 0 - \dfrac{Pa^2b}{L^2} = -\dfrac{Pa^2b}{L^2}$

21.4 STIFFNESS AND CARRY OVER FACTORS DETERMINED BY METHOD OF COLUMN ANALOGY

In simple words, stiffness is the value of moment to be applied at an end to cause slope of one radian and carry over factor is the ratio of moment generated at the far end due to the applied moment at the joint under investigation. Let us consider a fixed beam of rigidity EI as shown in Figure 21.4. The loading on the analogous column will be θ_A at A and width of the column will be $1/EI$ as shown in Figure 21.4. For the clockwise slope at A, the loading will be positive, and for clockwise slope at B, loading will be negative.

So, for this analogous column, the area of cross section will be $l \times 1/EI$. So, we get:

$$f = \frac{P}{A} + \frac{P \times e \times y}{I_y}$$

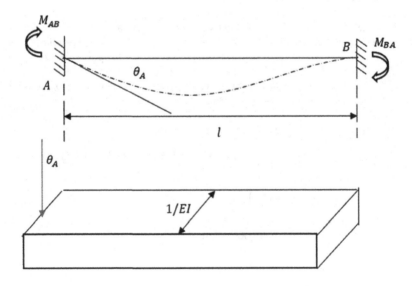

FIGURE 21.4 Fixed beam and corresponding analogous column model.

In the case of our analogous column model, we have $P = \theta_A$, $A = l/EI$, $e = l/2$, $y = l/2$, $I_y = l^3/12EI$.

So, substituting these values in the above equation, we get:

$$f_A = M_{AB} = \frac{\theta_A}{l/EI} + \frac{\theta_A \times (l/2) \times (l/2)}{l^3/12EI} = \frac{4EI\theta_A}{l}$$

So,

$$M_{AB} = M_s + M_i = 0 + f_A = 0 + \frac{4EI\theta_A}{l}$$

Remembering the definition of stiffness, putting $\theta_A = 1$, we get:

$$M_{AB} = \frac{4EI}{l}$$

This is in perfect agreement with the earlier derived stiffness factor for fixed beam found in moment distribution chapter (Chapter 19).

Also, let us check the stress at support B:

$$f_B = \frac{P}{A} - P \times e \times \frac{y}{I_y}$$

or,

$$f_B = \frac{\theta_A}{l/EI} - \frac{3EI\theta_A}{l} = -\frac{2EI\theta_A}{l}$$

So,

$$M_{BA} = M_s + M_i = 0 + f_B = -\frac{2EI\theta_A}{l} = -\frac{1}{2} M_{AB}$$

So, the carry over factor is 1/2 as we have already seen in the Chapter 19. So, all the results by earlier methods as discussed can be derived by constructing an imaginary column with geometric and loading properties as explained.

21.5 FIXED END MOMENTS DUE TO SUPPORT SETTLEMENT

Consider the same prismatic beam as drawn in Figure 21.5 with one exception. Let the support at B settles by an amount Δ with respect to support A. The slope of the chord joining the supports A and B, at this settled condition, will be $\theta = \Delta/l$.

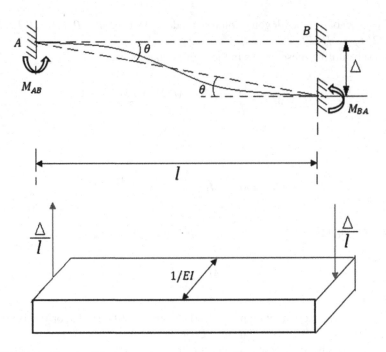

FIGURE 21.5   Fixed end moments due to settlement of supports.

So, following the earlier explained logic, the loading in the analogous column model will be Δ/l at B in downward direction and at A Δ/l in the upward direction.

As usual, the width of this analogous column will be $1/EI$. The net vertical load acting on the column $= P = 0$.

So,

$$M_y = \frac{\Delta}{l} \times l = \Delta$$

$$f_B = \frac{P}{A} + M_y \times \frac{x}{I_y} = 0 + \frac{\Delta \times (l/2)}{l^3/12EI} = +\frac{6EI\Delta}{l^2}$$

$$M_{BA} = 0 + f_B = \frac{6EI\Delta}{l^2}$$

Similarly,

$$f_A = 0 + \frac{\Delta \times (l/2)}{l^3/12EI} = \frac{6EI\Delta}{l^2}$$

So,

$$M_{AB} = 0 + f_A = \frac{6EI\Delta}{l^2}$$

So, these are also in perfect agreement with the earlier derived result in Chapters 18 and 19.

21.6 ANALYSIS OF PORTAL FRAMES

Continuing from the previous section, we will apply the method of column analogy to analyze the portal frames. We will show this method through Example 21.2. Please note this frame is not symmetrical due to loading and not due to its geometry.

Example 21.2: Analyze the fixed portal frame shown in Figure 21.6.

SOLUTION: Let us obtain a basic determinate structure as shown in Figure 21.7, of the given portal frame with A as the roller end and D as the hinged end.

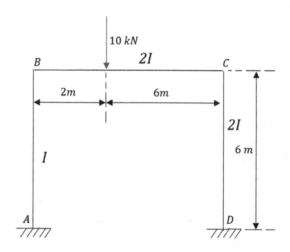

FIGURE 21.6 Portal frame example problem.

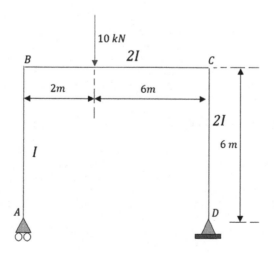

FIGURE 21.7 Determinate frame after removing the redundants.

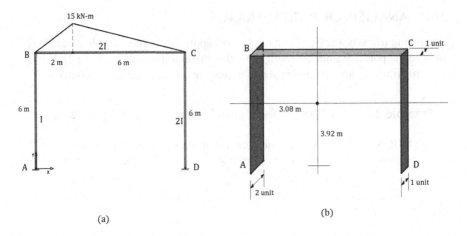

FIGURE 21.8   (a) M_s diagram and (b) analogous column of the example problem.

Figure 21.8 (a) shows the M_s diagram, and Figure 21.8 (b) shows the analogous column section. For AB the thickness of the section $\left(\frac{1}{EI}\right)$ is taken equal to 2. For BC and CD, the thickness of the section $\left(\frac{1}{2EI}\right)$ is taken equal to 1.

Load on the analogous column = Volume of the M_s diagram = $\frac{1}{2} \times 8 \times 15 \times 1 = 60$ units.

This acts on BC at a distance $\frac{10}{3}$ m from B.

Total area of the analogous column section, $= (2 \times 6) + (8 \times 1) + (6 \times 1) = 26$ units

Let us take A as the origin. Let $(\bar{x}, \bar{y})$ be the coordinates of the centroid, with respect to A.

$$\bar{x} = \frac{(8 \times 4) + (6 \times 8)}{26} = 3.08 \text{ m}$$

$$\bar{y} = \frac{(12 \times 3) + (8 \times 6) + (6 \times 3)}{26} = 3.92 \text{ m}$$

$$I_{BC} = \frac{2 \times 6^3}{3} + \frac{1 \times 6^3}{3} = 216 \text{ units}$$

$$I_{xx} = I_{BC} - A \times 2.08^2 = 216 - 26 \times (6 - 3.92)^2 = 103.51 \text{ units}$$

$$I_{AB} = \frac{1 \times 8^3}{3} + 6 \times 8^2 = 554.67 \text{ units}$$

$$I_{yy} = I_{AB} - A \times 3.08^2 = 554.67 - 26 \times 3.08^2 = 308.02 \text{ units}$$

$$I_{xy} = A_1 x_1 y_1 + A_2 x_2 y_2 + A_3 x_3 y_3$$

or,

$$I_{xy} = 12 \times (-3.08) \times (-0.92) + 8 \times 0.92 \times 2.08 + 6 \times 4.92 \times (-0.92)$$
$$= 34.0 + 15.3 - 27.16 = 22.14 \text{ units}$$

$$M_{xx} = P \times e_y = 60 \times 2.08 = 124.8 \text{ units}$$

$$M_{yy} = P \times e_x = 60 \times \left(4 - \frac{10}{3}\right) = 60 \times 0.67 = 40 \text{ units}$$

$$b = \frac{M_{yy}I_{xy} - M_{xx}I_{xy}}{I_{yy}I_{xx} - I_{xy}^2} = \frac{(40 \times 103.51) - (124.8 \times 22.14)}{(308.02 \times 103.51) - 22.14^2} = \frac{1377.33}{31392.97} = 0.044$$

$$c = \frac{M_{xx}I_{yy} - M_{yy}I_{xy}}{I_{yy}I_{xx} - I_{xy}^2} = \frac{(124.8 \times 308.02) - (40 \times 27.16)}{(308.02 \times 103.51) - 22.14^2} = \frac{37354.5}{31392.97} = 1.19$$

Now, the stress at any point of the analogous column section,

$$M_i = \frac{P}{A} + bx + cy = \frac{60}{26} + 0.044x + 1.19y = 2.31 + 0.044x + 1.19y$$

Therefore,
   Stress at A, (–3.08, –3.92)

$$M_{iA} = 2.31 + 0.044(-3.08) + 1.19(-3.92) = -2.49 \text{ kNm}$$

Stress at D, (4.92, –3.92)

$$M_{iD} = 2.31 + 0.044(4.92) + 1.19(-3.92) = -2.14 \text{ kNm}$$

Stress at B, (–3.08, 2.08)

$$M_{iB} = 2.31 + 0.044(-3.08) + 1.19(2.08) = +4.65 \text{ kNm}$$

Stress at C, (4.92, 2.08)

$$M_{iB} = 2.31 + 0.044(4.92) + 1.19(2.08) = +5.0 \text{ kNm}$$

Final Moments:

$$M_A = M_{sA} - M_{iA} = 0 - (-2.49) = +2.49 \text{ kNm}$$

$$M_D = M_{sD} - M_{iD} = 0 - (-2.14) = +2.14 \text{ kNm}$$

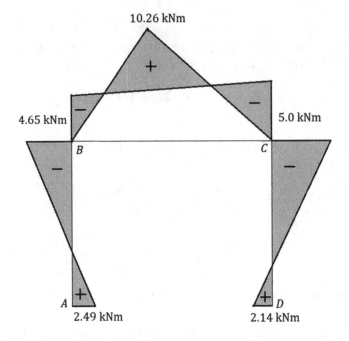

FIGURE 21.9   Final bending moment diagram of the portal frame.

$$M_B = M_{sB} - M_{iB} = 0 - (4.65) = -4.65 \text{ kNm}$$

$$M_C = M_{sC} - M_{iC} = 0 - (5.0) = -5.0 \text{ kNm}$$

Figure 21.9 shows the bending moment diagram of the portal frame.

22 Beams and Frames Having Nonprismatic Members

22.1 INTRODUCTION

In this chapter, we will apply previously acquired knowledge of slope deflection and moment distribution methods to analyze beams and frames composed of nonprismatic members. At first, how the necessary carry-over factors, stiffness factors, and fixed end moments are obtained will be discussed in detail. Finally, the analysis of statically indeterminate structures using the slope deflection and moment distribution methods will be discussed in detail.

Nonprismatic members are being used in recent large span structures like factory shades and large ceremony hall canopies. Modern days pre-engineered buildings are mostly used for storage sheds and various other industrial applications. In these structures, nonprismatic sections are used instead of making large truss members for holding the roof structure. Applications of these nonprismatic sections provide more aesthetic views and open spaces for other internal work installations.

22.2 DEFLECTIONS AND LOADING PROPERTIES OF NONPRISMATIC MEMBERS

As stated above, for economic design, girders used for long spans (in bridges or industrial sheds) are designed to be nonprismatic, that is, to have a variable moment of inertia. The most common form of nonprismatic member is the tapered section that is having more depth near the support and lesser depth at the crown portion. The depth of section is decided based on the detailed analysis and bending moment and shear force acting on the member at critical sections. If we can express the moment of inertia of these sections as function of x coordinates along the length of the member then we can use virtual work or Castigliano's methods to calculate the deflections. For recollection of earlier learned equations, deflection is calculated by applying the following:

$$\Delta = \int_0^l \frac{\partial M}{\partial P} \frac{M}{EI} dx$$

In many cases due to irregular geometry, the exact solution of the above integral is not possible. In those cases, various numerical methods and approximation techniques may be adopted to calculate the approximate value of deflection.

DOI: 10.1201/9781003081227-25

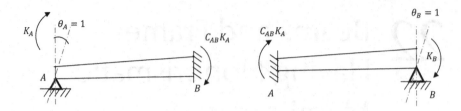

FIGURE 22.1 Stiffness coefficients and angular displacements.

To apply slope deflection or moment distribution methods, we need to quantify the following structural properties:

1. *Fixed end moments* – Assuming that supports are fixed, fixed end moments need to be calculated first for the given load applied on the same.
2. *Stiffness factor (K)* – The moment that is needed to be applied at the end of a member to make unit rotation at the end is called stiffness factor.
3. *Carry over factor* – It gives the amount of moment transferred from pin supported end to the fixed end of a structure.

There exists an important relationship between carry over factor and stiffness of structural elements. To understand the relationship, see Figure 22.1.

From the above diagram, we can apply Maxwell-Betti reciprocal theorem that stipulates that work done by loads in the first diagram with the displacement in the second diagram should be equal to the work done by the second diagram forces with the displacements in the first diagram. So, in short,

$$U_{AB} = U_{BA}$$

or,

$$K_A(0) + C_{AB}K_A(1) = C_{BA}K_B(1) + K_B(0)$$

or,

$$C_{AB}K_A = C_{BA}K_B$$

Although the relationship has been formed quite comfortably, determining numerical values of the above factors often involves considerable labor and efforts. To overcome these computational issues, design tables and graphs are often available in many standard books (see 'Bibliography') for ready reference. Most commonly used charts and tables are available in the *Handbook of Frame Constants* published by Portland Cement Association.+

22.3 MOMENT DISTRIBUTION FOR STRUCTURES HAVING NONPRISMATIC MEMBERS

After determining the fixed end moments, stiffness, and carry over factors for the nonprismatic element of a structure, application of the moment distribution method can be applied following the same procedure as outlined in Chapter 19. In this regard, remember that the distribution of moments may be shortened if a member stiffness factor is modified due to conditions of end-span pin support and structure symmetry or antisymmetry. Similar modifications are also required to be made to nonprismatic members.

22.3.1 BEAM PIN SUPPORTED AT FAR END

Let us consider the beam in Figure 22.2, which is pinned at its far end B. The absolute stiffness factor is the moment applied at A such that it rotates the beam at A, this can be calculated as follows. At first, let us assume that B is fixed end support, and a moment is applied at support A, as shown in Figure 22.2 (b). The moment generated at B is the carry over factor from A to B. Since B is not actually fixed, application of the opposite moment to the beam, Figure 22.2 (c), will generate a moment at support A. By superposition, the results of these two applications of moment produce the beam loaded as shown in Figure 22.2 (a). So, the absolute stiffness factor of the beam at support A is given by:

$$K'_A = K_A\left(1 - C_{AB}C_{BA}\right)$$

In this case, K_A is the absolute stiffness factor of the beam by assuming the far end B of the beam is fixed. For the sake of confirmation, in the case of prismatic beams, we know that $K_A = 4EI/4EI$ and $C_{AB} = C_{BA} = 1/2$. Substituting these values in the above equation yields:

$$K'_A = \frac{3EI}{l}$$

which is the well-known result from our earlier analysis of beams with far end pinned.

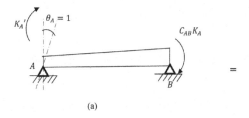

(a)

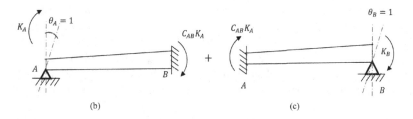

(b)   (c)

FIGURE 22.2 Principle of superposition for combined stiffness calculation.

22.3.2   Symmetric Beam and loading

In this case, we need to determine K_A' that is needed to rotate the beam at support A, $\theta_A = 1$ radian, while $\theta_B = -1$. In this case, we first assume that support B is fixed, and in this situation, we apply a moment K_A at support. Next, we apply negative moment K_B at support B by assuming support A is fixed. This induces a moment $C_{BA}K_B$ at end A. Now we apply superposition principle to get:

$$K_A' = K_A - C_{BA}K_B = K_A\left(1 - C_{AB}\right)$$

In the case of prismatic beams, $K_A = 4EI/l$ and $C_{AB} = 1/2$, so that:

$$K_A' = \frac{2EI}{l}$$

which is the same as the earlier derived value.

22.3.3   Symmetric Beam with Antisymmetric Loading

Now we will derive the relationship for symmetric beams with antisymmetric loading. To do this, we first take fixed support at B as fixed support and apply moment K_A at A. Similarly, applying moment K_B at support B by keeping A as fixed will yield another result. Combining these two, we get:

$$K_A' = K_A + C_{BA}K_B = K_A\left(1 + C_{AB}\right)$$

Substituting the values for prismatic beams, we get $K_A = 4EI/l$ and $C_{AB} = 1/2$, that gives:

$$K_A' = \frac{6EI}{l}$$

Hence, we have derived the earlier results once again by substituting the appropriate values for different parameters for prismatic beams.

22.3.4   Support Settlement

Now we will deal with the case of support settlements. In the case of support settlements, fixed end moments are developed at the joints, where the settlement has occurred. This we have already seen in Chapter 18. Now, to derive the result, we first take both supports of the beam as pin supported and apply a settlement at support B of the beam by an amount Δ. So, the rotation at the supports will be $\theta_A = \theta_B = \Delta/l$. Then, assuming B is fixed, we apply a moment $M_A' = -K_A(\Delta/l)$ to the support A. Following this, we take A end as fixed and apply a moment $M_B' = -K_B(\Delta/l)$, so that end rotates by an angle $\theta_B = -\Delta/l$. Thus, fixed end moment at A is given by:

$$(FEM)_{AB} = -K_A\frac{\Delta}{l} - \frac{C_{BA}K_B\Delta}{l}$$

Applying the earlier relationship,

$$C_{AB}K_A = C_{BA}K_B$$

we get,

$$(FEM)_{AB} = -K_A \frac{\Delta}{l}(1 + C_{AB})$$

For prismatic member, we have:

$$K_A = \frac{4EI}{l}, \ C_{AB} = \frac{1}{2}$$

Thus,

$$(FEM)_{AB} = -\frac{3EI\Delta}{l^2}$$

which is in exact agreement with the earlier derived values in slope deflection equation and moment distribution method chapters, i.e., Chapters 18 and 19, respectively.

22.4 SLOPE DEFLECTION EQUATION FOR STRUCTURES HAVING NONPRISMATIC MEMBERS

We have already learned slope deflection equations for prismatic members in Chapter 18. In this section, we will develop a generalized form of slope deflection equations so that they can be applied on nonprismatic members. To be able to do this, we take help of the results mentioned in the previous section and will formulate the equations in the same manner as discussed in Chapter 18, that is, considering the impacts caused by the imposed loads, relative joint displacement or support settlement, and each joint rotation separately, and then superimposing the results thus obtained.

22.4.1 LOADS

Loads are transformed into fixed end moments that are acting at the ends A and B of the span. Our convention is that positive moments act clockwise and negative moment acts anticlockwise in this chapter.

22.4.2 RELATIVE JOINT TRANSLATION

When a relative displacement (we commonly call this as support settlement) between the joints occurs, the induced moments are determined from the following equations developed in Section 22.3.4 at the two support ends A and B:

$$FEM_{AB} = -\left[\frac{K_A\Delta}{l}\right](1 + C_{AB})$$

$$FEM_{BA} = -\left[\frac{K_B\Delta}{l}\right](1 + C_{BA})$$

22.4.3 ROTATION AT *A*

If chord at support point A rotates by θ_A, the required moment in the span at A will be $K_A\theta_A$. Also, this induces a moment of $C_{AB}K_A\theta_A = C_{BA}K_B\theta_A$ at other support end B.

22.4.4 ROTATION AT *B*

If chord at support point B rotates by θ_B, the required moment in the span at B will be $K_B\theta_B$. Also, this induces a moment of $C_{BA}K_B\theta_B = C_{AB}K_A\theta_B$ at other support end A.

So, the total moment produced due to this effect will give us the generalized slope deflection equation that can be written down as follows:

$$M_{AB} = K_A[\theta_A + C_{AB}\theta_B - \frac{\Delta}{l}(1+C_{AB}) + \text{FEM}_{AB}$$

$$M_{BA} = K_B[\theta_B + C_{BA}\theta_A - \frac{\Delta}{l}(1+C_{BA}) + \text{FEM}_{BA}$$

By applying these equations for nonprismatic beams, we will get the final results in a fashion already explained in Chapter 18, slope deflection equation for prismatic beams.

23 Introduction to Matrix Structural Analysis

23.1 INTRODUCTION

In this chapter, the principles of using the stiffness method for analyzing structures will be explained. Although the procedure outlined here may seem very useful for manual calculation, but this is a quite tedious method for large structures. However, the repeated steps can be programmed in a computer as software package for solving large structures comprising of many structural elements. Modern computer programs systematically apply this procedure toward solving the structure with minor input parameters. Few examples will be provided as we move on, which will help us to develop the basics of matrix analysis techniques using stiffness matrix formulation. Also, it is to be noted that bold-faced upper- and lowercase letters in this chapter will represent matrix unless otherwise mentioned in the text.

23.2 ANALYTICAL MODEL

Matrix stiffness method requires following key steps to start the analysis process:

1. Subdividing the structure into a series of discrete members called elements.
2. Denoting their end points as nodes. For trusses, elements are represented by each of the members that compose the truss, and the nodes represent the joints.
3. The force-displacement equilibrium equations of each element are determined, and each member is combined in a way they are connected at the respective nodes.
4. These relationships, expressed in the form of matrix for the entire structure, are then grouped together into a single matrix known as the structure stiffness matrix K.
5. Once a complete stiffness matrix is established, the unknown displacements of the nodes can then be determined for any given loading on the structure.

The steps mentioned above are pretty tedious for manual calculations as for each element in the structures, one stiffness matrix needs to be formed. Then this stiffness matrix needs to be transformed into global stiffness matrix using transformation matrix. Once the global stiffness matrix of each element is formed, all these matrices need to be combined in a logical way, depending on their connectivity and arrangement in the whole structure. Thus, the global complete stiffness matrix of the entire structure is formed. For large structures having many elements, these steps are quite tedious in nature for manual calculations. However, many computer programs are

DOI: 10.1201/9781003081227-26

available these days to form the stiffness matrices and subsequently form the global complete stiffness matrix. We will provide few examples with structures having a small number of elements to provide the necessary feel toward the above mentioned steps. Students using software in the near future will have more confidence toward analyzing the output results by doing some elementary matrix operations.

We will explain each step mentioned above by taking a model two-dimensional (2D) truss. At the end of the chapter, we will provide sufficient insight toward forming stiffness matrices for more generalized frame elements.

23.3 MEMBER STIFFNESS RELATIONS IN LOCAL COORDINATES FOR 2D TRUSS

We will establish the stiffness matrix for a single truss member here. Let us consider a typical truss element as shown in Figure 23.1.

The axis passing through the member and aligned along the member is the local x axis of the member denoted by x. Similarly, y axis is also shown. These axes attached to the member or element is called local axis. The axial force acting at the two ends of the member has also been shown by respective arrows. Now due to the positive displacement d_N at the near end node N of the member, an axial force q'_N will be developed. The force-displacement equation for this member can be written as:

$$q'_N = \frac{AE}{l} d_N$$

And at the far end, since it is pinned, maintaining force equilibrium, we will get:

$$q'_F = -\frac{AE}{l} d_N$$

Now, if the N node is pinned and F node is free, the above equations will be just reversed, which can be written as:

$$q''_N = -\frac{AE}{l} d_F$$

$$q''_F = \frac{AE}{l} d_F$$

Thus, superimposing the above equations, for both type of displacement occurring simultaneously, we will get:

$$q_N = \frac{AE}{l} d_N - \frac{AE}{l} d_F$$

$$q_F = \frac{AE}{l} d_F - \frac{AE}{l} d_N$$

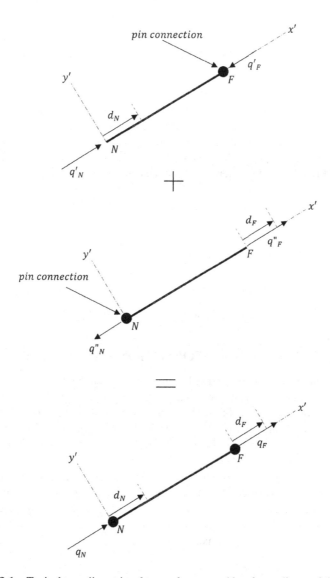

FIGURE 23.1   Typical two-dimensional truss element and local coordinate of the element.

Now, these two load displacement equations can be expressed in a one matrix equation as written below:

$$\begin{bmatrix} q_N \\ q_F \end{bmatrix} = \frac{AE}{l} \begin{bmatrix} 1 & -1 \\ -1 & 1 \end{bmatrix} \begin{bmatrix} d_N \\ d_F \end{bmatrix}$$

or,

$$q = k'd$$

where,

$$k' = \frac{AE}{l}\begin{bmatrix} 1 & -1 \\ -1 & 1 \end{bmatrix}$$

This is the stiffness matrix for the truss element NF in local coordinate system. The four elements that comprise the matrix are known as the member stiffness influence coefficients k'_{ij}. Physically, this means, force at joint i, when unit displacement is imposed at joint j. For example, $i = j = 1$, then k'_{11} is the force at joint 1, when far end joint is held fixed, and displacement at the near joint is $d_N = 1$. So,

$$q_N = k'_{11} = \frac{AE}{l}$$

Similarly, for $i = 2$, $j = 1$, we will get:

$$q_F = k'_{21} = -\frac{AE}{l}$$

These two are the first column, entered in the above stiffness matrix k'.

23.4 COORDINATE TRANSFORMATION FOR 2D TRUSS

A truss is composed of many individual members or elements oriented in different directions. In this section, we will learn the matrix method of transformation of coordinates from local to global scale. This is of utmost importance since all member local stiffness matrices need to be transformed into global stiffness matrix prior to assembling the same to form the final single global stiffness matrix for the entire structure.

To be able to do that, we need to consider the member orientation and cast on the global coordinate system as shown in Figure 23.2. As already stated in the previous section, member local axis is the axis, which is passing through the member. Depending upon the orientation of the member in the actual truss, there will be an angle between local and global axes. The angle of inclination of the member with

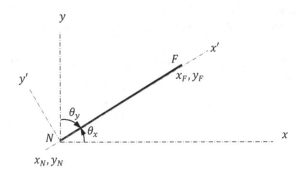

FIGURE 23.2 Local and global coordinate system.

respect to global axis needs to be determined first. The angles between the positive global x, y axes and the positive local x' axis are defined as θ_x and θ_y. Let us denote the global coordinates of the two ends of the member as (x_N, y_N) and (x_F, y_F) for N and F, respectively.

Let us make two more simplifying assumptions related to cosine of two angles θ_x and θ_y:

$$\delta_x = \cos \theta_x$$

$$\delta_y = \cos \theta_y$$

$$\delta_x = \cos \theta_x = \frac{x_F - x_N}{l} = \frac{x_F - x_N}{\sqrt{(x_F - x_N)^2 + (y_F - y_N)^2}}$$

$$\delta_y = \cos \theta_y = \frac{y_F - y_N}{l} = \frac{y_F - y_N}{\sqrt{(x_F - x_N)^2 + (y_F - y_N)^2}}$$

We will use the above relationships in the next section to develop the displacement transformation matrix.

23.5 DISPLACEMENT TRANSFORMATION MATRIX FOR 2D TRUSS

In global coordinate, the member displacement along local axes and corresponding components in global axis need to be connected. Otherwise, there will be no way to form the global matrix for the entire structure. We can form this relationship with the help of the direction cosines δ_x and δ_y and noting the displacement components as shown in Figure 23.3. Each joint of a truss element is having two degrees of freedom, i.e., Δ_{N_x}, Δ_{N_y} at node N, and Δ_{F_x}, Δ_{F_y} at node F in the direction of global coordinates.

From components of vectors, we can write:

$$d_N = \Delta_{N_x} \cos \theta_x + \Delta_{N_y} \cos \theta_y$$

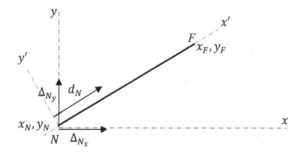

FIGURE 23.3 Local and global displacement coefficients.

Similarly, for the displacement at the other end F, we can write:

$$d_F = \Delta_{F_x} \cos \theta_x + \Delta_{F_y} \cos \theta_y$$

In terms of δ_x and δ_y, we can rewrite the above two equations in the following matrix form:

$$\begin{bmatrix} d_N \\ d_F \end{bmatrix} = \begin{bmatrix} \delta_x & \delta_y & 0 & 0 \\ 0 & 0 & \delta_x & \delta_y \end{bmatrix} \begin{bmatrix} \Delta_{N_x} \\ \Delta_{N_y} \\ \Delta_{F_x} \\ \Delta_{F_y} \end{bmatrix}$$

or in the compact matrix equation form, we can write:

$$d = T\Delta$$

where, T is the transformation matrix of displacement, which is known as displacement transformation matrix.

$$T = \begin{bmatrix} \delta_x & \delta_y & 0 & 0 \\ 0 & 0 & \delta_x & \delta_y \end{bmatrix}$$

23.6 FORCE TRANSFORMATION MATRIX

Next, we will investigate how the local and global force components can be related to the transformation matrix T already developed in the previous section. To do the same, refer to Figure 23.4 for global and local force components at the node N of the truss member.

$$F_{N_x} = q_N \cos \theta_x = q_N \delta_x$$

$$F_{N_y} = q_N \cos \theta_y = q_N \delta_y$$

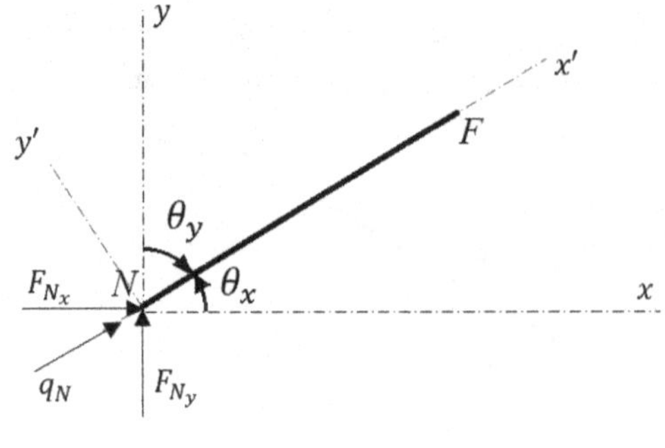

FIGURE 23.4 Local and global force coefficients.

Similarly, for the other node F, the above two equations can be written as:

$$F_{F_x} = q_F \cos \theta_x = q_F \delta_x$$

$$F_{F_y} = q_F \cos \theta_y = q_F \delta_y$$

In matrix form, the above four equations can be combined in the following elegant form:

$$\begin{bmatrix} F_{N_x} \\ F_{N_y} \\ F_{F_x} \\ F_{F_y} \end{bmatrix} = \begin{bmatrix} \delta_x & 0 \\ \delta_y & 0 \\ 0 & \delta_x \\ 0 & \delta_y \end{bmatrix} \begin{bmatrix} q_N \\ q_F \end{bmatrix}$$

or, it can be written in the compact matrix equation form as:

$$Q = T^T q$$

where:

$$T^T = \begin{bmatrix} \delta_x & 0 \\ \delta_y & 0 \\ 0 & \delta_x \\ 0 & \delta_y \end{bmatrix}$$

is the transpose matrix of the original transformation matrix T. So, for force transformation, we need to take transpose matrix of displacement transformation matrix.

23.7   MEMBER GLOBAL STIFFNESS MATRIX FOR 2D TRUSS

Now we have arrived at a point where we can form the member global stiffness matrix. Please be cautious that this is not the complete global matrix of the entire structure. We will form only a one-member global stiffness matrix that we were discussing in all previous sections. If we substitute, $d = T\Delta$ into the equation $q = k'd$, we will get:

$$q = k' T\Delta$$

Now substituting this into equation $Q = T^T q$ we will get:

$$Q = T^T k' T\Delta$$

or,

$$Q = K\Delta$$

where:

$$K = T^T k' T$$

is the global stiffness matrix of the member NF as discussed in all the sections. Substituting the values of T^T, k', and T, one can perform the matrix multiplication to obtain:

$$K = AE/l \begin{bmatrix} \delta_x^2 & \delta_x\delta_y & -\delta_x^2 & -\delta_x\delta_y \\ \delta_x\delta_y & \delta_y^2 & -\delta_x\delta_y & -\delta_y^2 \\ -\delta_x^2 & -\delta_x\delta_y & \delta_x^2 & \delta_x\delta_y \\ -\delta_x\delta_y & -\delta_y^2 & \delta_x\delta_y & \delta_y^2 \end{bmatrix} \begin{matrix} N_x \\ N_y \\ F_x \\ F_y \end{matrix}$$

with column headers $N_x \quad N_y \quad F_x \quad F_y$

Since 2D truss element has two degrees of freedom per node, this 4×4 symmetric matrix is referenced with each global degree of freedom related to near end N followed by far end F. All the global force and displacement components in the row and the column are provided at the top and right side of the global stiffness matrix for ease of understanding. Let us determine the structure stiffness matrix for a truss given in Example 23.1 below.

Example 23.1: Determine the structure stiffness matrix for the two-member truss as shown in Figure 23.5 (a). Consider AE as constant for all members.

SOLUTION: From the above configuration of the truss, it is clear that we will have two unknown displacements at node 2 only since nodes 1, and 3 are constrained in both the directions at supports. At node 2, we will have movement along x and y axis globally. Locally, the movement will be the resultant of this global displacement.

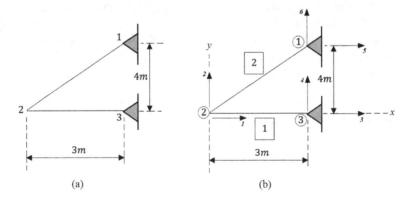

FIGURE 23.5 Two-member truss with node and member numbering.

Since 2 is the near end and 3 is the far end of the truss, hence,

$$\delta_x = \frac{3-0}{3} = 1$$

$$\delta_y = \frac{0-0}{3} = 0$$

Using the above global matrix and dividing each matrix element by the length of the member (i.e., by 3.0 m), we will get,

$$K_1 = AE \begin{array}{c} \\ \\ \end{array} \begin{array}{cccc} 1 & 2 & 3 & 4 \\ \left[\begin{array}{cccc} 0.333 & 0 & -0.333 & 0 \\ 0 & 0 & 0 & 0 \\ -0.333 & 0 & 0.333 & 0 \\ 0 & 0 & 0 & 0 \end{array}\right] & \begin{array}{c} 1 \\ 2 \\ 3 \\ 4 \end{array} \end{array}$$

Now the above matrix is the global stiffness matrix for member 1. For member 1, there are two nodes 2 and 3 of which 2 is free to move and 3 is constrained. Also, since this is a 2D truss, hence, each node can have degrees of freedom. For node 2, which is near end, is represented by 1 and 2, whereas for node 3, it is 3 and 4. Writing displacement of DOFs at the top and side of the stiffness matrix helps us to assemble the matrices easily to form the global stiffness matrix of the entire truss.

Now for next member 2, we have two nodes 2 and 1. Here also 2 is the near node and 1 is the far end node. Also, for this member, we have again four DOFs at two nodes 1, 2, 5, and 6 where 5 and 6 are the DOFs related to node 1.

So, for member 2, we have:

$$\delta_x = \frac{3-0}{5} = 0.6$$

$$\delta_y = \frac{4-0}{5} = 0.8$$

$$K_2 = AE \begin{array}{c} \\ \\ \end{array} \begin{array}{cccc} 1 & 2 & 5 & 6 \\ \left[\begin{array}{cccc} 0.072 & 0.096 & -0.072 & -0.096 \\ 0.096 & 0.128 & -0.096 & -0.128 \\ -0.072 & -0.096 & 0.072 & 0.096 \\ -0.096 & -0.128 & 0.096 & 0.128 \end{array}\right] & \begin{array}{c} 1 \\ 2 \\ 5 \\ 6 \end{array} \end{array}$$

Thus, the global stiffness matrix for member 2 becomes:

Since the structure consists of three nodes, so the total number of DOFs will be $3 \times 2 = 6$. Hence, the final matrix will be 6×6 matrix. Now, two matrices are added algebraically with entries in each individual matrix with zero in rows and columns, as shown below:

$$K = K_1 + K_2$$

or,

$$
K = AE
\begin{array}{c}
1 \quad2 \quad3 \quad456 \\
\begin{bmatrix}
0.333 & 0 & -0.333 & 0 & 0 & 0 \\
0 & 0 & 0 & 0 & 0 & 0 \\
-0.333 & 0 & 0.333 & 0 & 0 & 0 \\
0 & 0 & 0 & 0 & 0 & 0 \\
0 & 0 & 0 & 0 & 0 & 0 \\
0 & 0 & 0 & 0 & 0 & 0
\end{bmatrix}
\begin{array}{c}1\\2\\3\\4\\5\\6\end{array}
\end{array}
$$

$$
+ AE
\begin{array}{c}
1 \quad2 \quad3 \quad4 \quad5 \quad6 \\
\begin{bmatrix}
0.072 & 0 & 0.096 & 0 & -0.072 & -0.096 \\
0.096 & 0 & 0.128 & 0 & -0.096 & -0.128 \\
0 & 0 & 0 & 0 & 0 & 0 \\
0 & 0 & 0 & 0 & 0 & 0 \\
-0.072 & 0 & -0.096 & 0 & 0.072 & 0.096 \\
-0.096 & 0 & -0.128 & 0 & 0.096 & 0.128
\end{bmatrix}
\begin{array}{c}1\\2\\3\\4\\5\\6\end{array}
\end{array}
$$

or,

$$
K = AE
\begin{bmatrix}
0.405 & 0.096 & -0.033 & 0 & -0.072 & -0.096 \\
0.096 & 0.128 & 0 & 0 & -0.096 & -0.128 \\
-0.333 & 0 & 0.333 & 0 & 0 & 0 \\
0 & 0 & 0 & 0 & 0 & 0 \\
-0.072 & -0.096 & 0 & 0 & 0.072 & 0.096 \\
-0.096 & -0.128 & 0 & 0 & 0.096 & 0.128
\end{bmatrix}
$$

When the computer is called for analysis, generally, the software and/or algorithm of the analysis software starts with all zero elements in all cells of the 6 × 6 matrix. As the individual structural element global stiffness matrices are formed, the same are placed directly into their respective element positions in the global overall **K** matrix, instead of formulating the individual element stiffness matrices, computing and storing them, and finally assembling them.

23.8   APPLICATION OF STIFFNESS METHOD FOR TRUSS ANALYSIS

Once the structure global stiffness matrix is generated, the global force components Q acting on the truss can then be related to its global displacements D using:

$$Q = KD$$

The above equation is known as the structure stiffness equation. We can partition the above matrix equation to represent the known and unknown components of each matrix. Refer to the following matrix partition into known and unknown components as shown:

$$
\begin{bmatrix} Q_k \\ Q_u \end{bmatrix} =
\begin{bmatrix} K_{11} & K_{12} \\ K_{21} & K_{22} \end{bmatrix}
\begin{bmatrix} D_k \\ D_u \end{bmatrix}
$$

Here the subscripts k, u represent known and unknown components.

Expanding the above matrix equation further, we get:

$$Q_k = K_{11}D_u + K_{12}D_k$$

$$Q_u = K_{21}D_u + K_{22}D_k$$

In most of the cases, $D_k = 0$ since there are no movements at the support points. In such cases, we have:

$$Q_k = K_{11}D_u$$

Since the elements in matrix K_{11} represent the total resistance at a joint due to unit displacement at that joint in either x or in y direction, the equation is the matrix representation of the force equilibrium equation of the entire truss. Since the external applied load are either all known or are zero, hence, inverting the above matrix equation, we get:

$$D_u = K_{11}^{-1}Q_k$$

So, once the above equation is solved for D_u and noting that $D_k = 0,$ from the second equation to solve for unknown components of matrix Q_u

$$Q_u = K_{21}D_u$$

The member forces can be determined from the already developed equation in the previous section, as reproduced below:

$$q = k'TD$$

Expanding this equation yields:

$$\begin{bmatrix} q_N \\ q_F \end{bmatrix} = \frac{AE}{l} \begin{bmatrix} 1 & -1 \\ -1 & 1 \end{bmatrix} \begin{bmatrix} \delta_x & \delta_y & 0 & 0 \\ 0 & 0 & \delta_x & \delta_y \end{bmatrix} \begin{bmatrix} D_{Nx} \\ D_{Ny} \\ D_{Fx} \\ D_{Fy} \end{bmatrix}$$

We know that $q_N = -q_F$ following equilibrium condition, hence, only one force among these two needs to be calculated. Say, for example,

$$q_F = \frac{AE}{l} \begin{bmatrix} -\delta_x & -\delta_y & \delta_x & \delta_y \end{bmatrix} \begin{bmatrix} D_{Nx} \\ D_{Ny} \\ D_{Fx} \\ D_{Fy} \end{bmatrix}$$

23.9   APPLICATION OF STIFFNESS METHOD
FOR SPACE TRUSS ANALYSIS

The analysis for space truss also follows the same procedure as that of 2D truss. Only in the case of coordinate displacement, there will be three direction cosines:

$$\delta_x = \cos \theta_x$$
$$\delta_y = \cos \theta_y$$
$$\delta_z = \cos \theta_z$$

Due to this additional term, the transformation matrix becomes:

$$T = \begin{bmatrix} \delta_x & \delta_y & \delta_z & 0 & 0 & 0 \\ 0 & 0 & 0 & \delta_x & \delta_y & \delta_z \end{bmatrix}$$

Substituting this in the equation, $k = T^T k'T$, we get:

$$k = \begin{bmatrix} \delta_x & 0 \\ \delta_y & 0 \\ \delta_z & 0 \\ 0 & \delta_x \\ 0 & \delta_y \\ 0 & \delta_z \end{bmatrix} \frac{AE}{l} \begin{bmatrix} 1 & -1 \\ -1 & 1 \end{bmatrix} \begin{bmatrix} \delta_x & \delta_y & \delta_z & 0 & 0 & 0 \\ 0 & 0 & 0 & \delta_x & \delta_y & \delta_z \end{bmatrix}$$

Completing the matrix computation, the stiffness matrix yields:

$$k = AE/l \begin{bmatrix} \delta_x^2 & \delta_x\delta_y & \delta_x\delta_z & -\delta_x^2 & -\delta_x\delta_y & -\delta_x\delta_z \\ \delta_x\delta_y & \delta_y^2 & \delta_y\delta_z & -\delta_x\delta_y & -\delta_y^2 & -\delta_y\delta_z \\ \delta_z\delta_x & \delta_z\delta_y & \delta_z^2 & -\delta_z\delta_x & -\delta_z\delta_y & -\delta_z^2 \\ -\delta_x^2 & -\delta_x\delta_y & -\delta_z\delta_x & \delta_x^2 & \delta_x\delta_y & \delta_z\delta_x \\ -\delta_x\delta_y & -\delta_y^2 & -\delta_z\delta_y & \delta_x\delta_y & \delta_y^2 & \delta_z\delta_y \\ -\delta_z\delta_x & -\delta_z\delta_y & -\delta_z^2 & \delta_z\delta_x & \delta_z\delta_y & \delta_z^2 \end{bmatrix}$$

Once this global matrix has been formed, we can proceed to analyze the truss as per the procedures explained in Section 23.7 for 2D truss.

23.10   APPLICATION OF STIFFNESS METHOD
FOR BEAM ANALYSIS

In this section, we will develop stiffness matrix for a prismatic beam element. Necessary formulas for finding the stiffness matrix have already been developed in the slope deflection equation chapter. We will use those equations very frequently and

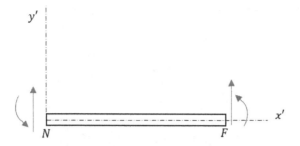

FIGURE 23.6 Beam element with local coordinate system and node numbering.

recall them as and when required in this section. Students and readers are advised to go through the slope deflection equation chapter in detail once again before progressing far from this point to learn the basic tools for the stiffness coefficient calculation for beams.

We have learnt from the slope deflection equation that if a support sinks by an amount Δ, there will be forces and moments induced at the supports due to this effect. We first define the local axes of beam elements as shown in Figure 23.6.

The positive direction of force and moments is also drawn in Figure 23.6 for understanding the local coordinate and force momentum vector conventions. Let the displacements at the nodes be denoted by δ'_y and rotation through an angle be represented by δ'_z. Linear force and moment at supports are represented by F'_y and M'_z, respectively. Now, due to linear and angular displacement occurring independently at two nodes, we will have the following force and moments induced at the support:

$$F'_{Ny} = \frac{6EI}{l^2} \delta'_{Ny}$$

$$F'_{Fy} = \frac{6EI}{l^2} \delta'_{Fy}$$

$$M'_{Ny} = \frac{12EI}{l^3} \delta'_{Nz}$$

$$M'_{Fy} = \frac{12EI}{l^3} \delta'_{Fz}$$

$$R'_{Nz} = \frac{4EI}{l} \delta'_{Nz}$$

$$R'_{Fz} = \frac{2EI}{l} \delta'_{Fz}$$

The above load displacement relationship can be expressed in a matrix form as follows:

$$
\begin{bmatrix} M'_{Ny} \\ F'_{Ny} \\ M'_{Fy} \\ F'_{Fy} \end{bmatrix} = \begin{bmatrix} \dfrac{12EI}{l^3} & \dfrac{6EI}{l^2} & -\dfrac{12EI}{l^3} & \dfrac{6EI}{l^2} \\ \dfrac{6EI}{l^2} & \dfrac{4EI}{l} & -\dfrac{6EI}{l^2} & \dfrac{2EI}{l} \\ -\dfrac{12EI}{l^3} & -\dfrac{6EI}{l^2} & \dfrac{12EI}{l^3} & -\dfrac{6EI}{l^2} \\ \dfrac{6EI}{l^2} & \dfrac{2EI}{l} & -\dfrac{6EI}{l^2} & \dfrac{4EI}{l} \end{bmatrix} \begin{bmatrix} \delta'_{Ny} \\ \delta'_{Nz} \\ \delta'_{Fy} \\ \delta'_{Fz} \end{bmatrix}
$$

The above matrix relationship can be expressed in a matrix equation form as follows:

$$Q = k\delta$$

The symmetric matrix k is called the member stiffness matrix for beam elements. Physically, all these components in the stiffness matrix represent the amount of force required for a unit displacement in the given positive sense of the member. Also, it is customary to note that the local and global axes for this beam element are the same since all these coordinates are parallel to each other. So, the stiffness matrix for beam element will retain its form in both coordinate systems.

23.11 BEAM STRUCTURE COMPLETE GLOBAL STIFFNESS MATRIX

When all the beam member stiffness matrices for an entire structure have been formed, one must combine them into the complete structure global stiffness matrix K. This computation first relies on determining the location of each member in the member stiffness matrix. In this case, the rows and columns of each k matrix (like the above matrix for each member) are denoted by the two code numbers at the near end of the member F_{Ny}, F_{Nz} followed by those at the far end F_{Fy}, F_{Fz}. So, when combining the matrices, each element must be placed in the same location of the complete global K matrix. Following this procedure, K matrix attains an order that will be equal to the highest code number assigned to the beam element, since this indicates the total number of degrees of freedom. Also, in the case of several members connected to a node, their member stiffness coefficients will have the same position in the global K matrix and so must be algebraically added to determine the nodal stiffness influence coefficient for the entire structure. This step is important since each coefficient represents the nodal resistance of the structure in a particular sense when a unit displacement occurs either at the same or another node. We will understand this concept of assembling each elemental matrix to form global stiffness matrix through Example 23.2.

Example 23.2: Determine the reaction at the supports of the continuous beam as shown in Figure 23.7 using the matrix method of analysis.

SOLUTION: The beam has two elements or members and three nodes. Out of these, 1–4 numbers are taken to indicate unconstrained DOFs, and 5 and 6 numbers are used to indicate constrained degrees of freedom.

The known load and displacement matrices are as follows:

$$F_k = \begin{bmatrix} 0 \\ -5 \\ 0 \\ 0 \end{bmatrix} \begin{matrix} 1 \\ 2 \\ 3 \\ 4 \end{matrix}$$

$$\delta = \begin{bmatrix} 0 \\ 0 \end{bmatrix} \begin{matrix} 5 \\ 6 \end{matrix}$$

Each DOF is written at the side of each element of the above matrices for ease of understanding.

Now member stiffness matrices will be prepared directly from the earlier developed **k matrix.**

Now we will assemble these two matrices to form the global stiffness matrix of the entire structure:

$$Q = K\delta$$

$$k_1 = EI \begin{array}{cccc} 6 & 4 & 5 & 3 \\ \begin{bmatrix} 1.5 & 1.5 & -1.5 & 1.5 \\ 1.5 & 2 & -1.5 & 1 \\ -1.5 & -1.5 & 1.5 & -1.5 \\ 1.5 & 1 & -1.5 & 2 \end{bmatrix} & \begin{matrix} 6 \\ 4 \\ 5 \\ 3 \end{matrix} \end{array}$$

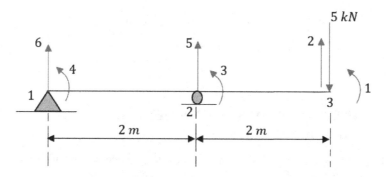

FIGURE 23.7   Example beam problem with node number, beam number, DOFs.

$$k_2 = EI \begin{array}{cccc} 5 & 3 & 2 & 1 \\ \begin{bmatrix} 1.5 & 1.5 & -1.5 & 1.5 \\ 1.5 & 2 & -1.5 & 1 \\ -1.5 & -1.5 & 1.5 & -1.5 \\ 1.5 & 1 & -1.5 & 2 \end{bmatrix} & \begin{array}{c} 5 \\ 3 \\ 2 \\ 1 \end{array} \end{array}$$

$$\begin{bmatrix} 0 \\ 0 \\ 0 \\ 0 \\ \delta_5 \\ \delta_6 \end{bmatrix} = \begin{bmatrix} 2 & -1.5 & 1 & 0 & 1.5 & 0 \\ -1.5 & 1.5 & -1.5 & 0 & -1.5 & 0 \\ 1 & -1.5 & -1.5 & 0 & -1.5 & 0 \\ 1 & -1.5 & 4 & 1 & 0 & 1.5 \\ 0 & 0 & 1 & 2 & -1.5 & 1.5 \\ 1.5 & -1.5 & 0 & -1.5 & 3 & -1.5 \\ 0 & 0 & 1.5 & 1.5 & -1.5 & 1.5 \end{bmatrix} \begin{bmatrix} \delta_1 \\ \delta_2 \\ \delta_3 \\ \delta_4 \\ 0 \\ 0 \end{bmatrix}$$

Now, carrying out the multiplication, the equations that can be formed from the above matrix are:

$$2\delta_1 - 1.5\delta_2 + \delta_3 = 0 = 0$$

$$-1.5\delta_1 + 1.5\delta_2 - 1.5\delta_3 + 0 = -\frac{5}{EI}$$

$$\delta_1 - 1.5\delta_2 + 4\delta_3 + \delta_4 = 0$$

$$0 + 0 + \delta_3 + 2\delta_4 = 0$$

Upon solution, we get:

$$\delta_1 = -\frac{16.67}{EI}$$

$$\delta_2 = -\frac{26.67}{EI}$$

$$\delta_3 = -\frac{6.67}{EI}$$

$$\delta_4 = \frac{3.33}{EI}$$

With this calculated value, we can determine the unknown forces as:

$$Q_5 = 10 \text{ kN}$$

$$Q_6 = -5 \text{ kN}$$

23.12 APPLICATION OF STIFFNESS METHOD FOR FRAME ANALYSIS

Having gained a fair amount of knowledge of truss and beam elements, we can apply these for the analysis of frames also. In case of frames, at each node, there are three degrees of freedom available in 2-D since the node can displace linearly in x and y

direction as well as there can be a rotation. So, for a frame element, there will be total $2 \times 3 = 6$ DOFs per element in two dimensions. Thus, for the stiffness, transformation, force, etc., matrices will be 6×6 matrix.

Also, unlike the beams, the frame elements can be oriented at different angles; hence, we also need transformation matrix to transform from local to global coordinate system. We will provide the required formulations in matrix forms in case of frame elements for reference. These matrices can be formed by the same procedure as that explained for truss and beams in previous sections of this chapter.

Typical stiffness matrix for a frame element in local coordinate system:

$$k' = \begin{bmatrix} AE/l & 0 & 0 & -AE/l & 0 & 0 \\ 0 & 12EI/l^3 & 6EI/l^2 & 0 & -12EI/l^3 & 6EI/l^2 \\ 0 & 6EI/l^2 & 4EI/l & 0 & -6EI/l^2 & 2EI/l \\ -AE/l & 0 & 0 & AE/l & 0 & 0 \\ 0 & -12EI/l^3 & -6EI/l^2 & 0 & 12EI/l^3 & -6EI/l^2 \\ 0 & 6EI/l^2 & 2EI/l & 0 & -6EI/l^2 & 4EI/l \end{bmatrix}$$

To transform the above local stiffness matrix into global, we need the following transformation matrix to operate:

$$T = \begin{bmatrix} \delta_x & \delta_y & 0 & 0 & 0 & 0 \\ -\delta_y & \delta_x & 0 & 0 & 0 & 0 \\ 0 & 0 & 1 & 0 & 0 & 0 \\ 0 & 0 & 0 & \delta_x & \delta_y & 0 \\ 0 & 0 & 0 & -\delta_y & \delta_x & 0 \\ 0 & 0 & 0 & 0 & 0 & 1 \end{bmatrix}$$

Similarly, the force transformation matrix will be of the following form:

$$T^T = \begin{bmatrix} \delta_x & -\delta_y & 0 & 0 & 0 & 0 \\ \delta_y & \delta_x & 0 & 0 & 0 & 0 \\ 0 & 0 & 1 & 0 & 0 & 0 \\ 0 & 0 & 0 & \delta_x & -\delta_y & 0 \\ 0 & 0 & 0 & \delta_y & \delta_x & 0 \\ 0 & 0 & 0 & 0 & 0 & 1 \end{bmatrix}$$

Frame member global stiffness matrix can be formed by carrying out the following operation:

$$K = T^T k' T$$

The global stiffness matrix for each frame element needs to be calculated by applying the above equation and the result is left as an exercise for readers. This is nothing but carrying out the stepwise row by column matrix multiplication that we have already done in previous sections.

After forming each member global stiffness matrix, we need to carry out the assembly process as per the member orientation and member numbering of the entire frame to form the overall global stiffness matrix. This method is quite tedious for large frames, and that is why, computer programs are called for to carry out the same automatically as per the algorithms set to do the operation. However, for small frame elements, we can attempt to do it manually and compare the result from computer analysis so that a concrete understanding of the underlying process remains at our confidence.

24 Introduction to Plastic Analysis of Structure

24.1 INTRODUCTION

In this chapter, plastic analysis of structure is introduced. Plastic analysis has advantages over elastic analysis in a way that members provide much more resistance and practical bending features. Material can utilize its reserve strength that remains unutilized in the elastic analysis if plastic analysis is adopted. We can define our limit load more realistically in plastic analysis, and thus, our design will be more economical. After completing this chapter, we will learn to calculate the plastic section modulus of different sections. Different types of frames using this analysis method will also be discussed in this chapter. A thorough understanding of this method of analysis will provide readers a very solid foundation to work in design analysis field.

24.2 STRESS-STRAIN CURVE OF A DUCTILE MATERIAL

For a ductile material like steel, the stress strain diagram under tension looks similar to the one shown in Figure 24.1.

Ideal ductile material is defined as the one that has a defined elastic range and then plastic range follows. In the above stress-strain curve, from O to A, the line is straight line, and in this portion the stress is proportional to strain. If you unload the material from point A, it will reach its origin O. There will be no residual strain left due to unloading. Point A is called the proportional limit. Once the load is little increased, and the stress-strain curve reaches point A', the material remains still elastic, but if the material is unloaded from this point, a very small residual strain remains. This point A' is called the elastic limit of the material. Once stress increases more than this, the elastic range is exceeded, and the material reaches yield point, namely upper yield point B, and lower yield point C. The portion between C and D is known to be the plastic range of the material, here strain increases at constant stress. Beyond point D, the material strain hardens up to point E, where the ultimate stress reaches. After point E, necking starts to develop, and upon further increase in load, the material gets fractured at point F. F' is the true fracture stress, where the change in cross section is considered. Plastic analysis is based on this elastic plastic stress-strain curve. In most situations, the portion CD is approximated into a flatten line, and an ideal zone for plastic behavior is taken for analysis and design purpose. Thus, the ideal stress-strain curve considered for plastic analysis will look similar to the one shown in Figure 24.2.

For ductile material, upper yield points generally not considered for calculating strength of material. It is considered that the material is elastic stage up to lower yield point and then it enters plastic state.

DOI: 10.1201/9781003081227-27

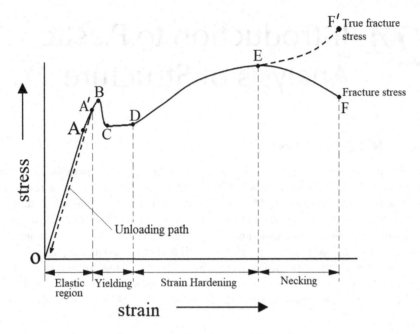

FIGURE 24.1   Stress strain diagram of ductile material.

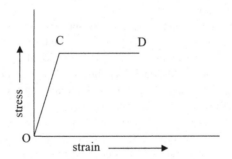

FIGURE 24.2   Ideal stress strain diagram of ductile material for plastic analysis.

24.3   PLASTIC MOMENT

Let us consider a prismatic cantilever beam with an applied increasing moment M at the end of the beam as shown in Figure 24.3. When moment intensity is such that bending stress in the extreme fiber does not attain yield stress, the beam will remain within the elastic zone of deformation. In this zone, the variation of stress will be zero at neutral axis up to the maximum at the extreme fiber level. With further increase in moment, the extreme fiber will reach the yield point, but the internal fibers of the beam will still have less stress intensity than the yield stress. So, the bending stress diagram will be triangular with zero intensity at the neutral axis.

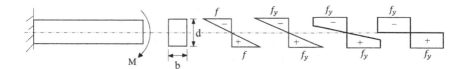

FIGURE 24.3 Progressive development of the plastic state for a rectangular beam under the application of increasing bending stress.

In this situation, the moment will be a yield moment, and the same is related to the yield stress of the material by the following equation:

$$M_y = f_y \times \frac{bd^2}{6}$$

where f_y is the yield stress of the beam material, b is breadth, and d is the depth of the beam.

With further increase in moment, the stress in extreme fiber will be unchanged while the stress in the inner fibers will be increased and the fiber at the immediate vicinity of the extreme fiber will attain the yield stress. So, the stress diagram will be slightly rectangular in nature and then there will be triangular portion reaching to zero at the neutral axis. With further increase in moment, all the fibers of the beam above and below the neutral axis will attain yield stress resulting in rectangular stress distribution above and below the neutral axis as shown in the last stress diagram in Figure 24.3. When all the fibers attain yield stress f_y, then the beam is said to attain its full plastic state. In Figure 24.4, the stress distribution of a beam having rectangular cross section in partially yielded (Figure 24.4 (b)) condition and fully yielded condition (Figure 24.4 (c)) is shown. In this same figure (Figure 24.4 (d)), the stress resultant at the fully yielded condition is also shown. Now, in equilibrium condition, the net force acting in the cross section will be zero. Let us consider the fully yielded condition,

$$F_C = F_t$$

or,

$$f_y A_c = f_y A_t$$

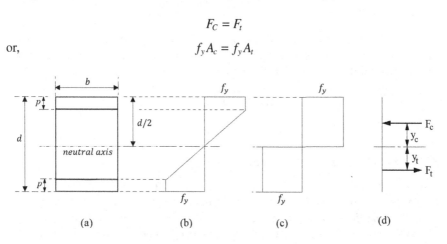

FIGURE 24.4 Stress distribution for rectangular section in partially and fully yielded condition.

where A_c and A_t are the areas under compression and tension, respectively. So, from the above expression, we can conclude that, in the fully yielded condition, the neutral axis divides the area exactly into two parts. Now, let us find out the moment of resistance (M.R.) in a partially yielded condition, where p is the depth of penetration of the yield zone as shown in Figure 24.4 (a) and (b).

M.R. of the section in partially yielded condition = M.R. of the yielded zone + M.R. of the elastic zone.

Therefore,

$$\text{M.R.} = M = \left[2 \times f_y bp \left(\frac{d}{2} - \frac{p}{2} \right) \right] + \left[f_y \times \frac{1}{6} \times b (d - 2p)^2 \right]$$

or,

$$\left[f_y bp (d - p) \right] + \left[f_y \times \frac{1}{6} \times b (d - 2p)^2 \right]$$

Let us consider, $p = \alpha d$; $M = \left[f_y b \alpha d (d - \alpha d) \right] + \left[f_y \times \frac{1}{6} \times b (d - 2\alpha d)^2 \right]$
or,

$$M = \frac{1}{6} f_y bd^2 \left(1 + 2\alpha - 2\alpha^2 \right)$$

The above expression can be written as,

$$M = \frac{1}{4} f_y bd^2 \times \frac{2}{3} \left(1 + 2\alpha - 2\alpha^2 \right)$$

For fully yielded condition, $p = 0.5d$, i.e., $\alpha = 0.5$; In that condition,

$$M = M_P = \frac{1}{4} f_y bd^2 = f_y S$$

where S is the plastic modulus. The ratio of fully plastic moment to elastic moment capacity of any section is called *shape factor* of the section. So, for our case, the shape factor for this rectangular beam will be:

$$\frac{M_P}{M_y} = \frac{f_y bd^2 / 4}{f_y bd^2 / 6} = 1.5$$

24.4 METHODS OF ANALYSIS

If we keep on increasing the load, a structure will attain fully plastic state in some regions at some point of time. The resisting bending moment cannot be increased at these fully plastic sections anymore, but the structure can rotate around these. Thus, we model this by placing a hinge at that point where structure has attained fully plastic state. So, a fully plastic section of structure forms a hinge, and it is called *plastic hinge*. With the plastic hinge formation, the structure attains a mechanism, and it

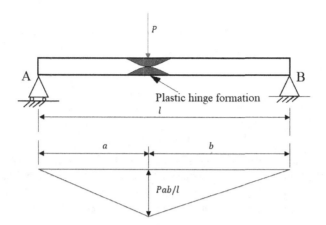

FIGURE 24.5 Plastic hinge formation and mechanism for a simply supported beam.

reaches on the verge of collapse. If the degree of static indeterminacy of a structure is D_S, plastic hinge required (n) to form mechanism is,

$$n = D_S + 1$$

a. *Simply supported beam loaded with concentrated load:* To understand this concept, let us take the example of a simply supported beam with a concentrated load P applied at a distance a from the support A as shown in Figure 24.5. The degree of static indeterminacy (D_S) in this case will be zero. Then number of plastic hinge (n) required to form mechanism is, $n = 0 + 1 = 1$.

As the applied point load intensity increases and reach βP, plastic hinge forms lead to a mechanism at the location of the applied load at some point in time, and the structure will collapse. At this situation, the structure can undergo any rotation about the plastic hinge, and the acting moment at that location will be fully plastic moment M_p. This can be related to initial bending moment at the same section with the following equation:

$$M_P = \beta \frac{Pab}{l}$$

or,

$$\beta = \frac{M_P l}{Pab}$$

This factor β is called the *collapse load factor* of the beam we are dealing with. In case the point load is acting at the midpoint of the beam, then collapse load factor will be ($a = b = l/2$):

$$\beta = \frac{4M_P}{Pl}$$

Thus, from the above equation, we can determine the collapse load factors under varying intensity of applied loads. If M_y is the moment when first yielding appears in the structure due to some load intensity of $\beta'P$, we can write:

$$\beta' = \frac{4M_y}{Pl}$$

Now, the ratio $\frac{\beta}{\beta'} = \frac{M_P}{M_y}$ is called as the shape factor.

b. *Propped cantilever loaded with concentrated load*: Let us consider a propped cantilever with a concentrated load acting on its span as shown in Figure 24.6. The degree of static indeterminacy (D_S) in this case will be one. Then number of plastic hinge (n) required to form mechanism is, $n = 1 + 1 = 2$.

In this case, maximum negative bending moment occurs at the fixed end, and maximum positive bending moment occurs under the load. If we keep on increasing the load, the first plastic hinge will occur either at the fixed end or under the applied load depending on where the numerical

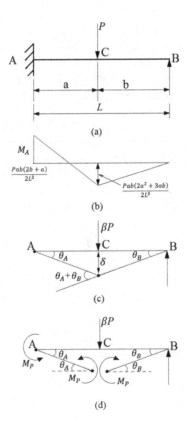

FIGURE 24.6   Plastic hinge formation and mechanism for a propped cantilever.

value of bending moment will be maximum. The beam will not still collapse because we need two plastic hinges to form, in this case, to create the mechanism. So, at the collapse state, the beam will form a mechanism as shown in Figure 24.6 (c) under the load magnitude of βP. In this condition, the internal work done by the plastic moments, M_P should be equal to the external work done by the applied force βP. We can write,

$$M_P\theta_A + M_P(\theta_A + \theta_B) = \beta P \times \delta$$

or,

$$\beta = \frac{M_P}{P} \times \frac{(2\theta_A + \theta_B)}{\delta}$$

Here, $\delta = a\theta_A = b\theta_B$; So, we can rewrite the earlier expression as,
$\beta = \frac{M_P}{P} \times \left[\frac{2\theta_A}{a\theta_A} + \frac{\theta_B}{b\theta_B}\right] = \frac{M_P(2b+a)}{Pab}$.
    If the point load is applied at the middle, $a = b = \frac{L}{2}$
Therefore,

$$\beta = \frac{M_P(2b+a)}{Pab} = \frac{M_P\left(L+\dfrac{L}{2}\right)}{P\left(\dfrac{L}{2}\times\dfrac{L}{2}\right)} = \frac{6M_P}{PL}.$$

Let yielding first occur for the load magnitude of $\beta'P$. From Figure 24.6 (b), we can understand that yielding will first commence in the fixed end A, for $a = b = \frac{L}{2}$. So, in the first yield condition, we can write,

$$M_y = \frac{\beta'P\left(\dfrac{L}{2}\times\dfrac{L}{2}\right)\times\left(L+\dfrac{L}{2}\right)}{2L^2} = \frac{\beta'\times 3PL}{16}$$

or,

$$\beta' = \frac{16}{3} \times \frac{M_y}{PL}$$

So,

$$\frac{\beta}{\beta'} = \frac{\dfrac{6M_P}{PL}}{\dfrac{16}{3}\times\dfrac{M_y}{PL}} = \frac{9}{8} \times \frac{M_P}{M_y} = \frac{9}{8} \times \text{Shape factor}$$

As it is clear from the above discussion, that plastic hinge formation may happen anywhere over the span of a structural element depending upon the intensity of loading at that region. Thus, possible collapse mechanism will vary in number, particularly if the number of indeterminacy is more, and it is practically impossible to carry out all types of mechanisms and determine the minimum load under which the collapse may occur. To identify the correct load factor in plastic analysis, there are three important criteria as below:

1. *Mechanism:* Mechanism is formed when the ultimate load or collapse load is reached in the structure. To form a mechanism, the number of plastic hinges developed should be just sufficient.
2. *Equilibrium:* The internal bending moment must be in equilibrium with the external loading. The conditions of equilibrium for 2D case are: $\sum F_x = 0$, $\sum F_y = 0$, $\sum M_z = 0$.
3. *Yield criteria:* The bending moment at a section should not exceed the plastic bending moment at that section of the structure.

To overcome the analysis hurdle and based on these three criteria, we have the following theorems that need to be studied and understood well.

24.4.1 THE MAXIMUM PRINCIPLE, STATIC THEOREM, LOWER BOUND THEOREM, OR SAFE THEOREM

If there exists a bending moment distribution (M) for a structure and loading which is safe $(M < M_P)$ and statically admissible (equations of statics are applicable) with a set of loads W then,

$$W \leq W_C$$

where W_C is the collapse load of the structure.

This theorem is safe, because the load factor will be less than or equal to the collapse load factor once the equilibrium and yield criteria are met. As the load factor comes out to be less than the actual collapse load, this theorem either gives a wrong and safe result or gives a right and safe result. Since the elastic analysis always meets the equilibrium and yield conditions, an elastic analysis will always be safe. As the load obtained by this theorem is always to be lesser than or equal to the collapse load, this theorem is also known as lower bound theorem.

24.4.2 THE MINIMUM PRINCIPLE, KINEMATC THEOREM, UPPER BOUND THEOREM, OR UNSAFE THEOREM

This theorem states that of all the mechanisms formed by assuming plastic hinge locations, the correct collapse mechanism is the one that requires the minimum loads. So, any choice of plastic hinge location will provide collapse load that is greater than or equal to the correct collapse load.

For a given structure subjected to a set of loads W, the value of W found corresponding to any assumed mechanism must be either greater or equal to the collapse load.
i.e.,

$$W \geq W_C$$

In this theorem, we always assume a mechanism is forming. This theorem is unsafe because as the load factor from this theorem comes out to be precisely correct or larger than the actual collapse load, it may lead the designer to think that the structure is capable of taking more load than the actual, which may lead to a dangerous situation. As the load obtained by this theorem is always to be greater than or equal to the collapse load, this theorem is also known as upper bound theorem.

24.4.3 The Uniqueness Theorem

This theorem states that the bending moment distribution, which fulfills all the three conditions mentioned above, i.e., mechanism, conditions of equilibrium, and yield, is unique. In other words, we can say if we obtain a bending moment distribution satisfying static theorem, in this distribution, the bending moment is equal to the plastic bending moment at sufficient sections to form mechanism then that load is equal to the collapse load.

i.e.,

$$W = W_C$$

Using these theorems, we can find the collapse load by the following two methods namely,

    a. Static method
    b. Kinematic method

Some examples on both these methods are discussed in the following sections.

24.5 STATIC METHOD FOR DETERMINING COLLAPSE LOAD

In this method, collapse load is computed based on the geometry of an assumed equilibrium bending moment diagram. Here the bending moment of a section is not greater than the plastic moment (less than or equal to) of that section. Some illustrative examples are provided below to clarify the concept.

Example 24.1: Determine the collapse load for the beam and loading condition shown in Figure 24.7 by static method.

SOLUTION: In this case, corresponding to the collapse condition, plastic hinge should develop at the center of the beam under the applied load. The bending moment diagram corresponding to the collapse condition is shown in Figure 24.8.

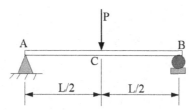

FIGURE 24.7   Example problem 1 on static method.

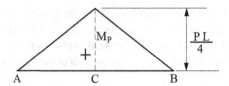

FIGURE 24.8   Bending moment diagram of Example 24.1.

From geometry of the collapse bending moment diagram, we have,

$$M_P = \frac{PL}{4}$$

Therefore, collapse load, $P_C = \dfrac{4M_P}{L}$

Example 24.2: Determine the collapse load for the beam and loading condition shown in Figure 24.9 by static method.

SOLUTION: In this case, corresponding to the collapse condition, plastic hinges should develop at the center of the beam under the applied load and at the fixed ends. The bending moment diagram corresponding to the collapse condition is shown in Figure 24.10.

From geometry of the collapse bending moment diagram, we have,

$$M_P + M_P = \frac{PL}{4}$$

Therefore, collapse load, $P_C = \dfrac{8M_P}{L}$

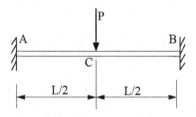

FIGURE 24.9   Example problem-2 on static method.

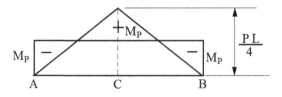

FIGURE 24.10   Bending moment diagram of Example 24.2.

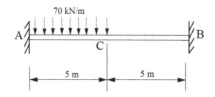

FIGURE 24.11   Example problem-3 on static method.

Example 24.3: A beam of span 8 m is fixed at both ends and carries a uniformly distributed load of 70 kN/m over the left half of the span as shown in Figure 24.11. Design the section considering static method of plasticity. Allow a load factor of 1.5. Yield stress for steel is 250 N/mm².

SOLUTION: Collapse load = load factor × safe load = 1.5 × 70 kN/m = 105 kN/m Considering the beam as simply supported and carrying a distributed load of 105 kN/m on the left half of the span, right end reaction,

$$R_B = \frac{105 \times 5 \times 2.5}{10} = 131.25 \text{ kN}$$

∴ left end reaction,

$$R_A = 105 \times 5 - 131.25 = 393.75 \text{ kN}$$

At any section in AC distant x meters from A, free bending moment, $M_x = 393.75x - 105\frac{x^2}{2}$
For the condition M_x is maximum, $\frac{dM_x}{dx} = 0$

$$\therefore \frac{dM_x}{dx} = 393.75 - 105x = 0$$

Therefore,   x = 3.75 m

∴ Maximum free bending moment = $393.75 \times 3.75 - 105 \times \frac{3.75^2}{2} = 738.28$ kNm
At the collapse condition, plastic hinges will be developed at A, B and the point of maximum sagging bending moment. The bending moment diagram corresponding to collapse condition is shown in Figure 24.12.
From the geometry of this diagram, we find

$$M_P + M_P = 738.28 \text{ kNm}$$

$$\therefore M_P = 369.13 \text{ kNm}$$

Again, $M_P = f_y S$

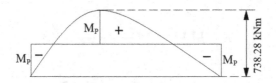

FIGURE 24.12 Bending moment diagram of Example 24.3.

or,

$$369.13 \times 10^6 = 250\ S$$

Therefore, plastic modulus required $= S = \frac{369.13 \times 10^6}{250}$ mm^3 = 1476000 mm^3
Assuming a shape factor of 1.15,
Section modulus required $= \frac{1476000}{1.15}$ mm^3 = 1283.5 $\times$ 10^3 mm^3.

Example 24.4: Determine the collapse load for the beam and loading condition shown in Figure 24.13 by static method.

SOLUTION: Figure 24.12 shows the bending moment diagram corresponding to the collapse condition. From the geometry of the collapse bending moment diagram, we have,

$$\frac{Pab}{L} = EF + FC = M_P + \frac{b}{L}M_P = M_P\left(1 + \frac{b}{L}\right) = \frac{b+L}{L}M_P$$

Therefore, collapse load, $P_C = \frac{b+L}{ab}M_P$

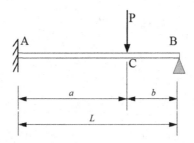

FIGURE 24.13 Example problem-4 on static method.

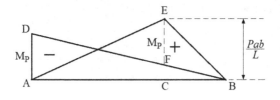

FIGURE 24.14 Bending moment diagram of Example 24.4.

24.6 KINEMATIC METHOD FOR DETERMINING COLLAPSE LOAD

The static method is suitable for simple structures, where the bending moment diagram is a simple one. But if the structure gets complicated, kinematic method of analysis is more suitable. In this method, work performed by the external loads is equated to the internal work absorbed by the plastic hinges to find out the collapse load. As this method is based on the assumed mechanisms of the structure, a load computed based on this method is always greater than or equal to the actual ultimate load. Some illustrative examples are provided below to clarify the concept.

Example 24.5: Determine the collapse load for the beam and loading condition shown in Figure 24.7 as given earlier by kinematic method.

SOLUTION: Figure 24.15 shows the collapse mechanism of the simply supported beam shown in Example 24.1. Let us now provide a virtual displacement δ at point C. Under the influence of this small displacement the beam will rotate about the plastic hinge. The internal plastic moment M_P will try to oppose this rotation. On the verge of collapse, the beam will be in equilibrium. In this equilibrium condition, work done by the external force will be equal to the work done by the internal force.

The external virtual work = $P\delta = P\frac{L\theta}{2}$ (since the displacement is small).
Internal virtual work = $2M_P\theta$
Equating both we get,

$$P\frac{L\theta}{2} = 2M_P\theta$$

or, collapse load, $P_C = \frac{4M_P}{L}$; It can be noted that the amplitude of the collapse load in this method is same as the earlier static method.

Example 24.6: Determine the collapse load for the beam and loading condition shown in Figure 24.9 as given earlier by kinematic method.

SOLUTION: Figure 24.16 shows the collapse mechanism of a fixed beam shown in Example 24.2. Let us now provide a virtual displacement δ at point C. Under the

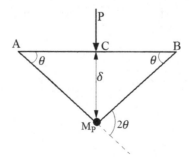

FIGURE 24.15 Collapse mechanism of beam in Figure 24.7.

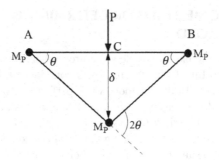

FIGURE 24.16 Collapse mechanism of beam in Figure 24.9.

influence of this small displacement, the beam will rotate about the plastic hinge. The internal plastic moment M_P will try to oppose this rotation. As the degree of indeterminacy is two, a total of three hinges need to form for the collapse mechanism. On the verge of collapse, the beam will be in equilibrium. In this equilibrium condition, work done by the external force will be equal to the work done by the internal force.

The external virtual work $= P\delta = P\frac{L\theta}{2}$ (since the displacement is small).

Internal virtual work $= 2M_P\theta + M_P\theta + M_P\theta = 4M_P\theta$

Equating both we get,

$$P\frac{L\theta}{2} = 4M_P\theta$$

or, collapse load, $P_C = \frac{8M_P}{L}$; it can be noted that the amplitude of the collapse load in this method is same as the earlier static method.

Example 24.7: Determine the collapse load for the beam and loading condition shown in Figure 24.17 as given earlier by kinematic method.

SOLUTION: Figure 24.18 shows the collapse mechanism of the fixed beam under uniformly distributed load. As the degree of indeterminacy is two, a total of three hinges, as shown in Figure 24.18, need to form for the collapse mechanism.

Total internal work $= 2M_P\theta + M_P\theta + M_P\theta = 4M_P\theta$

Total external work $= 2\int_0^{\frac{L}{2}} w\,dx\,y = 2\int_0^{\frac{L}{2}} w\,dx\,x\theta = 2w\theta\int_0^{\frac{L}{2}} x\,dx = \frac{w\theta L^2}{4}$

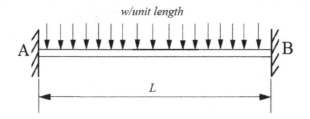

w/unit length

FIGURE 24.17 Example problem-3 on kinematic method.

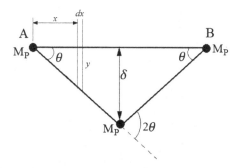

FIGURE 24.18   Collapse mechanism of the beam.

Equating both we get,

$$\frac{w\theta L^2}{4} = 4M_P\theta$$

or, $w = \frac{16M_P}{L^2}$

Total collapse load will be, $W_C = wL = \frac{16M_P}{L}$

Example 24.8: A beam AB of span L fixed at both ends has to carry a point load at a distance L/4 from the left end as shown in Figure 24.19 (a). Find the value of the load at the collapse condition if the plastic moment of resistance of the left half of the beam is $3M_P$ while the plastic moment of resistance of the right half of the beam is M_P.

SOLUTION: Figure 24.19 (a) shows the fixed beam subjected to collapse load P. There are two possible collapse mechanisms, shown in Figure 24.19 (b) and (c). In the first case shown in Figure 24.19 (b), plastic hinges are formed at A, B and under the load. Let a small virtual displacement, $CD = \delta$ be given.

Total external virtual work $= P\delta = P\dfrac{3L}{4}\theta$

Total internal virtual work $= 3M_P(3\theta) + 3M_P(3\theta) + 3M_P(\theta) + M_P(\theta) = 22M_P\theta$

Equating both we get, $P\dfrac{3L}{4}\theta = 22M_P\theta$

Therefore, $P = \dfrac{29.33M_P}{L}$

In the second case shown in Figure 24.19 (c), plastic hinges are developed at A, B, and E. Let a small virtual displacement, $EF = \delta$ be given.

Here, $\delta = \dfrac{L}{2}\theta$

Total external virtual work $= P\dfrac{L}{4}\theta$

Total internal virtual work $= 3M_P\theta + M_P\theta + M_P\theta + M_P\theta = 6M_P\theta$

Equating both we get, $P\dfrac{L}{4}\theta = 6M_P\theta$

Therefore, $P = \dfrac{24M_P}{L}$

Therefore, actual collapse load, $P_C = \dfrac{24M_P}{L}$

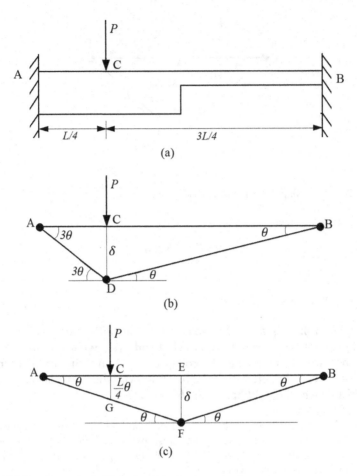

FIGURE 24.19 Example problem-3 on kinematic method.

24.7 PLASTIC ANALYSIS OF PORTAL FRAMES

A single portal frame can fail in three types of mechanisms, namely, beam mechanism, sway mechanism, and combined mechanism, as shown in Figure 24.20 (a), (b), and (c), respectively. The portal frame shown in this figure is indeterminate to degree three. For the total collapse of the structure, we need four plastic hinges to form. But in the case of beam mechanism, only partial collapse takes place. In this case, three plastic hinges are developed for the partial collapse of this member. Sway mechanism is a complete collapse of the frame. So, four hinges need to form for the total collapse. In the case of a sway mechanism, no plastic hinges are created due to vertical load. Only the horizontal load creates a mechanism here. The combined mechanism also led to the total collapse of the structure. So, four plastic hinges need to form here also. Here, the plastic hinge will not form in the left-hand side beam-column joint as both the column and the beam will rotate in the clockwise direction

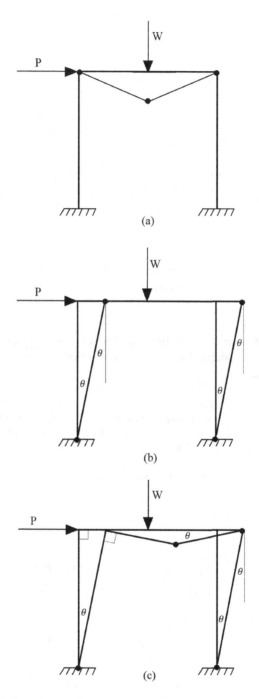

FIGURE 24.20 Collapse mechanism of portal frame.

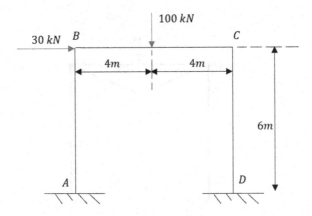

FIGURE 24.21   Frame example problem.

under the influence of applied load, as shown. So, orthogonality of this joint will remain unaffected. Some illustrative examples are provided below to clarify the concept.

Example 24.9: Find the plastic moment required for the portal frame subjected to the collapse load system shown in Figure 24.21. All members are of the same section.

SOLUTION: As discussed earlier, there can be three types of failure mechanism for these frames: beam, sway, and combined mechanisms. Three types of failure mechanisms are drawn one by one below for ease of understanding.

Let the load factor be β. Then from Figure 24.22, we get,

$$100\beta \times 4\theta - M_p\theta - M_p\theta - 2M_p\theta = 0$$

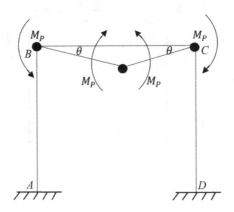

FIGURE 24.22   Beam mechanism.

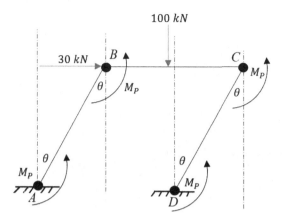

FIGURE 24.23   Sway mechanism.

or,

$$\beta = \frac{M_P}{100} = 0.01M_P$$

Then we will deal with sway mechanism as shown in Figure 24.23. As per Figure 24.23, the equation of sway mechanism will be:

$$30\beta \times 6\theta = M_P\theta + M_P\theta + M_P\theta + M_P\theta$$

or,

$$\beta = \frac{2}{90}M_P = 0.022M_P$$

Finally, we will draw the combined mechanism as shown in Figure 24.12.

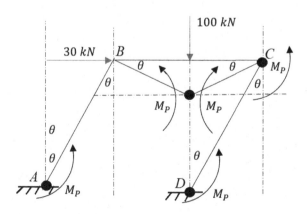

FIGURE 24.24   Combined mechanism.

The equation as per this mechanism will be:

$$100\beta \times 4\theta + 30\beta \times 6\theta - M_p \times \theta - M_p \times \theta - M_p \times \theta - M_p \times 2\theta - M_p \times \theta = 0$$

or,

$$\beta = \frac{3M_p}{290} = 0.01034M_p$$

So, comparing among all three cases, final collapse factor will be $\beta = 0.01M_p$.

Appendix A
Areas and Centroids of Geometric Shapes

Shape	Area	Centroid
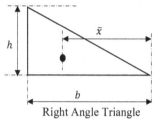 Right Angle Triangle	$A = \dfrac{bh}{2}$	$\bar{x} = \dfrac{2b}{3}$
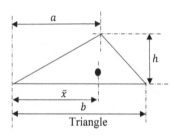 Triangle	$A = \dfrac{2h}{2}$	$\bar{x} = \dfrac{a+b}{3}$
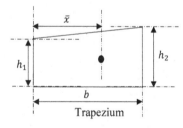 Trapezium	$A = \dfrac{b(h_1 + h_2)}{2}$	$\bar{x} = \dfrac{b(h_1 + 2h_2)}{3(h_1 + h_2)}$
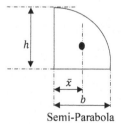 Semi-Parabola	$A = \dfrac{2bh}{3}$	$\bar{x} = \dfrac{3b}{8}$

(Continued)

Shape	Area	Centroid

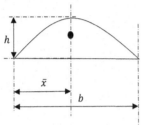

Parabolic Spandrel

$$A = \frac{bh}{3}$$

$$\bar{x} = \frac{3b}{4}$$

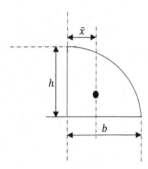

Parabolic Segment

$$A = \frac{2bh}{3}$$

$$\bar{x} = \frac{b}{2}$$

Cubic Parabola

$$A = \frac{3bh}{4}$$

$$\bar{x} = \frac{2b}{5}$$

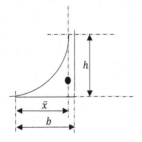

Cubic Spandrel

$$A = \frac{bh}{4}$$

$$\bar{x} = \frac{4b}{5}$$

Appendix B
Review of Matrix Algebra

B.1 INTRODUCTION TO MATRIX ALGEBRA

A matrix is a rectangular array of figures arranged in rows and columns. Algebraically, matrix A is represented as a_{ij} in which i represents the row number and j represents the column number. In tabular form of matrix, the same can be expressed as:

$$a_{ij} = \begin{bmatrix} a_{11} & \cdots & a_{1j} \\ \vdots & \ddots & \vdots \\ a_{i1} & \cdots & a_{ij} \end{bmatrix}$$

Sometimes, it is easier to understand the matrix operation using the index notation, rather than using the cumbersome tabular forms. A matrix can be only column or row matrix. So, a matrix having a single row is called row matrix, and a matrix having only one column is called column matrix.

A matrix is said to be a square matrix when it has an equal number of rows and columns.

Two matrices are said to be equal if and only if all the elements at each row are same. Two matrices can only be added together when both have the same rows and columns. In index notation, we can write:

$$A + B = D$$

i.e.,

$$a_{ij} + b_{ij} = d_{ij}$$

In tabular form, the abovementioned matrix operation can be expressed as follows:

$$\begin{bmatrix} a_{11} & \cdots & a_{1j} \\ \vdots & \ddots & \vdots \\ a_{i1} & \cdots & a_{ij} \end{bmatrix} + \begin{bmatrix} b_{11} & \cdots & b_{1j} \\ \vdots & \ddots & \vdots \\ b_{i1} & \cdots & b_{ij} \end{bmatrix} = \begin{bmatrix} d_{11} & \cdots & d_{1j} \\ \vdots & \ddots & \vdots \\ d_{i1} & \cdots & d_{ij} \end{bmatrix}$$

where $d_{11} = a_{11} + b_{11}$ etc.

If a matrix has all the entries as 0, only then it is called null matrix and is represented by 0.

If we add any matrix with null matrix, then the matrix remains unchanged:

$$a_{ij} + 0 = a_{ij}$$

i.e.,

$$A + 0 = A$$

Two matrices can be multiplied together by row and column multiplication rule. It is customary to note that two matrices can only be multiplied when number of columns in the first matrix is same as the number of rows of the second matrix.

$$C = A \times B = a_{ij} \times b_{jk} = c_{ik}$$

Matrix addition is commutative, i.e.,

$$A + B = B + A$$

$$a_{ij} + b_{ij} = b_{ij} + a_{ij}$$

Matrix multiplication is noncommutative, i.e.,

$$A \times B \neq B \times A$$

Transpose of a matrix is the process in which rows of the matrix are transformed into column or vice versa. It is as follows:

$$A = a_{ij} \ \textit{implies} \ A^T = a_{ji}$$

Example:
    If,

$$A = \begin{bmatrix} 1 & 2 \\ 3 & 4 \end{bmatrix}$$

Then,

$$A^T = \begin{bmatrix} 1 & 3 \\ 2 & 4 \end{bmatrix}$$

Example of matrix multiplication:
    If,

$$A = \begin{bmatrix} 1 & 2 \\ 3 & 4 \end{bmatrix} \text{ and } B = \begin{bmatrix} 5 & 7 \\ 6 & 8 \end{bmatrix} \text{ then } A \times B \text{ is given by,}$$

$$A \times B = \begin{bmatrix} 1 \times 5 + 2 \times 6 & 1 \times 7 + 2 \times 8 \\ 3 \times 5 + 4 \times 6 & 3 \times 7 + 4 \times 8 \end{bmatrix} = \begin{bmatrix} 17 & 23 \\ 39 & 83 \end{bmatrix}$$

It is instructed to readers that they should carry out the product $B \times A$ and check that:

$$A \times B \neq B \times A$$

If we multiply matrix by a scalar, then each element of the matrix is multiplied by the same scalar. In index notation, this can be written as:

$$\alpha A = \alpha a_{ij}$$

A matrix is said to be a symmetric matrix if it is symmetric about its diagonal:

$$a_{ij} = a_{ji}$$

A matrix is said to be a diagonal matrix if all the elements except its principal diagonal are 0. In index notation, this can be written as:

$$a_{ij} = 0 \text{ iff } i \neq j$$

Unit matrix is the one that has 1 along its main diagonal and all other elements are 0. Index notation for the same is as follows:

$$a_{ij} = 1 \text{ when } i = j$$

$$a_{ij} = 0 \text{ when } i \neq j$$

This matrix is also called identity matrix, because if we multiply any matrix with unit matrix, then the original matrix remains unchanged after multiplication. Identity or unit matrices are denoted as I.

B.2 MATRIX INVERSE

The inverse of a square matrix is the one that when multiplied with the original matrix gives unit or identity matrix as an outcome of multiplication. Mathematically it is expressed as:

$$A \times A^{-1} = A^{-1} \times A = I$$

B.3 PARTITIONING OF MATRICES

Sometimes, larger matrices are partitioned into smaller sub-matrices to perform operation in a much easier way. To understand this, the partitioning can be done in many ways. One of the partitioning methods is shown next to understand this concept.

$$A = \begin{bmatrix} a & d & g & j \\ b & e & h & k \\ c & f & i & l \end{bmatrix}$$

now, by carrying the partitioning as per the dotted line, we get:

$$A = \begin{bmatrix} C_{11} & C_{21} \\ C_{12} & C_{22} \end{bmatrix}$$

where $C_{11} = \begin{bmatrix} a & d & g \\ b & e & h \end{bmatrix}$, $C_{12} = \begin{bmatrix} c & f & i \end{bmatrix}$, $C_{21} = \begin{bmatrix} j \\ k \end{bmatrix}$, and $C_{22} = [l]$

B.4   MATRIX INVERSION BY GAUSS-JORDAN ELIMINATION PROCESS

The Gauss-Jordan Elimination process necessarily involves multiple row and column operation within the same matrix. In a matrix, we can add, multiply, and divide each element of row or column elements and add or subtract between rows and columns without causing any changes in the matrix. We need to carry out this type of operations multiple times so that the original matrix can be converted into a diagonal form. Then the augmented matrix written at the side of the main matrix will produce the inverse of the main matrix. To understand this process fully, see Example B.1.

Example B.1: Calculate the inverse of the following matrix.

$$A = \begin{bmatrix} 2 & -5 & 4 \\ 3 & 1 & 8 \\ 4 & -7 & -1 \end{bmatrix}$$

SOLUTION: To apply Gauss-Jordan Elimination process, we first form the augmented matrix by taking the 3×3-unit matrix at the side of the main matrix as shown next:

$$\left[\begin{array}{ccc|ccc} 2 & -5 & 4 & 1 & 0 & 0 \\ 3 & 1 & 8 & 0 & 1 & 0 \\ 4 & -7 & -1 & 0 & 0 & 1 \end{array} \right]$$

Now, our main task is to convert the main matrix into diagonal unit matrix by doing fundamental operations on rows of the total augmented matrix. Let us first divide the first row by 2, and we will get:

$$\left[\begin{array}{ccc|ccc} 1 & -2.5 & 2 & 0.5 & 0 & 0 \\ 3 & 1 & 8 & 0 & 1 & 0 \\ 4 & -7 & -1 & 0 & 0 & 1 \end{array} \right]$$

Next, we multiply row 1 by 3 and subtract it from row 2, we will get:

$$\begin{bmatrix} 1 & -2.5 & 2 \\ -3\times1+3 & -3\times2.5-1 & -2\times3-8 \\ 4 & -7 & -1 \end{bmatrix} \begin{matrix} 0.5 & 0 & 0 \\ 3\times0.5-0 & -3\times0+1 & 3\times0-0 \\ 0 & 0 & 1 \end{matrix}$$

which gives:
$$\begin{bmatrix} 1 & -2.5 & 2 \\ 0 & 8.5 & -14 \\ 4 & -7 & -1 \end{bmatrix} \begin{matrix} 0.5 & 0 & 0 \\ 1.5 & 1 & 0 \\ 0 & 0 & 1 \end{matrix}$$

Then we multiply row 1 by 4 and subtract it from row 3 to get:

$$\begin{bmatrix} 1 & -2.5 & 2 \\ 0 & 8.5 & -14 \\ 0 & 3 & -9 \end{bmatrix} \begin{matrix} 0.5 & 0 & 0 \\ 1.5 & 1 & 0 \\ -1.5 & 0 & 1 \end{matrix}$$

Now, dividing row 2 by 8.5, we get:

$$\begin{bmatrix} 1 & -2.5 & 2 \\ 0 & 1 & -1.647 \\ 0 & 3 & -9 \end{bmatrix} \begin{matrix} 0.5 & 0 & 0 \\ 0.176 & 0.118 & 0 \\ -1.5 & 0 & 1 \end{matrix}$$

Now, multiplying row 2 by −2.5 and subtracting it from row 1, we get:

$$\begin{bmatrix} 1 & 0 & 2.118 \\ 0 & 1 & -1.647 \\ 0 & 3 & -9 \end{bmatrix} \begin{matrix} -0.94 & 0 & 0 \\ 0.176 & 0.118 & 0 \\ -1.5 & 0 & 1 \end{matrix}$$

Then multiply row 2 by 3 and subtracting it from 3, we get:

$$\begin{bmatrix} 1 & 0 & 2.118 \\ 0 & 1 & -1.647 \\ 0 & 0 & -4.059 \end{bmatrix} \begin{matrix} -0.94 & 0 & 0 \\ 0.176 & 0.118 & 0 \\ -3.528 & -0.354 & 1 \end{matrix}$$

Now, dividing row 3 by −4.059, we get,

$$\begin{bmatrix} 1 & 0 & 2.118 \\ 0 & 1 & -1.647 \\ 0 & 0 & 1 \end{bmatrix} \begin{matrix} -0.94 & 0 & 0 \\ 0.176 & 0.118 & 0 \\ -0.869 & -0.09 & 1 \end{matrix}$$

Then, on multiplying row 3 by −2.118 and subtracting it from row 1, we get:

$$\begin{bmatrix} 1 & 0 & 0 \\ 0 & 1 & -1.647 \\ 0 & 0 & 1 \end{bmatrix} \begin{matrix} 0.9 & -0.017 & 2.118 \\ 0.176 & 0.118 & 0 \\ -0.869 & -0.008 & 1 \end{matrix}$$

Now, multiply row 3 by –1.647 and subtract it from row 2 to get:

$$\begin{bmatrix} 1 & 0 & 0 \\ 0 & 1 & 0 \\ 0 & 0 & 1 \end{bmatrix} \begin{array}{ccc} 0.9 & -0.017 & 2.118 \\ -1.255 & 0.104 & 1.647 \\ -0.869 & -0.008 & 1 \end{array}$$

Thus, we have converted the main matrix into the diagonal form by carrying out several fundamental row operations. So, the 3×3 obtained at the right side of the augmented matrix will be inverse of main matrix A.

$$\text{Hence, } A^{-1} = \begin{bmatrix} 0.9 & -0.017 & 2.118 \\ -1.255 & 0.104 & 1.647 \\ -0.869 & -0.008 & 1 \end{bmatrix}$$

It is customary to check that,

$$AA^{-1} = A^{-1}A = I$$

Appendix C
Three-Moment Equation

Three-moment equation was developed by Clapeyron in 1857. This method provides a very convenient way for analyzing continuous beams. The three-moment equation produces, in its general form, the compatibility condition that the slope of the elastic curve be continuous at an interior support of the continuous beam. Since the equation involves three moments – the bending moments at the support under our observation and at the two supports adjacent to it – is commonly referred to as three-moment equation. While applying this method, the bending moments at the interior (and any fixed) supports of the continuous beam are taken as the redundants. The three-moment equation is then applied at the location of each redundant locations to obtain a set of compatibility equations that can be solved for the unknown redundant moments. We will learn this method of analysis with the help of Example C.1.

Example C.1: Analyze the following continuous beam as shown in Figure C.1 using three-moment equation.

SOLUTION: The beam shown in Figure C.1 has one degree of redundancy. The moment at support B is the redundant moment. Hence, we will start at B and its adjacent two supports, i.e., A and C as shown previously.

The general form of three-moment equation is given by,

$$\frac{M_{AB}L_{AB}}{I_{AB}} + 2M_B \left(\frac{L_{AB}}{I_{AB}} + \frac{L_{CB}}{I_{CB}} \right) + \frac{M_{CB}l_{CB}}{I_{CB}}$$

$$= -\sum \frac{P_A L_{AB}^2 k_{AB}}{I_{AB}} \left(1 - k_{AB}^2\right) - \sum \frac{P_C l_{CB}^2 k_{CB}}{I_{CB}} \left(1 - k_{CB}^2\right) - \frac{w_{AB}L_{AB}^3}{4I_{AB}}$$

$$- \frac{w_{CB}l_{CB}^3}{4I_{CB}} - 6E \left(\frac{\Delta_A - \Delta_B}{L_{AB}} + \frac{\Delta_C - \Delta_B}{L_{CB}} \right)$$

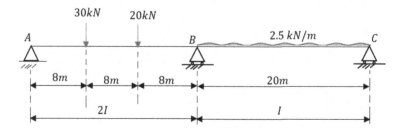

FIGURE C.1   Example problem of three-moment equation.

It is to be noted that the redundant moments at support B have been written in the abovementioned equation as M_B. k_{AB} and k_{CB} are the ratios of distances of point loads from left and right supports, respectively. All the other terms are self-explanatory. Since in span AB we have two-point loads and along span CB, only one uniformly distributed load is acting, we have,

$$L_{AB} = 24 \text{ m}$$

$$L_{CB} = 20 \text{ m}$$

$$I_{AB} = 2I$$

$$I_{CB} = I$$

$$P_{AB1} = 30, \ k_{AB1} = \frac{1}{3}$$

$$P_{AB2} = 20, \ k_{AB2} = \frac{2}{3}$$

$$w_{AB} = 0$$

$$w_{CB} = 2.5 \text{ kN/m}$$

All other parameters in the abovementioned equation are zero.

Hence, once we substitute the above values in the abovementioned equation, we will get:

$$\frac{M_{AB}(24)}{2I} + 2M_B\left(\frac{24}{2I} + \frac{20}{I}\right) + \frac{M_{CB}(20)}{I}$$

$$= -\frac{30(24)^2(1/3)}{2I}\left(1-(1/3)^2\right) - \frac{20(24)^2(2/3)}{I}\left(1-(2/3)^2\right) - \frac{2.5(20)^3}{4I}$$

Now, since A and C supports are hinge supports, by inspection we have:

$$M_{AB} = M_{CB} = 0$$

Substituting these values in the abovementioned equation, we have:

$$M_B = -151.5 \text{ kNm}$$

Once this redundant reaction is known, we can form the free-body diagrams of each span separately, and by applying equilibrium equations, we can determine the unknown support reactions at these three supports as well. Since we have carried out these steps several times in the text, the same is left as an exercise for the readers.

Appendix D
Solved Examples of Selected Problems

Here we will provide a few solved examples of different types of problems. After completing the texts in previous chapters it is suggested to students to do these problems on their own and then check the solutions provided here to bridge any gap between computational and conceptual issues.

Example D.1: Determine the reactions at the supports for the three-hinged arch as shown in Figure D.1.

SOLUTION: Free-body diagram: See Figure D.1(b)
Static Determinacy: The arch is internally unstable. It is composed of two rigid portions, *AB* and *BC*, connected by an internal hinge at *B*.
The arch has,
No. of support reactions, r = 4; Equation of condition due to internal hinge, $e_c = 1$
i.e., Degree of external indeterminacy, $i_e = r - (3 + e_c) = 4 - (3 + 1) = 0$
The arch is statically determinate.
Support reactions:

$$\circlearrowleft + \sum M_C = 0$$

$$-A_y \times 20 + 4 \times 10 \times \frac{10}{2} + \frac{1}{2} \times 20 \times 12 \times \frac{2}{3} \times 20 = 0$$

$$A_y = 90 \text{ kN} \uparrow$$

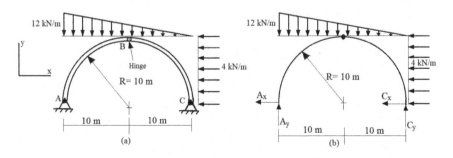

(a)   (b)

FIGURE D.1   Example problem of a three-hinged arch.

$$\uparrow + \sum F_y = 0$$

$$90 + C_y - \frac{1}{2} \times 20 \times 12 = 0$$

$$C_y = 30 \text{ kN} \uparrow$$

$$\circlearrowleft + \sum M_B^{BC} = 0$$

$$30 \times 10 - C_x \times 10 - 4 \times 10 \times \frac{10}{2} - \frac{1}{2} \times 10 \times 6 \times \frac{1}{3} \times 10 = 0$$

$$C_x = 0 \text{ kN}$$

$$\rightarrow + \sum F_x = 0$$

$$-A_x - 0 - 4 \times 10 = 0$$

$$A_x = -40 \text{ kN} \rightarrow$$

Check:

To check the computation, we apply the equilibrium $\circlearrowleft + \sum M_B = 0$ for the entire structure.

$$\circlearrowleft + \sum M_B = 0$$

$$40 \times 10 - 90 \times 10 + 30 \times 10 - 4 \times 10 \times \frac{10}{2} + \frac{1}{2} \times 20 \times 12 \times \left(10 - \frac{1}{3} \times 20 \right) = 0$$

Hence, o.k.

Example D.2: Analyze the complex truss as shown in Figure D.2.

SOLUTION: Inspect the truss properly. If a joint is reached, where there are three unknowns, remove one of the members at the joint and replace it with an imaginary member elsewhere in the truss. Here, in this case, let us remove the member *BE* and instead add the member *AC*. Now we can analyze the truss by the method of joints with actual loading present in the truss.

The member forces are obtained by this, let us name them as *F'* set of forces.

Calculation of F' set of forces:

Joint *E*: See Figure D.3 (a)

$$\uparrow + \sum F_y = 0$$

$$F_{ED} + 20 = 0$$

$$F_{ED} = -20 \text{ kN} \left(C \right)$$

$$\rightarrow + \sum F_x = 0$$

$$F_{EF} = 0$$

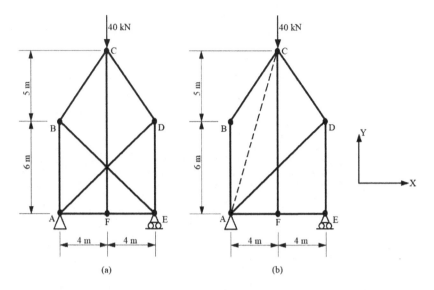

FIGURE D.2 Problem of a complex truss.

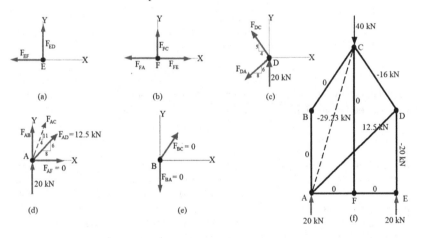

FIGURE D.3 Example problem of a complex truss: various joint equilibriums for F' set of forces.

Joint F: See Figure D.3 (b)

$$F_{FA} = 0$$
$$F_{FC} = 0$$

Joint D: See Figure D.3 (c)

$$\uparrow + \sum F_y = 0$$

$$20 + \frac{5}{6.4} \times F_{DC} - \frac{6}{10} \times F_{DA} = 0$$

$$\rightarrow + \sum F_x = 0$$

$$-F_{DA} \times \frac{8}{10} - F_{DC} \times \frac{4}{6.4} = 0$$

Solving the abovementioned two equations,

$$F_{DA} = 12.5 \text{ kN } (T)$$

$$F_{DC} = -16 \text{ kN } (C)$$

Joint A: See Figure D.3 (d)

$$\uparrow + \sum F_y = 0$$

$$20 + \frac{6}{10} \times 12.5 + \frac{11}{11.7} \times F_{AC} + F_{AB} = 0$$

$$\rightarrow + \sum F_x = 0$$

$$12.5 \times \frac{8}{10} + F_{AC} \times \frac{4}{11.7} = 0$$

$$F_{AC} = -29.23 \text{ kN } (C)$$

Replacing the value of F_{AC} in the abovementioned equation, we get $F_{AB} = 0$.
Joint A: See Figure D.3 (e)

$$F_{BC} = 0 \text{ and } F_{BA} = 0$$

Now, consider the simple truss without the external load of 40 kN. Place equal, but opposite collinear unit loads on the truss at the two joints from which the member was removed. Due to these unit loads at joint B and E, let's assume that a U set of forces develops in the truss members.

If these forces develop a force U_i in the ith truss member, then by proportion an unknown force X in the removed member would exert a force, XU_i in the ith member.

Calculation of U set of forces:
Joint E: See Figure D.4 (a)

$$\rightarrow + \sum F_x = 0$$

$$-1 \times \frac{8}{10} - F_{EF} = 0$$

$$F_{EF} = -0.8 \text{ kN } (C)$$

$$\uparrow + \sum F_y = 0$$

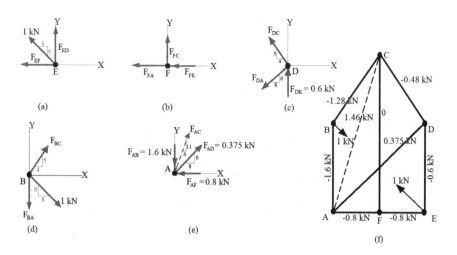

FIGURE D.4 Example problem of a complex truss: various joint equilibriums for U set of forces.

$$1 \times \frac{6}{10} + F_{ED} = 0$$

$$F_{ED} = -0.6 \text{ kN } (\text{C})$$

Joint F: See Figure D.4 (b)

$$\uparrow + \sum F_y = 0$$

$$F_{FC} = 0$$

$$\rightarrow + \sum F_x = 0$$

$$-F_{FE} - F_{FA} = 0$$

$$F_{FA} = -0.8 \text{ kN}$$

Joint D: See Figure D.4 (c)

$$\uparrow + \sum F_y = 0$$

$$F_{DC} \times \frac{5}{6.4} + 0.6 - \frac{6}{10} \times F_{DA} = 0$$

$$\rightarrow + \sum F_x = 0$$

$$-F_{DC} \times \frac{4}{6.4} - F_{DA} \times \frac{8}{10} = 0$$

Solving the abovementioned two equations, we get:

$$F_{DC} = -0.48 \text{ kN } (C)$$

$$F_{DA} = 0.375 \text{ kN } (T)$$

Joint D: See Fig. D.4 (d)

$$\rightarrow + \sum F_x = 0$$

$$F_{BC} \times \frac{4}{6.4} + 1 \times \frac{8}{10} = 0$$

$$F_{BC} = -1.28 \text{ kN } (C)$$

$$\uparrow + \sum F_y = 0$$

$$F_{BC} \times \frac{5}{6.4} - F_{BA} - 1 \times \frac{6}{10} = 0$$

Substituting the value of F_{BC}, we get, $F_{BA} = -1.6 \text{ kN } (C)$.
    Joint A: See Figure D.4 (e)

$$\rightarrow + \sum F_x = 0$$

$$-0.8 + 0.375 \times \frac{8}{10} + F_{AC} \times \frac{4}{11.7} = 0$$

$$F_{AC} = 1.46 \text{ kN } (T)$$

Now, if the effects of the abovementioned two loadings are considered, the forces in the ith member of the truss will be, $F_i = F_i' + XU_i$.
    The member AC does not exist in the actual truss. So, we have to choose the magnitude of X in such a way so that $F_i = 0$ in member AC.
    For member AC we get,

$$F_{AC}' + XU_{AC} = 0$$

$$-29.23 + 1.46 \times X = 0$$

$$X = 20.02$$

Finally, we get the value of the member forces in the actual truss from Table D.1.

TABLE D.1
Calculating Member Forces in the Actual Truss

Truss Members	F' Force (kN)	U Force (kN)	$F_i = F_i' + XU_i$ (kN)
AB	0	−1.6	−32.03
BC	0	−1.28	−25.62
CD	−16	−0.48	−25.61
DE	−20	−0.60	−32.01
EF	0	−0.80	−16.02
FA	0	−0.80	−16.02
AD	12.5	0.375	20.01
BE	0	1	20.02
FC	0	0	0

Example D.3: Analyze the following multistory portal frame as shown in Figure D.5 by moment distribution method.

SOLUTION: As all the applied loads are nodal loads only, there will be no fixed-end moments induced in the members. Only moments will be generated due to sidesway of the portal frame.

Distribution factor at C will be:

$$DF_{CD} = \frac{2I/6}{(I/6)+(2I/6)} = \frac{2}{3}$$

$$DF_{CB} = \frac{1}{3}$$

Distribution factor at D will be:

$$DF_{DC} = \frac{2}{3}$$

$$DF_{DE} = \frac{1}{3}$$

Distribution factor at B will be:

$$DF_{BE} = \frac{2I/6}{(I/6)+(2I/6)+(I/6)} = \frac{1}{2}$$

$$DF_{BC} = \frac{I/6}{(I/6)+(2I/6)+(I/6)} = \frac{1}{4}$$

$$DF_{BA} = \frac{I/6}{(I/6)+(2I/6)+(I/6)} = \frac{1}{4}$$

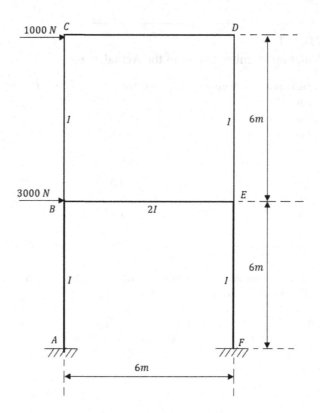

FIGURE D.5 Example on multistory portal frame.

Distribution factor at E will be:

$$DF_{EB} = \frac{2I/6}{(I/6) + (2I/6) + (I/6)} = \frac{1}{2}$$

$$DF_{EF} = \frac{I/6}{(I/6) + (2I/6) + (I/6)} = \frac{1}{4}$$

$$DF_{ED} = \frac{I/6}{(I/6) + (2I/6) + (I/6)} = \frac{1}{4}$$

Sway Correction for Top Story

Let us assume that δ be the sway at the top story toward right side. For this sway, the induced moments at joints will be:

$$\bar{M}_{BC} = \bar{M}_{CB} = -\frac{6EI\delta}{6^2}$$

$$\bar{M}_{DE} = \bar{M}_{ED} = -\frac{6EI\delta}{6^2}$$

$$\bar{M}_{BC} : \bar{M}_{DE} = 1:1$$

Let us assume that

$$\bar{M}_{BC} = \bar{M}_{CB} = \bar{M}_{DE} = \bar{M}_{ED} = -1000 \text{ Nm}$$

The moment distribution for top-story sway toward right is provided in the following tabular form:

Joints	A	B			C		D		E			F
Member	AB	BA	BE	BC	CB	CD	DC	DE	ED	EB	EF	FE
DF	–	1/4	1/2	1/4	1/3	2/3	2/3	1/3	1/4	1/2	1/4	–
FEM				−1000	−1000			−1000	−1000			
BAL		+250	+500	+250	+330	+667	+667	+333	+250	+500	+250	
COM	+125		+250	+107	+125	+334	+334	+125	+167	+250		+125
BAL		−104	−209	−104	−153	−306	−306	−153	−104	−209	−104	
COM	−52		−105	−77	−52	−153	−153	−52	−77	−105		−52
BAL		+46	+90	+46	+68	+137	+137	+68	+46	+90	+46	
COM	+23		+45	+34	+23	+69	+69	+23	+34	+45		+23
BAL		−20	−39	−20	−31	−61	−61	−31	−20	−39	−20	
COM	−10		−20	−16	−10	−31	−31	−10	−16	−20		+10
BAL		+9	+18	+9	+14	+27	+27	+14	+9	+18	+9	
COM	−5		+9	+7	+5	+14	+14	+5	+7	+9		−5
BAL		−4	−8	−4	−6	−13	−13	−6	−4	−8	−4	
COM	−2		−4	−3	−2	−7	−7	−2	−3	−4		−2
BAL		+2	+3	+2	+3	+6	+6	+3	+2	+3	+2	
COM	+1		+2	+2	+1	+3	+3	+1	+2	+2		+1
BAL		−1	−2	−1	−1	−3	−3	−1	−1	−2	−1	
FINAL MOMENT	+90	+178	+530	−708	−683	+683	+683	−683	−708	+0.530	−178	+90

Let the actual sway moment is x times the moment calculated for top-story sway toward right side.

Sway Correction for Bottom Story

Let δ_2 be the sway toward right of the bottom sway:

$$\bar{M}_{AB} = \bar{M}_{BA} = -\frac{6EI\delta_2}{6^2}$$

$$\bar{M}_{EF} = \bar{M}_{FE} = -\frac{6EI\delta_2}{6^2}$$

$$\bar{M}_{AB} : \bar{M}_{EF} = 1:1$$

Let us assume,

$$\bar{M}_{AB} = \bar{M}_{BA} = \bar{M}_{FE} = \bar{M}_{EF} = -1000 \text{ Nm}$$

The moment distribution of the assumed sway moments for the bottom story is provided in the below table:

Joints	A	B			C		D		E			F
Member	AB	BA	BE	BC	CB	CD	DC	DE	ED	EB	EF	FE
DF	–	1/4	1/2	1/4	1/3	2/3	2/3	1/3	1/4	1/2	1/4	–
FEM	−1000	−1000									−1000	−1000
BAL		+250	+100	+250					+250	+500	+250	
COM	+125		+250		+125			+125		+250		+125
BAL		−63	−124	−63	−42	−83	−83	−42	−63	−124	−63	
COM	−32		−62	−21	−32	−42	−42	−32	−21	−62		−32
BAL		+21	+41	+21	+25	+49	+49	+25	+21	+41	+21	
COM	+11		+21	+13	+11	+25	+25	+11	+13	+21		+11
BAL		−9	−16	−9	−12	−24	−24	−12	−9	−13	−9	
COM	−5		−8	−6	−5	−12	−12	−5	−6	−8		−5
BAL		+4	+6	+4	+6	+11	+11	+6	+4	+6	+4	
COM	+2		+3	+3	+2	+6	+6	+2	+3	+3		+2
BAL		2	−2	−2	−3	−4	−5	−3	−2	−2	−2	
COM	−1											−1
FINAL MOMENT	−900	−700	+609	+190	+75	−75	−75	+75	+190	+601	+799	−900

Let the actual moment be y times the moment obtained from assumed moment values. Thus, total moment in various members will be:

$$M_{AB} = 90x - 900y$$

$$M_{BC} = -708x - 190y$$

$$M_{CB} = -683x - 75y$$

$$M_{DC} = 683x - 75y$$

$$M_{ED} = -708x - 190y$$

$$M_{EF} = 178x - 799y$$

$$M_{BA} = 178x - 799y$$

$$M_{BE} = 530x + 600y$$

$$M_{CD} = 683x - 75y$$

$$M_{BD} = -683x + 75y$$

$$M_{DE} = 530X + 609y$$

$$M_{FE} = 90x - 900y$$

Now, shear force at the base of top story must be zero to maintain the equilibrium if forces,

$$H_B + H_E + P = 0$$

$$\frac{M_{BC} + M_{CB}}{6} + \frac{M_{DE} + M_{ED}}{6} + 1000 = 0$$

Also, total summation of moment will also need to be zero to maintain moment equilibrium conditions.

$$M_{BC} + M_{CB} + M_{DE} + M_{ED} = -6000$$

$$-708x + 190y - 683x + 75y - 683x + 75y - 708x + 190y = -6000$$

or,

$$-2782x + 530x = -6000$$

$$-x + 0.1905y = -2.157$$

Similarly, shear at bottom of the story needs to be zero to maintain force equilibrium conditions.

$$H_A + H_E + P = 0$$

or,

$$\frac{M_{AB} + M_{BA}}{6} + \frac{M_{EF} + M_{FE}}{6} + 1000 + 3000 = 0$$

$$M_{AB} + M_{BA} + M_{EF} + M_{FE} = -24,000$$

$$90x - 900y + 178x - 799y + 90x - 900y + 178x - 799y = -24,000$$

$$536x - 3398y = -24,000$$

$$x - 6.339y = -44.27$$

On addition of these final equations, we get:

$$-6.1485y = -46.927$$

or,

$$y = +7.633$$

$$x = -44.77 + 63.339 \times 7.633 = +3.63$$

Once we got the values of unknown x and y, we need to just substitute back them in the abovementioned total moment expressions in terms of these parameters. The final moments are provided in a table below:

Joints	A	B			C		D		E			F
Member	AB	BA	BE	BC	CB	CD	DC	DE	ED	EB	EF	FE
3.63 times sway moment for top story	+325.8	+645	+1918	−2563	−2473	+2473	+2473	−2473	−2563	+1918	+645	+325.8
7.633 times sway moment for bottom story	−6869.7	−6099	+4648	+1451	+572.5	−572.5	−572.5	+572.5	+1451	+4648	−6098	−6869.7
Final moment	−6543.9	−5454	+6566	−1112	−1900.5	+1900.5	+1900.5	−1900.5	−1112	+6566	−5454	−6543.9

Example D.4: Analyze the following frame as shown in Figure D.6 by strain energy method. Support A is hinged and support D is fixed.

SOLUTION: We have shown all the support reactions at the support points A and D in Figure D.6. Let M_D be the fixed-end moment at the support D. Taking moment about D, we get:

$$M_D + R_A \times 4 + H_A \times 3 - 4000 \times 2 - 6 \times 1000 \times 3 = 0$$

$$R_A = 6500 - \frac{3}{4}H - \frac{M_D}{4}$$

As per the support conditions, the supports A and D will not yield. Hence, partial derivative of strain energy with respect to H_A and M_D will be zero separately. Refer to the following table for complete analysis of the given frame by Castigliano's method.

Origin at	Portion	Bending Moment	$\dfrac{\partial M}{\partial H}$	$\dfrac{\partial M}{\partial M_D}$	Limits
A	AB	$H \times y$	y	y	$0 \to 3$
B	BE	$\left(6500 - \dfrac{3}{4}H - \dfrac{M_D}{4}\right)x - 3H$	$-\left(\dfrac{3}{4}x + 3\right)$	$-\dfrac{x}{4}$	$0 \to 2$
B	BC	$6500x - \left(\dfrac{3}{4}x + 3\right)H - \dfrac{M_D}{4}x - 4000(x-2)$	$-\left(\dfrac{x}{4} + 3\right)$	$-\dfrac{x}{4}$	$2 \to 4$
D	DC	$(6000 - H)y - M_D - 1000\dfrac{y^2}{2}$	$-y$	-1	$0 \to 6$

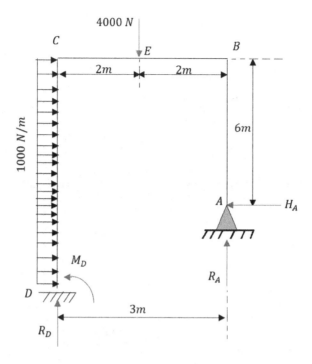

FIGURE D.6 Example of frame analysis problem.

Now, let us first take partial derivative of strain energy with respect to horizontal reaction H_A at A:

$$\frac{\partial U}{\partial H} = \frac{1}{EI} \int_0^6 Hy \times y\, dy$$

$$+ \frac{1}{2EI} \int_0^2 \left[6500x - \left(\frac{3}{4}x + 3 \right) H - \frac{M_D}{4} x \right] \left[-\left(\frac{3}{4}x + 3 \right) \right] dx$$

$$+ \frac{1}{2EI} \int_2^4 \left[6500x - \left(\frac{3}{4}x + 3 \right) H - \frac{M_D}{4} - 4000(x-2) \right] \left[-\left(\frac{3}{4}x + 3 \right) \right] dx$$

$$+ \frac{1}{EI} \int_0^6 \left[(6000 - H)y - M_D - 1000 \frac{y^2}{2} \right] [-y]\, dy = 0$$

So,

$$H \left[\frac{y^3}{3} \right]_0^3 - \frac{1}{2} \int_0^4 \left[6500 \left(\frac{3}{4}x^2 + 3x \right) - H \left(\left[\frac{3}{4}x + 3 \right]^2 \right) - \frac{M_D}{4} \left(\frac{3}{4}x^2 + 3x \right) \right] dx$$

$$+ 2000 \int_2^4 \left(\frac{3}{4}x^2 + \frac{3}{2}x - 6 \right) dx - (6000 - H) \left[\frac{y^3}{3} \right]_0^6 + M_D \left[\frac{y^2}{3} \right]_0^6 + 500 \left[\frac{y^4}{4} \right]_0^5 = 0$$

or,

$$9H - \frac{6500}{2}\left[\frac{x^3}{4} + \frac{3}{2}x^2\right]_0^4 + \frac{H}{2}\left[\frac{4}{9}\left(\frac{3}{4}x+3\right)^3\right]_0^4 + \frac{M_D}{8}\left[\frac{x^3}{4} + \frac{3}{2}x^2\right]_0^4$$

$$+ 2000\left[\frac{x^3}{4} + \frac{3x^2}{4} - 6x\right]_0^4 - (6000 - H)\times 72 + 18M_D + 162,000 = 0$$

So, on simplification we get:

$$9H - \frac{6500}{2}[16+24] + \frac{H}{2}\times\frac{4}{9}\left(6^3 - 3^2\right) + \frac{M_D}{8}(16+24)$$

$$+ 2000\left[\frac{\left(4^3 - 2^3\right)}{4} + \frac{3}{4}\left(4^2 - 2^2\right) - 6(4-2)\right]$$

$$- 432,000 + 72H + 18M_D + 162,000 = 0$$

or,

$$5.347H + M_D = 16,434.8$$

Second, we take partial derivative of strain energy with respect to M_D to get:

$$\frac{\partial U}{\partial M_D} = \frac{1}{2EI}\int_0^4\left[6500x - \left(\frac{3}{4}x+3\right)H - \frac{M_D}{4}x\right]\left(-\frac{x}{4}\right)dx$$

$$+ \frac{1}{2EI}\int_2^4\left[6500x - \left(\frac{3}{4}x+3\right)H - \frac{M_D}{4}x - 4000(x-2)\right]\left(-\frac{x}{4}\right)dx$$

$$+ \frac{1}{EI}\int_0^6\left[(6000-H)y - M_D - \frac{1000y^2}{2}\right](-1)dy = 0$$

So,

$$-\frac{1}{2}\int_0^4\left[\frac{6500}{4}x^2 - \frac{H}{4}\left(\frac{3}{4}x^2 + 3x\right) - \frac{M_Dx^2}{16}\right]dx + \frac{4800}{8}\int_0^4\left(x^2 - 2x\right)dx$$

$$- \int_0^6\left[(6000-H)y - M_D - 5000y^2\right]dy = 0$$

Upon carrying out the integrations and simplification, we finally get:

$$-\frac{52,000}{3} + 5H + \frac{2}{3}M_D + 500\left(\frac{56}{3} - 12\right) - 108,000 + 18H + 6M_D + 36,000 = 0$$

or,

$$5.347H + M_D = 16,434.8$$

Second, we take partial derivative of strain energy with respect to M_D to get:

$$\frac{\partial U}{\partial M_D} = \frac{1}{2EI} \int_0^4 \left[6500x - \left(\frac{3}{4}x + 3 \right)H - \frac{M_D}{4}x \right]\left(-\frac{x}{4} \right) dx$$

$$+ \frac{1}{2EI} \int_2^4 \left[6500x - \left(\frac{3}{4}x + 3 \right)H - \frac{M_D}{4}x - 4000(x-2) \right]\left(-\frac{x}{4} \right) dx$$

$$+ \frac{1}{EI} \int_0^6 \left[(6000 - H)y - M_D - \frac{1000y^2}{2} \right](-1) dy = 0$$

So,

$$-\frac{1}{2}\int_0^4 \left[\frac{6500}{4}x^2 - \frac{H}{4}\left(\frac{3}{4}x^2 + 3x \right) - \frac{M_D x^2}{16} \right] dx + \frac{4800}{8}\int_0^4 (x^2 - 2x) dx$$

$$- \int_0^6 \left[(6000 - H)y - M_D - 5000y^2 \right] dy = 0$$

Upon carrying out the integrations and simplification, we finally get:

$$-\frac{52,000}{3} + 5H + \frac{2}{3}M_D + 500\left(\frac{56}{3} - 12 \right) - 108,000 + 18H + 6M_D + 36,000 = 0$$

$$23H + \frac{20}{3}M_D = 86,000$$

$$3.45H + M_D = 12,900$$

Solving these two final equations, we get:

$$H = 1863 \text{ N}$$

$$M_D = 6472 \text{ Nm}$$

$$M_{BA} = -H \times 3 = -1863 \times 3 = -5589 \text{ Nm}$$

Taking moment about C, we get:

$$M_{CD} + (6000 - H) \times 6 - M_{DC} - 1000 \times 6 \times 3 = 0$$

or,

$$M_{CD} = 18,000 + 6472 - 4137 \times 6 = -340 \text{ Nm}$$

Example D.5: A Continuous beam *ABC* as shown in Figure D.7, having uniform cross section has two equal spans of *AB* and *BC* each of length *l*. During application of imposed loads, the support *B* sinks by an amount δ_1 and support *C* by an amount δ_2. Find the reaction at the supports at this situation in terms of settlement parameters and *EI*.

SOLUTION: Let M_B be the moment at *B*.
Taking moments about *B*,

$$R_A \times l + M_B = 0$$

or,

$$R_A = -\frac{M_B}{l}$$

Similarly, taking moment about *B* for *BC*, we will get:

$$R_C = -\frac{M_B}{l}$$

$$R_A = R_C = R$$

$$R_B = -2R$$

And for *AB*,

$$M = R \times x$$

Strain energy of beam *ABC*,

$$= 2\int_0^l \frac{M^2 dx}{2EI}$$

or,

$$2\int_0^l \frac{(Rx)^2 dx}{2EI}$$

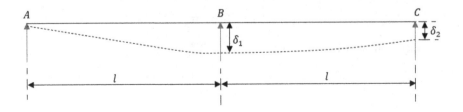

FIGURE D.7 Example problem of a continuous beam with support settlement.

or,

$$\frac{R^2}{EI}\left[\frac{x^3}{3}\right]_0^l$$

i.e.,

$$\frac{R^2 l^3}{EI}$$

So, the total work done on the structure,

$$W = \frac{R^2 l^3}{EI} - 2R \times \delta_1 + R \times \delta_2$$

For work to be maximum,

$$\frac{\partial W}{\partial R} = 0$$

Upon carrying out the partial derivative, we will get:

$$R = 3EI\frac{2\delta_1 - \delta_2}{2l^3}$$

Hence,

$$R_A = R_C = 3EI\frac{2\delta_1 - \delta_2}{2l^3}$$

And,

$$R_B = -2R = -3EI\frac{2\delta_1 - \delta_2}{2l^3}$$

Example D.6: All the members of the below frame have same cross section A and same EI. Find the force in each member due to the applied loading as shown in Figure D.8.

SOLUTION: The given frame is a redundant frame with more members than required for stability. Let us take member BC as the redundant member. We will remove this redundant member from the original frame to make it a deterministic structure. The same is shown in Figure D.9.

In Figure D.9, we have marked all the unknown support reactions at the supports A and D of the determinate frame. In the next step, we will apply unit loads at nodes B and C in the direction of the redundant member BC. The same is also shown in Figure D.9.

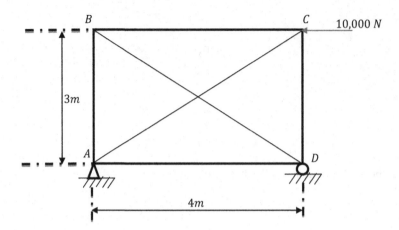

FIGURE D.8 Example problem of redundant frame with loading.

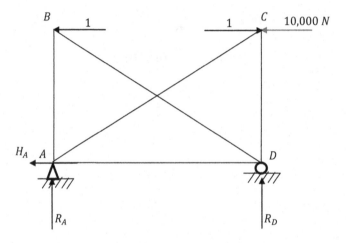

FIGURE D.9 Determinate frame with redundant member BC removed.

The force in various members due to this unit load is provided in the following tabular form.

Member	Length in cm.	Force F	k	k²	Fkl	k²l	Final Force F + kX
AB	300	0	+3/4	9/16	0	2700/16	−2568
CD	300	+7500	+3/4	9/16	+6,750,000/4	2700/16	4432
DA	400	0	+1	1	0	400	−3425
AC	500	−12,500	−5/4	25/16	+31,250,000/4	12,500/16	−8220
BD	500	0	−5/4	25/16	0	12,500/16	+4280
Σ					+38,000,000/4	+36,000/16	

where k is the force due to unit load in various members in the determinate frame after removing the redundant member from the original structure. And force F is the force in the determinate frame due to applied load of 10,000 N at joint C.
     We have:

$$X = \frac{\Sigma(Fkl/AE)}{\Sigma(k^2l/AE) + (l_0/A_0E)}$$

where l_0 and A_0 are the length and area of cross section of redundant member that has been removed from the original structure to convert it into a determinate structure.
     Since all the member having same cross-sectional area a and length of redundant member as per Figure D.8 is 400 cm, we have,

$$X = \frac{\Sigma Fkl}{\Sigma k^2l + l_0}$$

Putting the values, we will get:

$$X = -3425 \text{ N}$$

Final force in each member will be $F + kX$, and the same has already been provided in table's last column.

Example D.7: Find the fixed-end moment for the fixed beam as shown in Figure D.10 using column analogy method.

SOLUTION: The basic determinate structure of the above fixed beam will be a cantilever beam AB. Bending moment diagram and loading on analogous column of the same is drawn sequentially in Figure D.11 for reference. Please note that load acting in analogous column is $-M_s$, where M_s is the bending moment in determinate structure.

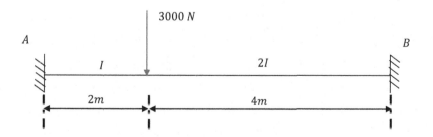

FIGURE D.10   Fixed beam with point load.

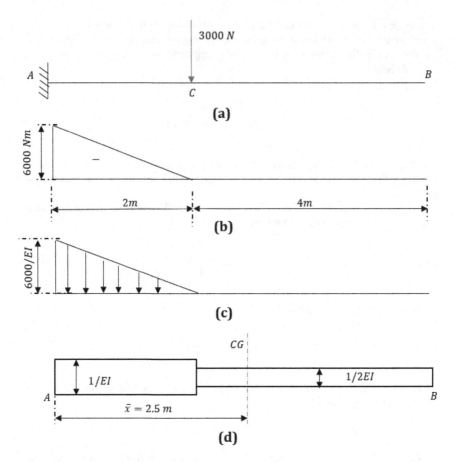

FIGURE D.11 (a) Determinate cantilever, (b) bending moment diagram, (c) loading on analogous column, and (d) section of analogous column.

Distance of *CG* from the left support point *A* is given by,

$$\bar{x} = \frac{2 \times (1/E) \times 1 + 4 \times (1/2EI) \times (2+2)}{2 \times (1/EI) + 4 \times (1/2EI)} = \frac{2+8}{2+2} = 2.5 \text{ m}$$

Moment of inertia of the analogous column is given by,

$$I_y = \frac{1}{12} \times \frac{1}{EI} \times 2^3 + \frac{2}{EI} \times 1.5^2 \times + \frac{1}{12} \times \frac{1}{2EI} \times 4^3 + \frac{4}{2EI} \times 1.5^2$$

or,

$$I_y = \frac{12.333}{EI}$$

Area of the analogous column is given by,

$$A = \frac{1}{EI} \times 2 + \frac{1}{2EI} \times 4 = \frac{4}{EI}$$

Net force acting on the analogous column is given by,

$$P = \frac{1}{2} \times \frac{6000}{EI} \times 2 = \frac{6000}{EI}$$

Eccentricity,

$$e = 2.5 - \frac{2}{3} = \frac{5.5}{3}$$

Hence, net stress acting in analogous column at A,

$$f_A = \frac{P}{A} + \frac{Pex}{I_y}$$

where in abovementioned equation of stress, $x = 2.5$ m
So, we have:

$$f_A = \frac{P}{A} + \frac{Pex}{I_y} = 3729.7$$

Hence, net stress acting in analogous column at B,

$$f_B = \frac{P}{A} + \frac{Pex}{I_y}$$

Here $x = 6 - 2.5 = 3.5$ m
Thus,

$$f_B = 1621.6$$

So, the final moments at A and B will be,

$$M_A = -M_s + f_A = -6000 + 3729.7 = -2270.3 \text{ Nm}$$

$$M_B = 0 - f_B = -1621.6 \text{ Nm}$$

Bibliography

1. ASCE Standard Minimum Design Loads for Buildings and Other Structures. (2010) ASCE/SEI 7-10, American Society of Civil Engineers, Virginia.
2. Arbabi, F. (1991) *Structural Analysis and Behavior.* McGraw-Hill, New York, NY.
3. Bathe, K.J., and Wilson, E.L. (1976) *Numerical Methods in Finite Element Analysis.* Prentice Hall, Englewood Cliffs, NJ.
4. Beer, F.P., and Johnston, E.R.., Jr. (1981) *Mechanics of Materials.* McGraw-Hill, New York, NY.
5. Betti, E. (1872) Il Nuovo Cimento. Series 2, Vols. 7–8.
6. Boggs, R.G. (1984) *Elementary Structural Analysis.* Holt, Rinehart & Winston, New York, NY.
7. BS 8110-1. (1997) Structural use of Concrete-Part 1: Code of practice for design and construction.
8. Chajes, A. (1990) *Structural Analysis*, 2nd ed. Prentice Hall, Englewood Cliffs, NJ.
9. Colloquim on History of Structures. (1982) *Proceedings, International Association for Bridge and Structural Engineering*, Cambridge, England.
10. Cross, H. (1930) "Analysis of Continuous Frames by Distributing Fixed-End Moments." *Proceedings of the American Society of Civil Engineers* 56, 919–928.
11. Elias, Z.M. (1986) *Theory and Methods of Structural Analysis.* Wiley, New York, NY.
12. Gere, J.M., and Weaver, W. Jr. (1965) *Matrix Algebra for Engineers.* Van Nostrand Reinhold, New York, NY.
13. Glockner, P.G. (1973) "Symmetry in Structural Mechanics." *Journal of the Structural Division, ASCE* 99, 71–89.
14. Goldstein, H., Poole, C.P., and Safko, J. (2011) *Classical Mechanics.* Pearson Education, India.
15. Hibbeler, R.C. (2012) *Structural Analysis*, 8th ed. Prentice Hall, Englewood Cliffs, NJ.
16. Holzer, S.M. (1985) *Computer Analysis of Structures.* Elsevier Science, New York, NY.
17. Hourani, M. (2002) *Mathematical Model of Influence Lines for Indeterminate Beams,* Proceeding of the 2002 American Society for Engineering Education Annual Conference & Exposition, American Society for Engineering Education.
18. International Building Code. (2012) International Code Council, Chicago, IL.
19. IS-875 Part – I (1997) Indian Standard Code of Practice for Design Loads (Other Than Earthquake) for Buildings and Structures Part 1 – Dead Loads: Unit Weights of Building Materials and Stored Materials, 2nd ed.
20. IS-875 Part – II, (1987) Indian Standard Code of Practice for Design Loads (Other Than Earthquake) for Buildings and Structures Part 2 – Imposed Loads, 2nd revision, Reaffirmed 2008.
21. IS 875 Part – III (2015) Indian Standard Code of Practice for Design Loads (Other Than Earthquake) for Buildings and Structures Part 3 – Wind Loads, 3rd revision.
22. IS1893 Part – I (2002) Indian Standard Criteria for Earthquake Resistant Design of Structures Part – 1 General Provisions and Buildings, 5th revision.
23. Kassimali, A. (2011) *Matrix Analysis of Structures*, 2nd ed. Cengage Learning, Stamford, CT.
24. Kassimali, A. (2014) *Structural Analysis*, 5th ed. Cengage Learning, Boston, MA.
25. Kennedy, J.B., and Madugula, M.K.S. (1990) *Elastic Analysis of Structures: Classical and Matrix Methods.* Harper & Row, New York, NY.
26. Laible, J.P. (1985) *Structural Analysis.* Holt, Rinehart & Winston, New York, NY.

27. Langhaar, H.L. (1962) *Energy Methods in Applied Mechanics.* Wiley, New York, NY.
28. Laursen, H.A. (1988) *Structural Analysis,* 3rd ed. McGraw-Hill, New York, NY.
29. Leet, K.M. (1988) *Fundamentals of Structural Analysis.* Macmillan, New York, NY.
30. McCormac, J. (1984) *Structural Analysis,* 4th ed. Harper & Row, New York, NY.
31. McCormac, J., and Elling, R.E. (1988) *Structural Analysis: A Classical and Matrix Approach.* Harper & Row, New York, NY.
32. McGuire, W., and Gallagher, R.H. (1979) *Matrix Structural Analysis.* Wiley, New York, NY.
33. Maney, G.A. (1915) *Studies in Engineering,* Bulletin 1. University of Minnesota, Minneapolis, MN.
34. Manual for Railway Engineering. (2011) American Railway Engineering and Maintenance of Way Association, Maryland, VA.
35. Maxwell, J.C. (1864) "On the Calculations of the Equilibrium and Stiffness of Frames." *Philosophical Magazine* 27, 294–299.
36. Marium, J.L., and Craig, L.G. (2017) *Engineering Mechanics (Part-1 – Statics).* Wiley, India.
37. Noble, B. (1969) *Applied Linear Algebra.* Prentice Hall, Englewood Cliffs, NJ.
38. Norris, C.H., Wilbur, J.B., and Utku, S. (1976) *Elementary Structural Analysis,* 3rd ed. McGraw-Hill, New York, NY.
39. Parcel, J.H., and Moorman, R.B.B. (1955) *Analysis of Statically Indeterminate Structures.* Wiley, New York, NY.
40. Petroski, H. (1985) *To Engineer Is Human—The Role of Failure in Successful Design.* St. Martin's Press, New York, NY.
41. Popov, E.P. (1968) *Introduction to Mechanics of Solids.* Prentice Hall, Englewood Cliffs, NJ.
42. Rao, D.S.P. (1997) *Graphical Methods in Structural Analysis.* University Press, India.
43. Ramamrutham, S., Narayan, R. (2003) *Theory of Structures.* Dhanpat Rai Publishing Company (P) LTD., New Delhi.
44. Timoshenko, S. (1930) Strength of Materials, D. Van Nostrand Company Inc., Toronto, NY, London.
45. Uniform building code. (1997), Vol. II.
46. Vazirani, V.N., Ratwani, M.M. (2002) *Analysis of Structures,* Vol. II. Khanna Publishers, New Delhi.
47. Wang, C.K. (1983) *Intermediate Structural Analysis.* McGraw-Hill, New York, NY.

Index

Note: Locators in *italics* represent figures and **bold** indicate tables in the text.

Printed in the United States
by Baker & Taylor Publisher Services